Waves, Atoms and Solids

Waves, Atoms and Solids

D. A. Davies
BSc, PhD, FInstP

Longman
London and New York

Longman Group Limited London

Associated companies, branches and representatives throughout the world

Published in the United States of America by Longman Inc., New York

First published 1978

Library of Congress Cataloging in Publication Data
Davies, D. A.
Waves, atoms and solids.
1. Solids—Electric properties. 2. Solids—Thermal properties. 3. Semiconductors. I. Title.
QC176.8.E35D38 530.4'1 77-2192
ISBN 0-582-44174-9

Printed in Great Britain by
Richard Clay (The Chaucer Press) Ltd, Bungay, Suffolk

Contents

Preface

This book is intended as an undergraduate text for students wishing to obtain a broad introduction to the electrical and thermal properties of solids and an understanding of the theory of semiconductors. In view of the use made in the theory of solids of basic concepts in wave theory and atomic theory I have started the text with a fairly detailed treatment of these topics. This has allowed development of concepts in the theory of electron orbitals, a topic more commonly included in chemistry texts on valence theory. In my view the inclusion of this material should prove to be of value in the better understanding of many of the topics in the text. I would also hope that this approach will make the text acceptable to a wider range of readers than would otherwise have been the case.

SI units are used throughout the text and I have attempted as far as possible to follow the recommendations of the Symbols Committee of the Royal Society (*Quantities, Units, and Symbols*, Second Edition, 1975). I have, however, retained the use of the electron volt (eV) as a unit of energy. This unit is recognised for continued use with SI and it does not appear to me that there is at present any suitable alternative unit (e.g. a designated multiple for 10^{-21} for use with the joule). The very occasional reference to other units should not cause difficulty.

I wish to thank my colleagues Dr J. S. Brooks, Mr T. R. Barrass, Mr L. S. D. Fuller and Dr M. Goldstein for reading and commenting upon different parts of the text, and also Dr D. M. Allen-Booth and Dr R. M. Wood for their assistance with specific points. The responsibility for any remaining errors, omissions or incorrect emphases is mine.

Finally I wish to thank my wife for the very many hours she has spent in assisting me with the preparation and finalisation of the typescript.

D. A. Davies

Acknowledgements

We are grateful to the following for permission to reproduce copyright material: Allyn and Bacon Inc. for Fig. 6.11 from *Elementary Modern Physics* 1968 by Weidner and Sells; American Institute of Physics for Fig. 8.10 by J. C. Slater and Fig. 11 by P. P. Debye and E. M. Conwell from *Phys. Rev.* Vol. 45; G. Bell & Sons Ltd. for a figure based on Fig. 1.6 from *Principles of Semiconductor Device Operation* by A. K. Jonscher; Butterworths & Co. (Publishers) Ltd. for Fig. 12 of *Progress in Applied Materials Research* by D. A. Davies, published by Iliffe Books Ltd; Cambridge University Press for Fig. 8.7A from an article by J. Friedell in *Physics of Metals* Vol. 1 by Ziman; *Chemistry in Britain* and the author for Fig. 10 from *Principles of Atomic Orbitals* 1969 by N. Greenwood; Heinemann Eductional Books Ltd for Fig. 10.10 from *Wave Mechanics For Chemists* by Cumper; McGraw-Hill Book Company for Figs. 1.5, 6.17, 17.7 and 17.5 by Azaroff from *Elements of X-Ray Crystallography* and Fig. 4 by Barrett from *Structure of Metals,* used with permission of McGraw-Hill Book Company; Methuen & Co. Ltd for Fig. 7.2 from *Atomic and Molecular Structure* 1960 by Barrett; Oxford University Press for Fig. 7.13 from *Bonds Between Atoms* by A. Holden, published by Oxford University Press 1971; W. B. Saunders Co. for Fig. 3.38 from *Solid State Physics* 1969 by Blakemore; Taylor & Francis Ltd for a figure from *Electrons in Metals* 1964 by J. M. Ziman; John Wiley & Sons Inc. for Figs. 2.9 and 2.12 from *Valence Theory* 1965 by J. N. Kettle and J. M. Tedder for Fig. 2.6 and Fig. 5.2 by D. Long and Fig. 7.33 from *Energy Bands in Semiconductors* 1968.

Values of physical constants[†]

Quantity	Symbol	Value	Unit
speed of light in a vacuum	c	2.998×10^{8}	$m\,s^{-1}$
Planck constant	h	6.626×10^{-34}	$J\,s$
	$\hbar = \frac{h}{2\pi}$	1.055×10^{-34}	$J\,s$
Boltzmann constant	k	1.381×10^{-23}	$J\,K^{-1}$
gas constant	R	8.314	$J\,K^{-1}\,mol^{-1}$
Avogadro constant	N_A	6.022×10^{23}	mol^{-1}
elementary charge (of proton)	e	1.602×10^{-19}	C
rest mass of electron	m_e	9.110×10^{-31}	kg
rest mass of proton	m_p	1.673×10^{-27}	kg
rest mass of neutron	m_n	1.675×10^{-27}	kg
unified atomic mass constant	m_u	1.661×10^{-27}	kg
permeability of a vacuum	μ_0	$4\pi \times 10^{-7}$	$H\,m^{-1}$
permittivity of a vacuum	ε_0	8.854×10^{-12}	$F\,m^{-1}$
Bohr radius	a_0	5.292×10^{-11}	m
Rydberg constant	R_∞	1.097×10^{7}	m^{-1}
fine structure constant	α	7.297×10^{-3}	
gravitational constant	G	6.672×10^{-11}	$N\,m^2\,kg^{-2}$
Bohr magneton	μ_B	9.274×10^{-24}	$J\,T^{-1}$

Conversion factors

electron volt	$1\,eV = 1.602 \times 10^{-19}\,J = 9.648 \times 10^{4}\,J\,mol^{-1}$
Ångstrom	$1\,Å = 10^{-10}\,m = 10^{-1}\,nm$

Prefixes

Multiple	Prefix	Symbol
10^{6}	mega	M
10^{3}	kilo	k
10^{-2}	centi	c
10^{-3}	milli	m
10^{-6}	micro	μ
10^{-9}	nano	n
10^{-12}	pico	p

Values of mathematical constants

$\pi = 3.142$
$e = 2.718$
$\ln 10 = 2.303$
$\ln 2 = 0.693$

[†] Data from *Quantities, Units, and Symbols*, Second Edition, The Royal Society, 1975.

1. Wave motion

1.1 Simple harmonic motion

Simple harmonic motion (SHM) is a phenomenon which is common to most branches of physics, but is perhaps most familiar at an introductory level in the vibration of simple mechanical systems. All systems performing SHM have the common property that the displacement of the system from its equilibrium position gives rise to a restoring force towards this point. The magnitude of the force is proportional to the displacement. Taking the equilibrium position as the origin the *displacement* ψ is therefore related to a restoring force F by

$$F = -b\psi, \qquad \ldots(1.1)$$

where b is a constant for the particular system. We note that the force is always directed towards the origin whatever the sign of the displacement ψ, i.e. the equation represents a restoring force.

Since the acceleration $\mathrm{d}^2\psi/\mathrm{d}t^2$ of a particle of mass m is F/m we may write

$$\frac{\mathrm{d}^2\psi}{\mathrm{d}t^2} = -\omega^2\psi, \qquad \ldots(1.2)$$

where $\omega = (b/m)^{\frac{1}{2}}$ is a constant termed the *angular frequency* of the motion. The reason for this nomenclature will become apparent shortly. The differential equation (1.2) defines the SHM and has for its solution the equation

$$\psi = A\cos(\omega t + \alpha). \qquad \ldots(1.3)$$

The maximum value of the displacement ψ is A, and this is termed the *amplitude* of the motion. The angle $(\omega t + \alpha)$ is called the *phase angle*, or simply the *phase*. Putting $t = 0$ the angle α is then the initial phase of the motion or its *epoch*. Its value is arbitrary since it is governed simply by the instant from which time is measured.

A graphical representation of equation (1.3) is given in Fig. 1.1. Since the function (1.3) repeats itself whenever t is increased by an integral number of $2\pi/\omega$, it follows that it represents a periodic motion of *periodic time* T given by

$$T = \frac{2\pi}{\omega}. \qquad \ldots(1.4)$$

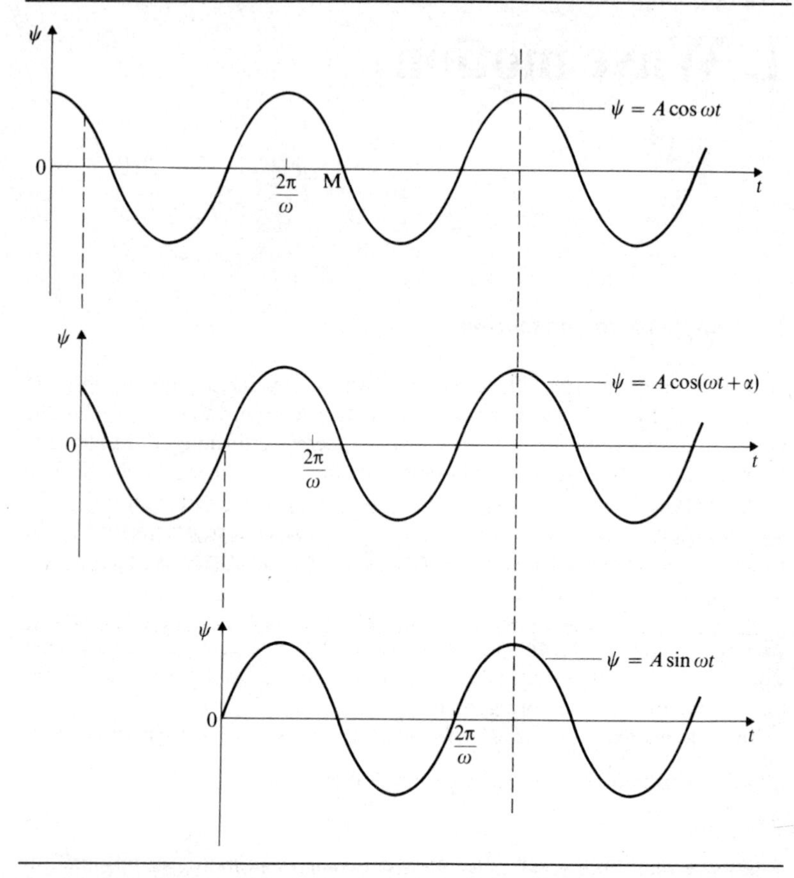

Fig. 1.1 The same simple harmonic motion is described by different functions by measuring time from a different instant.

The *frequency* ν of the motion is the number of oscillations of the system performed in unit time and is therefore given by

$$\nu = \frac{1}{T} = \frac{\omega}{2\pi}. \qquad \ldots(1.5)$$

A geometrical interpretation of the SHM represented by equation (1.3) is shown in Fig. 1.2. From this figure we observe that the projection on a coordinate axis of a point moving in a circle with uniform angular velocity ω performs a SHM in accordance with equation (1.3). As we have seen ω is referred to as the angular frequency of the SHM.

Since the value of the initial phase α is governed by the displacement at the

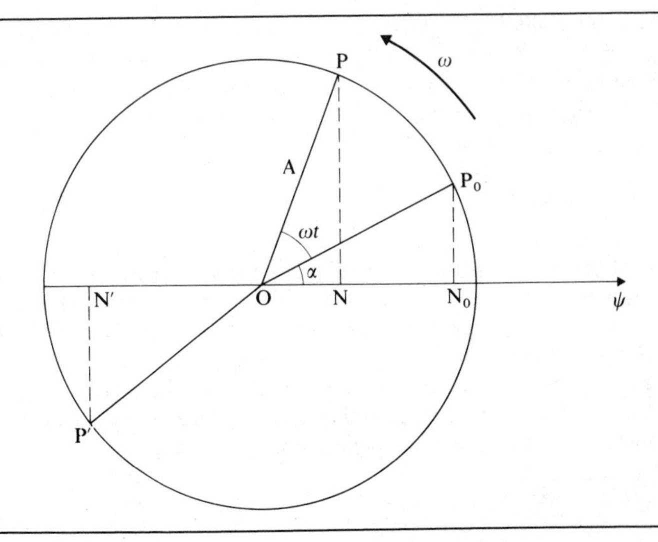

Fig. 1.2 The point P moves in a circular path at constant angular velocity ω and the motion of N is given by equation (1.3). P is initially at P_0 and the initial phase angle is α.

particular instant from which time is measured we may use any value of α we wish, e.g. we may put

$$\alpha' = \alpha - \frac{\pi}{2}$$

and thus represent the SHM of equation (1.3) by

$$\psi = A \sin(\omega t + \alpha'). \qquad \ldots(1.6)$$

Sine and cosine representations of SHMs are therefore totally equivalent.

Exponential functions representing the SHM may also be used as solutions of the differential equation (1.2). It is readily confirmed by differentiation that

$$\begin{aligned} \psi &= A \operatorname{expi}(\omega t + \alpha) \\ &= B \operatorname{expi} \omega t \end{aligned} \qquad \ldots(1.7)$$

is a solution. Except in the special case of $\alpha = 0$, the constant B here is complex and contains both the amplitude and phase components of the motion. The relationship of the exponential solution to the sine and cosine type of solutions may be seen by expanding equation (1.7) in the form

$$\psi = (C + \mathrm{i}D)(\cos \omega t + \mathrm{i} \sin \omega t). \qquad \ldots(1.8)$$

We may take either the real or imaginary parts of equation (1.8) and the two solutions obtained will correspond to the trigonometrical type of solution already discussed. It should also be emphasised that equation (1.7) as it stands is a solution of the differential equation (1.2) and a description of the system.

Although a discussion of damping in a system performing SHM will not be developed in this text it is of interest to note that solutions are then required which describe changes in the system over a period of time. The differential equation for such a system

$$\frac{d^2\psi}{dt^2}+2k\frac{d\psi}{dt}+\omega^2\psi = 0 \qquad \ldots(1.9)$$

has the additional damping term which gives this time-dependence. Solutions involving imaginary exponents are used to describe the damped oscillatory motion with the decaying amplitude described by a real negative exponent. The overdampened system, which returns slowly to equilibrium without oscillation, is described simply by the real negative exponent. We shall later come across many exponential solutions representing waves and will find these particularly useful in dealing with electron waves in solids.

1.2. Travelling waves

When a stone is thrown into a pond a pulse or group of waves is propagated in the surface of the water, each particle of the water near the surface being in turn set into oscillation as the waves pass through. In a similar manner if any part of an elastic or deformable medium is displaced and then released a travelling wave will be set up in the medium. If the displacement of the particle in the path of the wave is in a plane normal to the direction of propagation of the wave, as for example in a wave on a stretched string, the wave is termed transverse. Conversely when the displacement lies along the direction of propagation, such as for example in a wave travelling along a spiral spring, the wave is termed longitudinal. In other cases a more complex displacement may occur, e.g. in surface waves on water the particle motion is elliptical. Electromagnetic waves have transverse electric and magnetic vectors whose directions determine the direction of polarisation of the radiation. In contrast in a wave description of electrons we specify only the direction of motion of the waves and the wave displacement is then a scalar quantity.

1.2.1 *Representation of travelling waves*

Let us now consider the form of a function which can represent a travelling wave. This *wave-function* will be required to describe the system at any position and time. In order to simplify the approach let us consider the one-dimensional case of a wave travelling along the x direction. If we let the wave-function at $t = 0$ be given by

$$\Psi = f(x) \qquad \ldots(1.10)$$

then this equation gives the profile of the wave. This may, for example, be of sinusoidal form, or as in Fig. 1.3 be simply a single pulse. If this pulse is travelling forward unchanged in form with a velocity v the wave-function at any time t will be given by

$$\Psi = f(x - vt). \qquad \ldots(1.11)$$

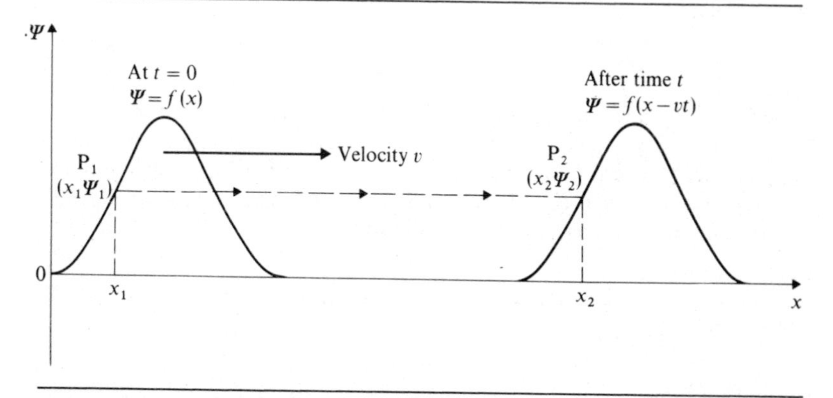

Fig. 1.3 The wave profile $\Psi = f(x)$ travels forward with velocity v and is represented at any time t by $\Psi = f(x - vt)$.

This is justified as follows—consider a point P_1 on the pulse whose coordinates at time $t = 0$ are (x_1, Ψ_1) so that for this point

$$\Psi_1 = f(x_1). \qquad \ldots(1.12)$$

Since the velocity of any point on the wave is v, after a time t, x_1 is increased to $(x_1 + vt)$ and, *assuming* that equation (1.11) holds, the ordinate Ψ_2 of the same point P_2 will be

$$\Psi_2 = f(x_2 - vt) = f(\overline{x_1 + vt} - vt) = f(x_1) = \Psi_1. \qquad \ldots(1.13)$$

Hence, *provided* equation (1.11) is obeyed, the ordinate remains unchanged as the wave moves forward. We can clearly repeat the argument for all points on the pulse and consequently equation (1.11) must be the appropriate form of equation to represent a wave propagated forward unchanged with a constant velocity v. It may similarly be verified that

$$\Psi = f(x + vt) \qquad \ldots(1.14)$$

represents a wave travelling in the negative direction of x.

Implicit in this discussion is that the velocity v, which is termed the *wave* or *phase velocity*, is the same for all points of the profile. There are some important cases in which this assumption is not valid and the pulses propagated suffer a progressive change of form. The phenomenon is termed dispersion and it will be discussed again in Section 1.5.2.

The most important type of wave whose propagation is of interest is that of harmonic waves. Let us take as an example the waves propagated along a stretched string from a source located at $x = 0$ and oscillating with SHM given by

$$\Psi = A\cos\left(\omega t + \frac{\pi}{2}\right) \qquad \ldots(1.15)$$

$$= A\sin(-\omega t).$$

The form chosen for this equation is such that time is measured from the point marked M in Fig. 1.1. A harmonic wave is propagated along the string and the displacement Ψ at any position of the string at any instant of time will be given by a function of the form of equation (1.11). Since the travelling wave-equation must also reduce to equation (1.15) at $x = 0$ it must be

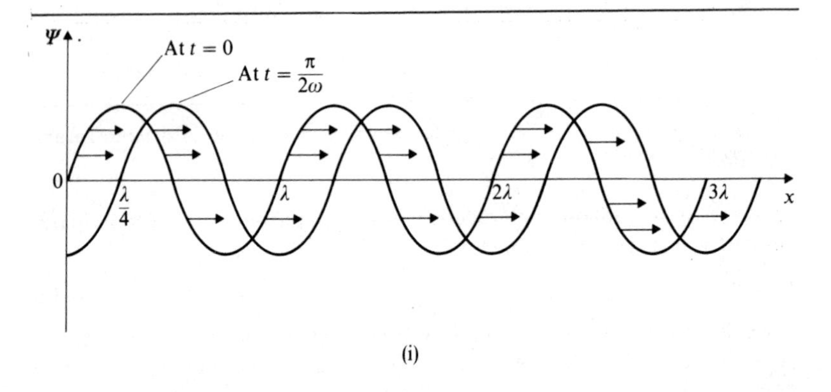

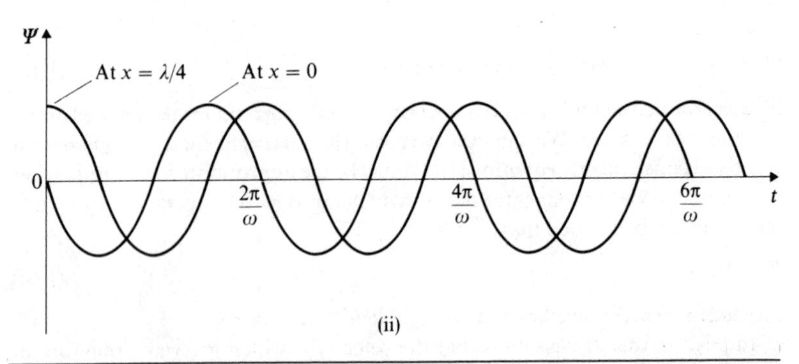

Fig. 1.4 A periodic wave profile shown in (i) is propagated forward unchanged. After a time $t = \pi/2\omega$ the profile has moved forward a distance $\lambda/4$. The displacement in (ii) varies sinusoidally at each point in the track of the wave. The phase of the wave at $x = \lambda/4$ lags that at the point $x = 0$ by $\pi/2\omega$.

$$\Psi = A \sin k(x - vt)$$

or ...(1.16)

$$\Psi = A \sin(kx - \omega t),$$

where

$$v = \frac{\omega}{k}. \qquad ...(1.17)$$

k is termed the propagation constant of the wave. The travelling wave is illustrated in Fig. 1.4.

The wave profile at $t = 0$ will be the function

$$\psi = A \sin kx \qquad ...(1.18)$$

from which we note that the wave repeats itself at repeated distances of

$$\lambda = \frac{2\pi}{k}. \qquad ...(1.19)$$

λ is termed the wavelength. It should be pointed out that k is sometimes defined as $1/\lambda$, in which case it represents the number of waves in unit distance, i.e. it is then a wavenumber. Inherent in equations (1.5), (1.17) and (1.19) is the fundamental relationship for phase velocity

$$v = \nu\lambda = \frac{\omega}{k}, \qquad ...(1.20)$$

which gives the rate at which a point on the wave of given phase moves forward.

The treatment above is applicable to all waves propagated without dispersion. In a dispersive medium another velocity, termed *group velocity*, is defined and the relationship between wave and group velocity will be discussed in Section 1.5.2.

The wave-function Ψ of equation (1.16) represents a harmonic wave and is periodic in space and time. It represents a wave travelling forward unchanged in form, each point in the track of the wave performing a simple harmonic displacement and the wave profile at any instant being sinusoidal in form.

As in the case of the SHM equation, we again note that alternative forms to equation (1.16) are equally valid representations of travelling waves. Thus if an initial phase angle ε is included, the general equation

$$\Psi = A \sin(kx \pm \omega t + \varepsilon) \qquad ...(1.21)$$

may be used to obtain the most appropriate form for any particular problem. Alternatively the exponential form

$$\begin{aligned} \Psi &= A \exp \mathrm{i}(kx - \omega t) \\ &= A \exp \mathrm{i}kx \exp(-\mathrm{i}\omega t) \end{aligned} \qquad ...(1.22)$$

is, in accordance with equation (1.11), suitable to represent a travelling wave. The comments made regarding the SHM equation (1.7) are equally valid here so that we may use this equation as it stands, or the trigonometrical solutions derived from it.

1.3 The wave equation in one dimension

The wave functions

$$\Psi = f(x - vt) \quad \ldots(1.11)$$

$$\Psi = f(x + vt) \quad \ldots(1.14)$$

and the specific forms they have taken, such as equation (1.22), are solutions to the *wave equation*

$$\frac{\partial^2 \Psi}{\partial x^2} = \frac{1}{v^2} \frac{\partial^2 \Psi}{\partial t^2}, \quad \ldots(1.23)$$

as can be verified by differentiation and substitution. This differential wave equation can be derived directly for specific examples such as waves propagated on strings (see, e.g., French, 1971).

The most general solution to the wave equation (1.23) is

$$\Psi = f(x - vt) + f(x + vt) \quad \ldots(1.24)$$

and we shall consider solutions of this form when dealing with stationary waves. It is important to note that the wave equation (1.23) in the form given is applicable only to systems in which the wave velocity v, given by ω/k, is constant and independent of the angular frequency ω, i.e. it is applicable provided there is no dispersion of the waves. This will be the case, for example, for sound waves and for the propagation of electromagnetic waves *in vacuo*. In such cases the wave profile is propagated unchanged. In the case of electron waves, which is of major interest to us, we will find that there is dispersion and that a differential wave equation different in form from equation (1.23) is valid instead. We will however find the harmonic solutions, as for example equation (1.22), to be equally valid for the new equation.

1.4 The wave equation in three dimensions

We can generalise the treatment in the previous section to the consideration of waves in three dimensions. If a plane wave is travelling with velocity v in a direction having direction cosines (l, m, n) it will be represented by

$$\Psi = f(lx + my + nz - vt). \quad \ldots(1.25)$$

At any instant of time all points having the same phase and displacement will lie on a plane. This plane is normal to the direction of propagation and travels

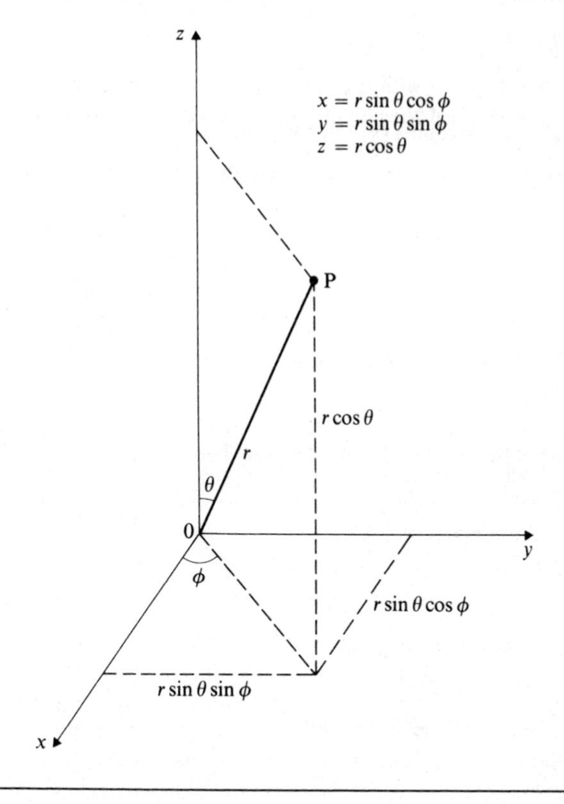

Fig. 1.5 The relationship between Cartesian and spherical polar coordinates.

forward with the velocity of the wave. The wave-function of equation (1.25) is a solution of the three-dimensional form of the wave equation

$$\frac{\partial^2 \Psi}{\partial x^2}+\frac{\partial^2 \Psi}{\partial y^2}+\frac{\partial^2 \Psi}{\partial z^2}=\frac{1}{v^2}\frac{\partial^2 \Psi}{\partial t^2}$$

or simply ...(1.26)

$$\nabla^2 \Psi=\frac{1}{v^2}\frac{\partial^2 \Psi}{\partial t^2},$$

where ∇^2 is Laplace's operator, and may be viewed as the particular mathematical operation carried out on the wave displacement Ψ. The equation is valid only for non-dispersive waves.

The solution to the wave equation (1.26) for harmonic waves is particuarly useful. If the complex exponential type of solution is used the wave-function will take the form

$$\Psi = A\exp[ik(lx+my+nz)]\exp(-i\omega t). \quad \ldots(1.27)$$

This should be compared with the one-dimensional form of the wave-function given by equation (1.22). A considerable simplification is obtained here by the use of vector notation.

We replace k by the wave-vector $\boldsymbol{k}$ whose components along the three coordinate axes are

$$\begin{aligned} k_x &= kl \\ k_y &= km \\ k_z &= kn. \end{aligned} \quad \ldots(1.28)$$

Bearing in mind that for direction cosines

$$l^2+m^2+n^2 = 1,$$

the magnitude of the wave-vector will be given by

$$\begin{aligned} |\boldsymbol{k}| &= (k_x^2+k_y^2+k_z^2)^{\frac{1}{2}} \\ &= [k^2(l^2+m^2+n^2)]^{\frac{1}{2}} \\ &= k = \frac{2\pi}{\lambda}, \end{aligned} \quad \ldots(1.29)$$

which confirms the procedure adopted.

The wave-function (1.27) then becomes

$$\begin{aligned} \Psi &= A\exp[i(k_x x+k_y y+k_z z)]\exp(-i\omega t) \\ &= A\exp i(\boldsymbol{k}\cdot\boldsymbol{r}), \end{aligned} \quad \ldots(1.30)$$

where

$$r = |\boldsymbol{r}| = (x^2+y^2+z^2)^{\frac{1}{2}} \quad \ldots(1.31)$$

and $\boldsymbol{k}\cdot\boldsymbol{r}$ is the usual scalar product of the two vectors.

This wave-function is also a solution to the differential equation representing electron waves and its use in this form, and other modified forms, is of considerable importance in studying the behaviour of electrons in solids.

In addition to plane wave solutions to the wave equation, one may also have solutions representing cylindrical waves spreading out from a line source, or those for spherical waves radiating from a point source. For dealing with this latter problem it is convenient to convert the wave equation (1.26) from Cartesian to spherical polar coordinates. From Fig. 1.5 we have the following relationships:

$$\begin{aligned} x &= r\sin\theta\cos\phi \\ y &= r\sin\theta\sin\phi \\ z &= r\cos\theta \end{aligned} \quad \ldots(1.32)$$

which yields the spherical polar form

$$\left[\frac{1}{r^2}\frac{\partial}{\partial r}\left(r^2\frac{\partial}{\partial r}\right)+\frac{1}{r^2\sin\theta}\frac{\partial}{\partial\theta}\left(\sin\theta\frac{\partial}{\partial\theta}\right)+\frac{1}{r^2\sin^2\theta}\frac{\partial^2}{\partial\phi^2}\right]\Psi=\frac{1}{v^2}\frac{\partial^2\Psi}{\partial t^2}. \qquad \ldots(1.33)$$

The transformation is rather lengthy and will not be reproduced here. It is given in many standard texts (see e.g. Speigel, 1959). For a spherically symmetrical wave, the solution to equation (1.33) must be independent of both θ and ϕ and the wave equation will then take on the simpler form

$$\frac{\partial^2\Psi}{\partial r^2}+\frac{2}{r}\frac{\partial\Psi}{\partial r}=\frac{1}{v^2}\frac{\partial^2\Psi}{\partial t^2}. \qquad \ldots(1.34)$$

The general solution for spherical waves then becomes

$$\Psi=\frac{1}{r}f(r-vt)+\frac{1}{r}g(r+vt), \qquad \ldots(1.35)$$

the first term representing a train of diverging waves and the second a train of converging waves.

1.5 Stationary waves

If at a given position two simple harmonic trains of waves are passing, the resultant wave displacement at that point is given by the sum of the displacements of the individual waves. This is termed the *superposition principle.* It is obeyed because of the direct proportionality between the wave displacement and the applied force which leads to a wave-function linear in Ψ.

Let us now consider the superposition of two waves of equal amplitudes and frequencies travelling in opposite directions with the same speeds. The two waves may be represented by

$$\Psi_1=a\sin(kx-\omega t) \qquad \ldots(1.36)$$

$$\Psi_2=a\sin(kx+\omega t) \qquad \ldots(1.37)$$

and we note for future reference that other forms of the waves, implying a different phase relationship between them, could alternatively have been chosen. In accordance with the superposition principle the resultant wave-function will be given by

$$\begin{aligned}\Psi&=\Psi_1+\Psi_2\\&=a\sin(kx-\omega t)+a\sin(kx+\omega t)\\&=2a\sin kx\cos\omega t.\end{aligned} \qquad \ldots(1.38)$$

The general form of this equation is shown in Fig. 1.6 and has several interesting features. First, we note that at a series of equally spaced points the

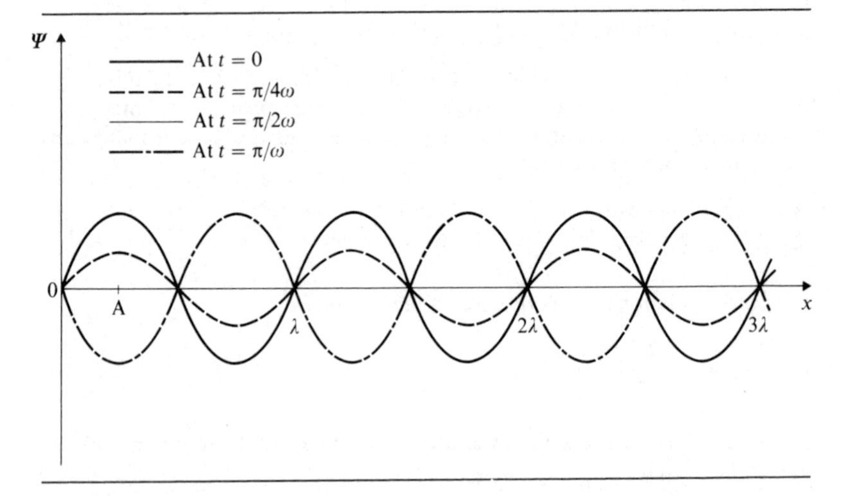

Fig. 1.6 The stationary wave $\Psi = 2a \sin kx \cos \omega t$. If the origin is moved to A the wave is represented by $\Psi = 2a \cos kx \cos \omega t$.

wave-function has zero displacement at all times. These points are termed nodes and the solution (1.38) represents a stationary or standing wave. At all other points the displacement is governed by the time factor $\cos \omega t$ and the wave is therefore harmonic. The space factor $2a \sin kx$ gives the amplitude of the wave at any point and imposes a sinusoidal wave profile at all times, whereas the time factor determines the scale of the profile at any instant. The wave-function has a spatial periodicity of $2\pi/k = \lambda$ since the phase angle kx is repeated whenever x increases by $2\pi/k$. The position of the nodes is given by

$$x = 0, \frac{\lambda}{2}, \lambda \ldots (n-1)\frac{\lambda}{2} \qquad (n = 1, 2, 3 \ldots) \qquad \ldots(1.39)$$

so that the nodes, and intermediate antinodes, are one half wavelength apart.

We have thus found a stationary wave to be formed by the superposition of identical waves travelling in opposite directions. In practice this superposition takes place between a wave incident on a discontinuity and a wave reflected back from the discontinuity. The conditions at the discontinuity determine the phase relationship between the waves. Typical examples of stationary waves are those on strings, the vibrations of air columns and standing electromagnetic waves in wave guides. We will also find that atomic states can be represented by stationary waves.

Let us extend the consideration of the specific example chosen. Inherent in the choice of the two waves in equations (1.36) and (1.37) is an implicit *boundary condition* of the form that

$$\Psi_1 = \Psi_2 = 0 \text{ at } x = 0, t = 0. \qquad \text{...(1.40)}$$

A different boundary condition could have been imposed and we will shortly return to this point. However let us first see the effect of setting an *additional* boundary condition

$$\Psi_1 = \Psi_2 = 0 \text{ at } x = L, t = 0, \qquad \text{...(1.41)}$$

which together with condition (1.40) would be appropriate for examining the modes of vibration of a string of length L attached at each end. Equation (1.38) will now only be satisfied provided that the additional condition

$$k = \frac{n\pi}{L} \quad \text{or} \quad \lambda = \frac{2L}{n}, \qquad \text{...(1.42)}$$

where n is integral, is obeyed. This condition therefore governs the possible values of wavelength and the corresponding allowed wave-functions for the stationary states of the system. This type of behaviour is quite general and we shall see in Chapter 5 how the boundary conditions for a model representing electron waves in a crystal will similarly yield a set of allowed wave-functions.

So far we have considered only one particular way of combining two wave-functions to form a standing wave. If instead the sine functions of the waves of equations (1.36) and (1.37) are replaced by cosine functions this will be equivalent to changing their relative phases and this will correspond to the boundary condition

$$\Psi_1 = \Psi_2 = a \text{ at } x = 0, t = 0. \qquad \text{...(1.43)}$$

It may be verified that the resultant wave-function will be

$$\Psi = 2a \cos kx \cos \omega t \qquad \text{...(1.44)}$$

which has antinodes at $x = 0, \lambda/2$, etc. instead of nodes, i.e. the standing wave has been displaced along the x-axis by one half wavelength.

It is of interest to note that by changing the phase of Ψ_1 in equation (1.36) to $(kx - \omega t + \pi)$ this is equivalent to changing Ψ_1 to $-\Psi_1$, i.e. a standing wave can be obtained as a *difference* between Ψ_2 and Ψ_1. This yields

$$\Psi = 2a \cos kx \sin \omega t. \qquad \text{...(1.45)}$$

When dealing with stationary waves one is primarily interested in the space factor so that equation (1.45) will represent the same physical situation as equation (1.44). Any other phase difference introduced between Ψ_1 and Ψ_2 will merely have the effect of transferring the standing wave along the x-axis. Thus by suitable choice of origin an equation such as (1.38) or (1.44) may conveniently be used to represent any system.

1.5.1 *Stationary waves by separation of variables*

The standing wave solutions we have found, such as for example equation (1.38), are all of the form

$$\Psi = \psi(x)T(t), \qquad \ldots(1.46)$$

where $\psi(x)$ is a function independent of time and $T(t)$ is independent of position. Wave-functions capable of being expressed in this form are termed stationary wave-functions and when the function $\psi(x)$ is real will represent standing waves.

Let us consider the one-dimensional wave equation

$$\frac{\partial^2 \Psi}{\partial x^2} = \frac{1}{v^2}\frac{\partial^2 \Psi}{\partial t^2} \qquad \ldots(1.23)$$

and assume a separable solution as in equation (1.46) of the form

$$\Psi = \psi(x)\exp(-\mathrm{i}\omega t), \qquad \ldots(1.47)$$

where the function $\psi(x)$ is independent of time. Differentiating with respect to time we have

$$\frac{\partial^2 \Psi}{\partial t^2} = -\omega^2 \Psi \qquad \ldots(1.48)$$

and on substituting in equation (1.23) it follows that

$$\frac{\partial^2 \Psi}{\partial x^2} + k^2 \Psi = 0$$

or

$$\frac{\mathrm{d}^2 \psi}{\mathrm{d}x^2} + k^2 \psi = 0. \qquad \ldots(1.49)$$

For three dimensions the corresponding equation

$$\nabla^2 \psi + k^2 \psi = 0 \qquad \ldots(1.50)$$

is obtained.

We note here the implicit assumption that there is no dispersion of the waves since we are putting $k = \omega/v$. Equations (1.49) and (1.50) have wider validity than might be inferred from this derivation. These derived equations are obeyed for wave propagation in dispersive media for which v is a function of ω and when the wave equation (1.23) must take on a different form.

It is readily verified that the solution to equation (1.49) for the wave-function ψ must be of the form of complex exponentials or of the sine and cosine functions derivable from them.

The general solution to equation (1.49) is

$$\psi = A\exp(\mathrm{i}kx) + B\exp(-\mathrm{i}kx) \qquad \ldots(1.51)$$

so that the complete standing wave equation is

$$\Psi = [A\exp(ikx) + B\exp(-ikx)]\exp(-i\omega t). \qquad \ldots(1.52)$$

In this form it is seen to represent the superposition of two waves travelling in opposite directions. If we put $A = B = a$ and take the real part we obtain

$$\Psi = 2a\cos kx\cos\omega t. \qquad \ldots(1.44)$$

Any of the other geometrical forms of solution can be obtained in a similar manner, noting that A and B may be complex and that either the real or imaginary part of equation (1.52) may be used. Since it is not necessary that the magnitudes of the constants A and B are equal, the general solution of equation (1.52) may also be interpreted as a wave having both a standing wave and a travelling wave component. In three dimensions the solutions are independent for each direction and we have the possibility of a travelling wave in one direction, a standing wave in another and zero displacement in the third.

The solution in the form of equation (1.52) with either A or $B = 0$ is a valid separable solution of the wave equation (1.23), and clearly represents a travelling wave. The ψ part of the wave, which is now complex, is independent of time so that the complete function Ψ still represents a stationary condition of the system. This may be contrasted with the time-dependence obtained when a non-periodic wave such as a pulse traverses a system. When we come to deal with the wave representation of electrons in crystals we will normally find that the factor $\exp(-i\omega t)$ is omitted both when dealing with standing waves and also for travelling waves.

1.5.2 *Phase velocity and group velocity*

Waves propagated without dispersion have a velocity which is independent of frequency given by

$$v = \frac{\omega}{k}. \qquad \ldots(1.20)$$

This gives the rate at which the energy of the wave is transmitted. In a dispersive medium velocity is now a function of frequency and the energy of the wave is propagated at a different velocity termed the *group velocity*.

Let us now consider how the group velocity arises. If we have a source emitting simultaneously two notes of nearly equal frequencies ν and $\nu + \delta\nu$, maxima of amplitude, termed beats, will occur $\delta\nu$ times every second and will be propagated from the source. If the phase velocities of the waves of frequencies ν and $\nu + \delta\nu$ are identical then the beat frequency will also have this velocity. The velocity of the *group* of waves is the same as the phase velocity. The wave received can be regarded as a single wave which is amplitude-modulated and for this particular example the modulation velocity is the same as that of the wave itself.

Consider next two waves W_1 and W_2 which now have different phase velocities v and $v + \delta v$ corresponding to these two different wavelengths λ and

$\lambda+\delta\lambda$. Referring to Fig. 1.7, the two waves are shown initially in phase at $t = 0$ at the plane A_1A_2. This will define a position of maximum displacement of the combined wave. Let the waves now move forward. Since the velocity of wave W_2

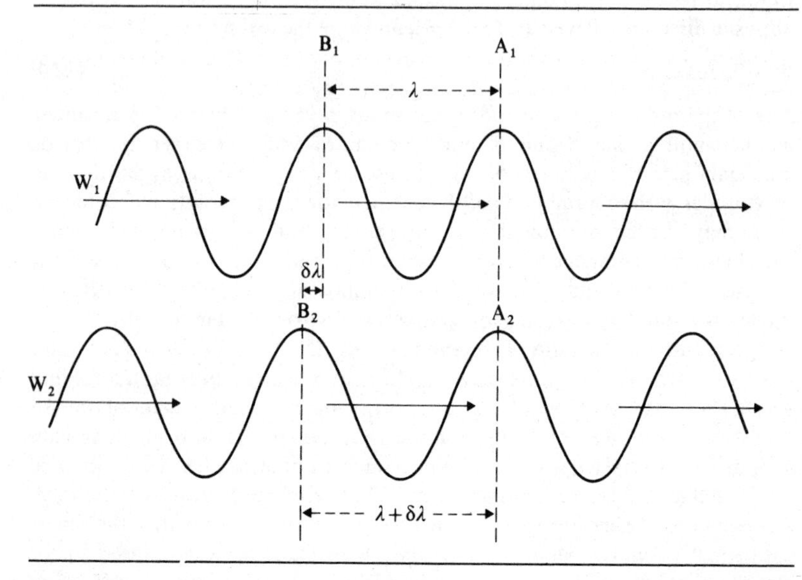

Fig. 1.7 The waves W_1 and W_2 have slightly different wavelengths and travel forwards with slightly different velocities. At $t = 0$ the two waves coincide in phase at plane A_1A_2. Due to the dispersion the wave coincidence will be occurring at a slightly later instant at the plane B_1B_2.

is greater than that of wave W_1 at some later instant of time t, instead of A_1 and A_2 coinciding we have B_1 and B_2 and these will now define the displacement maximum. The *relative* velocity of the two waves is δv so that the initial separation $\delta\lambda$ of the crests B_1 and B_2 must be

$$\delta\lambda = t\delta v. \qquad \text{...(1.53)}$$

In the same time, wave W_1 has moved forward a distance vt and the displacement maximum a distance one wavelength *less*, i.e. $vt-\lambda$. The velocity of the displacement maximum is thus

$$u = v - \frac{\lambda}{t}$$

$$= v - \lambda\frac{\delta v}{\delta\lambda} \qquad \text{...(1.54)}$$

by using equation (1.53).

If instead of two waves being superimposed we have a dispersive medium for which v is a function of λ then we can define a *group velocity* given by

$$u = v - \lambda \frac{dv}{d\lambda}, \qquad \ldots(1.55)$$

where $dv/d\lambda$ is now a measure of the dispersion of the waves.

Alternative forms of this relationship may be obtained. Thus by differentiating equation (1.20) and substituting in equation (1.55) we have

$$u = -\lambda^2 \frac{d\nu}{d\lambda} \qquad \ldots(1.56)$$

and using equations (1.5) and (1.19)

$$u = \frac{d\omega}{dk}. \qquad \ldots(1.57)$$

This latter form is particularly useful since dispersion relations are often in the form $\omega = \omega(k)$. The correspondence between this expression for group velocity and that in equation (1.20) for phase velocity is illustrated in Fig. 1.8. This is the

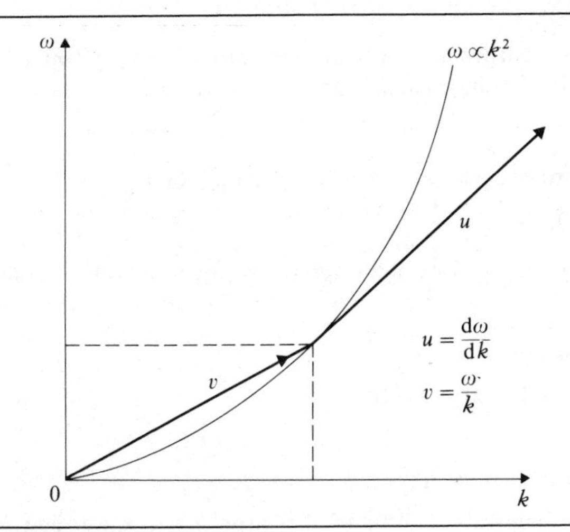

Fig. 1.8 Dispersion relation for electron waves on free electron theory.

type of dispersion relationship obtained for an electron in a crystal. It will be discussed again in Section 1.6.3.

An alternative method of arriving at the group velocity is to consider the superposition of two travelling waves:

$$\Psi_1 = a \exp i(kx - \omega t)$$
$$\Psi_2 = a \exp i[(k + \delta k)x - (\omega + \delta\omega)t] \qquad \ldots(1.58)$$

which give the wave

$$\Psi = \Psi_1 + \Psi_2 = a \exp \mathrm{i}(kx - \omega t)[1 + \exp \mathrm{i}(x\delta k - t\delta\omega)]. \qquad \ldots(1.59)$$

This represents a travelling wave whose amplitude is modulated by the second factor in the expression. By taking the real part of this factor we find that the amplitude varies between a maximum of $2a$ and a minimum of zero. The form of the factor shows that the modulation is propagated forward with a velocity $\delta\omega/\delta k$, which is in accord with the previous discussion.

We may return briefly to consider the form of the wave equation for dispersive media. The general wave equation is of the form

$$\nabla^2 \Psi + g^2(q)\Psi = a\frac{\partial \Psi}{\partial t} + b\frac{\partial^2 \Psi}{\partial t^2}. \qquad \ldots(1.60)$$

where the coefficients a and b are independent of space and time and one or other of which may vanish according to the system being described. The function $g^2(q)$ can be space-dependent and for electron waves we will find that it relates the wave function to the potential energy of the electron.

If we look for separable solutions these must be of the form

$$\Psi = \psi(\boldsymbol{r})T(t), \qquad \ldots(1.61)$$

where $\psi(\boldsymbol{r})$ is independent of time and $T(t)$ is independent of the space coordinates. Proceeding as in the one-dimensional case we assume

$$T(t) = \exp(-\mathrm{i}\omega t) \qquad \ldots(1.62)$$

so that the derived equation is then obtained for ψ in the form

$$\nabla^2 \psi + k^2 \psi = 0, \qquad \ldots(1.50)$$

i.e. of the same form as for a non-dispersive medium. It may be verified that k is given by

$$k = \pm(g^2 + \mathrm{i}a\omega + b\omega^2)^{\frac{1}{2}} \qquad \ldots(1.63)$$

so that the wave velocity v will be

$$v = \frac{\omega}{k} = \pm\frac{\omega}{(g^2 + \mathrm{i}a\omega + b\omega^2)^{\frac{1}{2}}}. \qquad \ldots(1.64)$$

Equation (1.63) gives the form of the dispersion relation and only when $g^2 = a = 0$ can we have non-dispersive waves such as electromagnetic waves in vacuum. In ionospheric propagation g is related to ω_p, the plasma oscillation frequency, and dispersion is observed.

1.6 The electron as a wave

1.6.1 *The Einstein photoelectric equation*

Effects such as interference and diffraction are characteristic phenomena exhibited by all types of waves and in the case of optics may be explained fully in terms of a wave theory of light. Other phenomena exhibited by light must however be explained in terms of a particle behaviour, i.e. in terms of the photon concept.

The most important is the photoelectric effect. This is concerned with the emission of electrons from a clean metal surface *in vacuo* when it is irradiated by light of a suitable wavelength. The main feature of the effect is that an increase in light intensity is accompanied by an increase in electron emission but *not* in electron energy. This observation cannot be explained in terms of classical theory. As illustrated in Fig. 1.9, the minimum retarding voltage U_0 necessary to

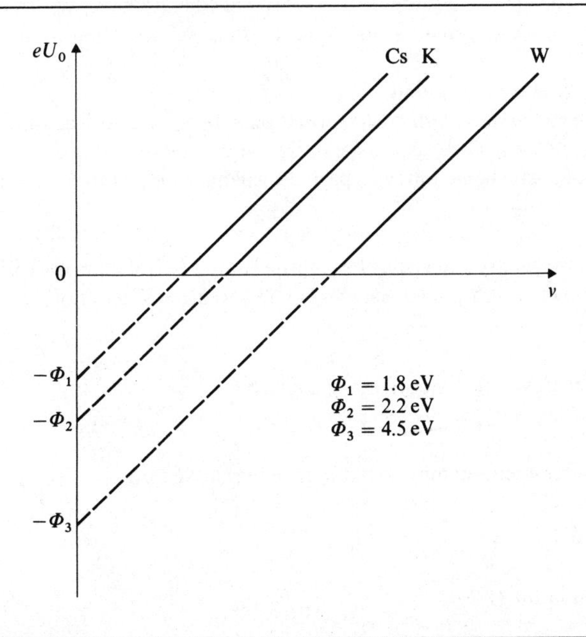

Fig. 1.9 The photoelectric effect. The different metals yield linear graphs of equal slope h, Planck's constant. The intercepts give values of the work functions of the metals.

prevent the collection of electrons is measured as a function of the light frequency ν for various metal emitters. Einstein postulated that the light consisted of a stream of photons each of energy E given by

$$E = h\nu, \qquad \ldots(1.65)$$

where h is Planck's constant and has a value of 6.626×10^{-34} Js. The kinetic energy of each emitted electron is thus given as a difference between the energy $h\nu$ of an incident photon and the energy necessary to remove the electron from the metal. The latter will vary according to the binding state of the electron and will be smaller for some exposed crystallographic planes than for others. There will, however, always be a minimum binding energy for each metal which is termed its work function Φ. Electrons bound with this energy will attain maximum kinetic energy on release and will require the retarding potential U_0 to prevent their collection. The photoelectric equation thus becomes

$$eU_0 = h\nu - \Phi \qquad \ldots(1.66)$$

and is in excellent agreement with experiment.

Planck initially postulated that radiation consisted of quanta of energies $h\nu$ in order to explain the form of the wavelength distribution of electromagnetic radiation from a heated body. This is the so-called black-body spectrum and the agreement with experiment formed an early success for the quantum theory.

1.6.2 *The de Broglie postulate*

We have seen that electromagnetic radiation may behave both as waves and as a stream of photons. These aspects of radiation may be summarised as follows.

A photon, in its behaviour as a particle, will have momentum p given by

$$p = mc, \qquad \ldots(1.67)$$

where c is its velocity and m may be regarded as its equivalent mass arising due to its kinetic energy. It has zero rest mass and its total (or kinetic) energy will be

$$E = mc^2. \qquad \ldots(1.68)$$

Since the energy of the photon is also given by

$$E = h\nu, \qquad \ldots(1.65)$$

the photon momentum may be put in the alternative form

$$p = \frac{h\nu}{c} \qquad \ldots(1.69)$$

or using equation (1.20)

$$p = \frac{h}{\lambda}. \qquad \ldots(1.70)$$

Summarising, electromagnetic radiation may behave both as waves of frequency ν and wavelength λ, or as photons having energy $h\nu$ and momentum h/λ.

De Broglie proposed that an analogous relationship to that of equation (1.70) would apply to actual particles, i.e. that the duality found in the case of radiation also extended to particles. Thus according to the postulate a particle of mass m and velocity u would have an associated wavelength λ given by

$$\lambda = \frac{h}{p} = \frac{h}{mu}. \qquad \ldots(1.71)$$

The correctness of the de Broglie relationship was first confirmed by Davisson and Germer who showed by experiment that electrons did indeed behave as waves. In the experiments the electrons were diffracted by the regular array of atoms in a single metal crystal, and the electron wavelengths measured were found to agree with that expected from the de Broglie relationship. Electron diffraction now forms an important and everyday technique for the investigation of the surfaces of a wide range of materials. The wave properties of atoms and molecules have similarly been confirmed, as have those of neutrons. The techniques of electron diffraction and neutron diffraction thus complement those of X-ray diffraction in the investigation of the structure of materials.

1.6.3 *The dispersion of electron waves*

In order to represent the electron by a wave, or group of waves, we require to be able to state whether the wave will show dispersion and if so what form the dispersion relation takes.

Let us consider a single electron of mass m_e moving with velocity u. As we shall see later there are fundamental difficulties concerning the simultaneous observation of both position and momentum of a single electron, but this will not affect the argument here. It is convenient to develop the theory in terms of $k(=2\pi/\lambda)$ rather than λ and to write

$$\hbar = \frac{h}{2\pi}. \qquad \ldots(1.72)$$

From the de Broglie relationship (1.71) we therefore have

$$p = m_e u = \hbar k \qquad \ldots(1.73)$$

and following a non-relativistic approach, the kinetic energy K of the electron is then

$$K = \tfrac{1}{2} m_e u^2 = \frac{\hbar^2 k^2}{2m_e}. \qquad \ldots(1.74)$$

We now put the total electron energy E as

$$E = K + V \qquad \ldots(1.75)$$

where V is the potential energy of the electron, and assign a frequency to the electron in accordance with

$$E = h\nu = \hbar\omega. \qquad \ldots(1.76)$$

From these last three relationships it follows that

$$E = \hbar\omega = \frac{\hbar^2 k^2}{2m_e} + V \qquad \ldots(1.77)$$

which is the dispersion relation for de Broglie waves.

A wave equation representation of electrons and associated solutions must be compatible with this relation. Since however there is no true zero of energy it is often convenient when dealing with electron waves in a region of constant potential to put $V = 0$ and use instead

$$E = \frac{\hbar^2 k^2}{2m_e}. \qquad \ldots(1.78)$$

The energy relationship in this form is extensively used in the free electron theory of metals.

In order to represent the electron as an identifiable single particle it is necessary to make up a wave packet consisting of a group of waves having k values distributed about a mean value of $k = k_0$. This group of waves will obey the dispersion relation (1.77) so that the group velocity is

$$\frac{d\omega}{dk} = \frac{\hbar k}{m_e}, \qquad \ldots(1.79)$$

where $d\omega/dk$ is evaluated at the centre k_0 of the packet. From equation (1.73) we can now confirm the association of the group velocity with the velocity u specified for the single electron.

A difficulty arises however when we come to calculate the phase velocity. If we assume $V = 0$, as for free electrons, the phase velocity v determined from the dispersion relation is

$$v = \frac{\omega}{k} = \frac{\hbar k}{2m_e} = \frac{u}{2}. \qquad \ldots(1.80)$$

This in itself would be perfectly feasible and would not be dissimilar to surface waves in water where the individual waves in this case travel forward through the group. However, the potential energy V can be assigned any value we wish and, in accordance with the dispersion relation, there is no single value of phase velocity. Consideration of equations (1.75) and (1.76) brings out a similar difficulty regarding the assigning of a definite frequency to an electron. As might therefore be expected we find that neither the phase velocity nor the frequency of an electron can be determined experimentally. We may also note here that a relativistic approach leads to a different dispersion relation from that of equation (1.77) and its form is such that the electron rest mass energy equivalent is included within the total energy E. This approach gives the same group velocity for the electron as obtained here but leads to a totally different phase velocity of c^2/u and a different electron frequency.

1.7 Heisenberg's uncertainty principle

Inherent in the dual concepts of the wave representation and the particle behaviour of electrons is a fundamental difficulty regarding the specification of both the position and momentum of an electron simultaneously. The *Heisenberg uncertainty principle* states that if Δx and Δp are the uncertainties in the simultaneous determinations of position and momentum of a particle then

$$\Delta x \cdot \Delta p \geqslant \frac{\hbar}{2}. \qquad \ldots(1.81)$$

This is sometimes referred to as the *indeterminacy principle.* The magnitude of Planck's constant determines that the principle is of importance when dealing with particles such as electrons and photons.

A simple example of the consequences of this principle is found in the possible determination of the exact location of an electron using a beam of photons. The resolving power of any system designed would be roughly limited to the wavelength of the radiation used. Thus, if a resolution of better than 0.01 nm is required γ-rays would need to be used. Under these conditions, however, although the position of the electron would be closely determined, a photon–electron interaction would occur and would result in a discontinuous change in the momentum of the electron at the instant its position is being determined (Compton effect). Similar arguments can be put forward against any other proposed methods.

The representation of an electron by a wave packet has already been briefly mentioned in the last section. This forms an interesting example of the consequences of the uncertainty principle and will now be developed. First of all, however, it will be necessary to review very briefly some of the basic concepts of Fourier analysis.

The theorem due to Fourier is concerned with the analysis of periodic functions which can be of any complexity and which may be either periodic in time (e.g.a violin note) or spatially periodic (e.g. waves on a string). It states that any single valued periodic function can be expressed as a summation of a number of harmonic terms of suitable amplitudes and of frequencies which are all multiples of the frequency of the periodic function. The development of the theory is similar for both time or spatial periodicity but will be summarised here only for spatial periodicity.

We let the wave to be analysed have periodicity λ so that

$$F(x+\lambda) = F(x) \qquad \ldots(1.82)$$

for all values of x. The Fourier theorem expressed in terms of k $(=2\pi/\lambda)$ then states that

$$F(x) = a_0 + \sum_{n=1}^{\infty} a_n \sin nkx + \sum_{n=1}^{\infty} b_n \cos nkx, \qquad \ldots(1.83)$$

where a_0, a_n and b_n are constants. These may be determined by integrating $F(x)$ and may be either positive or negative to allow for components differing in phase. The corresponding exponential form

$$F(x) = \sum_{-n}^{n} A_n \exp(inkx) \qquad \ldots(1.84)$$

is a very useful mathematical representation (although term-by-term interpretation might be taken to imply compex amplitudes A_n and negative propagation constants!). Provided $F(x)$ is real then the summation here is also real and is in every way equivalent to that in equation (1.83). An example of the use of the Fourier analysis theorem is given in Fig. 1.10.

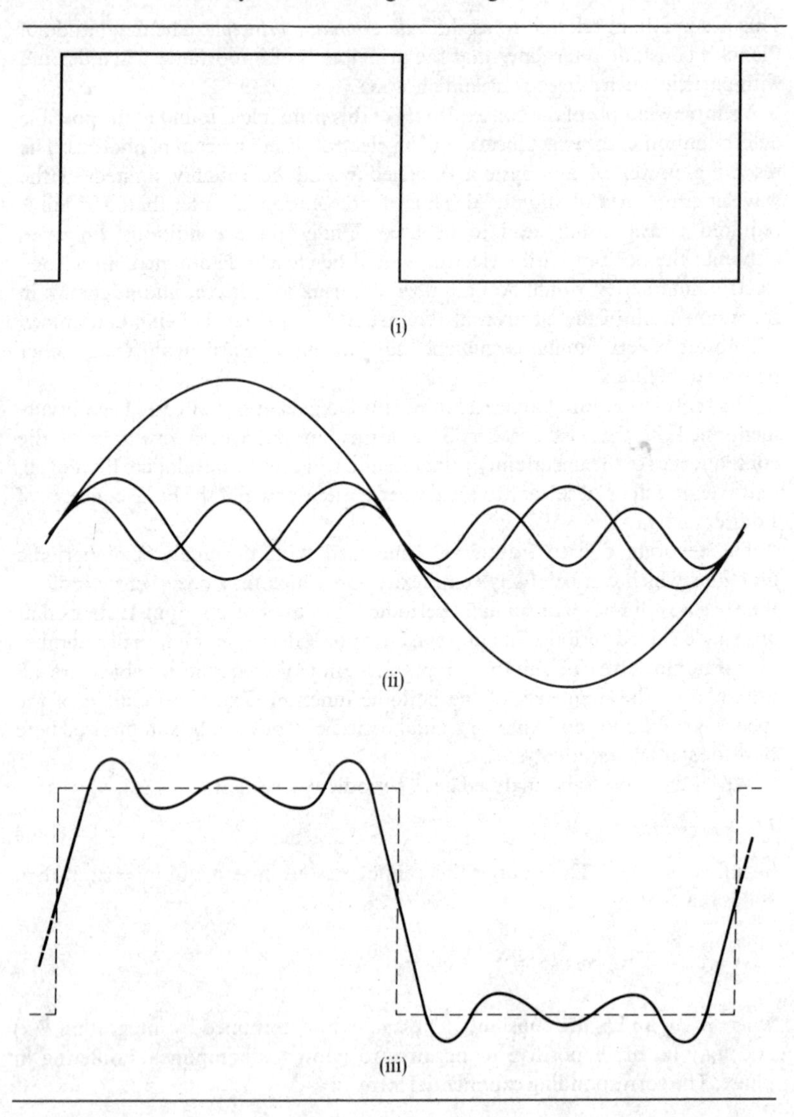

Fig. 1.10 Fourier Analysis of a square wave (as (i)). Fourier series is $F(x) = (4/n\pi)\sin kx$ ($n = 1, 3, 5 \ldots$) of which the first three terms are shown in (ii) and their superposition in (iii).

If now we let the spatial periodicity λ of $F(x)$ become very large, so that k is very small, the adjoining terms in the Fourier series become very close together. Consequently the summation is replaced by an integral of the form

$$F(x) = \int_{-\infty}^{+\infty} A(k) \exp ikx \, dk, \qquad \ldots(1.85)$$

where $A(k)$ is now a distribution function. $A(k)$ is termed the *Fourier transform* of $F(x)$ and may be shown to be given by

$$A(k) = \frac{1}{2\pi} \int_{-\infty}^{\infty} F(x) \exp(-ikx) \, dx. \qquad \ldots(1.86)$$

$F(x)$ is also the transform of $A(k)$ so that both form a Fourier transform pair.

The above is only a brief summary of Fourier analysis. Further details are given in many standard texts (see e.g. Braddick, 1965).

We can now return to our discussion of the representation of particles by waves and consider first the Gauss error function

$$F(x) = A \exp\left(\frac{-x^2}{a^2}\right) \qquad \ldots(1.87)$$

represented in Fig. 1.11. This can be regarded as a wave pulse and can be made into a travelling pulse simply by replacing x by $(x - vt)$ so that it becomes of the form of equation (1.11) for a travelling wave. In accordance with the Fourier analysis concept this pulse may be synthesised in the form of equation (1.85) where the distribution function $A(k)$ is determined using equation (1.86). By the use of the standard integral

$$\int_{-\infty}^{\infty} \exp(ax - bx^2) = \left(\frac{\pi}{b}\right)^{\frac{1}{2}} \exp\left(\frac{a^2}{4b}\right) \qquad \ldots(1.88)$$

the distribution function is found to be

$$A(k) = \frac{a}{2\pi^{\frac{1}{2}}} A \exp\left(-\frac{a^2 k^2}{4}\right). \qquad \ldots(1.89)$$

It is useful to compare the functions $F(x)$ and $A(k)$ both of which it may be noted are of the Gaussian form. The height of the function $F(x)$ is given by A and the parameter a determines its width. A half-width corresponding to the width at $1/e$ of the maximum height may be defined and is equal to a. The parameters of the transform function are clearly similarly defined and we have the interesting result that as the pulse half-width a is decreased the distribution function decreases in height and increases in width. The sharper we make the pulse the greater the range of k values necessary to represent it.

Let us now modify the distribution function so that it is centred about $k = k_0$ rather than $k = 0$. This is achieved by putting

$$A(k) = \frac{aA}{2(\pi)^{\frac{1}{2}}} \exp\left[-\frac{a^2(k - k_0)^2}{4}\right]. \qquad \ldots(1.90)$$

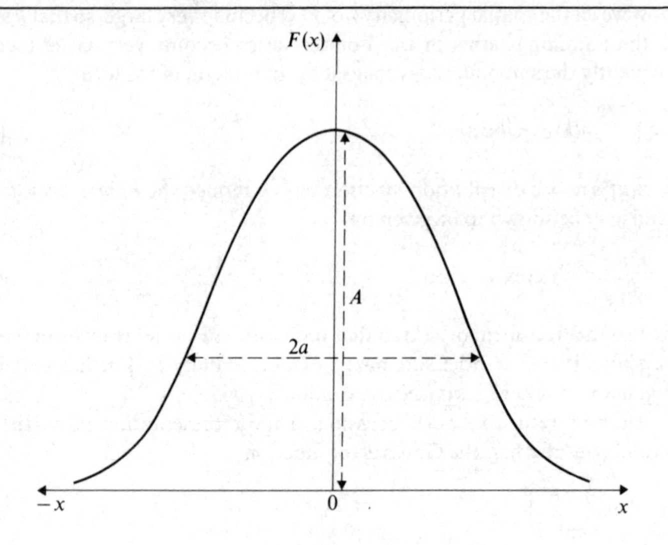

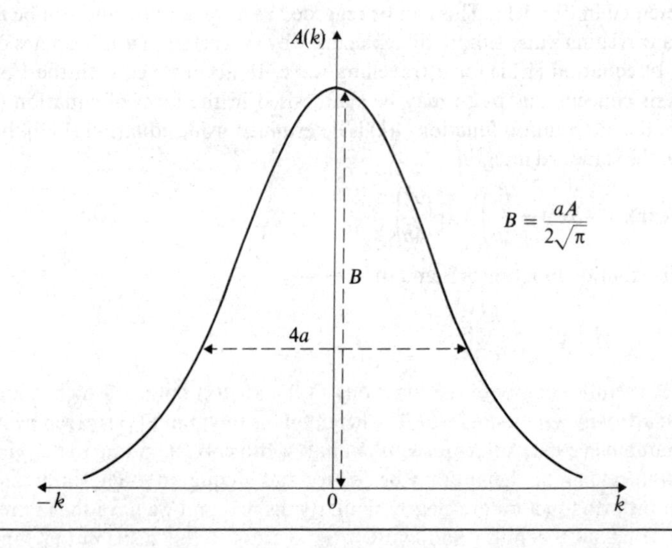

Fig. 1.11 The functions $F(x)$ and $A(k)$. Each forms the Fourier transform of the other.

The corresponding function $F(x)$ given by the Fourier transform of $A(k)$ is readily shown to be

$$F(x) = A\exp\left(-\frac{x^2}{a^2}\right)\exp ik_0x \qquad \ldots(1.91)$$

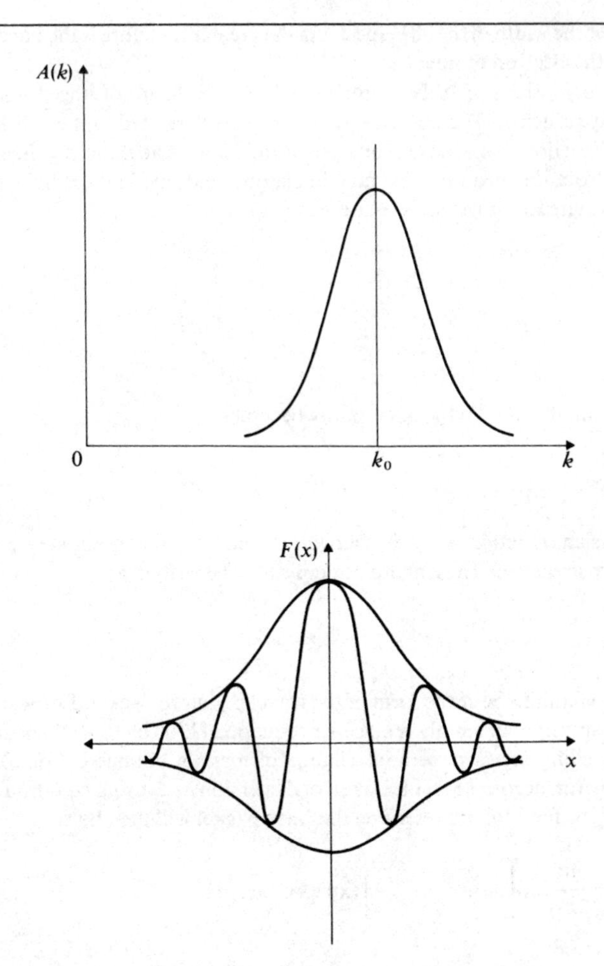

Fig. 1.12 Representation of an electron by a wave packet.

and is illustrated in Fig. 1.12. The Gaussian function now forms the envelope to a wave of wavelength $2\pi/k_0$ and a wave packet in this form may be used to represent a particle such as an electron. The uncertainty in the position of the electron is governed by the width of the wave packet, so that the narrower the packet the more closely will the position of the electron be determined. On the other hand, the $A(k)$ spectrum, whose width determines the sharpness of the wave packet, is related to the electron momentum through

$$p = \hbar k. \quad \ldots(1.73)$$

The greater the width of the $A(k)$ spectrum the greater therefore is the uncertainty in k_0 and the electron momentum.

These conclusions can be summarised in the form of the Heisenberg uncertainty principle. The uncertainty in each case is related to the width of the Gaussian function and an exact analysis using root-mean-square values of the deviation from the mean for this function shows that one half the half-width is the appropriate factor to use. We thus have

$$\Delta x = \frac{a}{2} \qquad \ldots(1.92)$$

$$\Delta k = \frac{1}{a}$$

and using equation (1.73) the uncertainty becomes

$$\Delta x \cdot \Delta p = \frac{\hbar}{2}. \qquad \ldots(1.93)$$

The Gaussian function may be shown to lead to the minimum possible uncertainty so that the Heisenberg principle may be written as

$$\Delta x \cdot \Delta p \geqslant \frac{\hbar}{2}. \qquad \ldots(1.94)$$

We may enquire about the form of the travelling wave packet. For waves *not* showing dispersion we simply replace x in equation (1.91) by $(x - vt)$ and a wave packet travelling forward without change of form is obtained. For electron waves, however, account must be taken of dispersion and, using equation (1.85), the form of the function representing the wave packet will then be

$$F(x) = \int_{-\infty}^{\infty} \frac{Aa}{2\pi^{\frac{1}{2}}} \exp\left[-\frac{a^2(k-k_0)^2}{4}\right] \exp \mathrm{i}(kx - \omega t)\,\mathrm{d}k, \qquad \ldots(1.95)$$

where

$$\omega = \frac{E}{\hbar} = \frac{\hbar k^2}{2m_e} \qquad \ldots(1.96)$$

in accordance with equation (1.78) for free electrons. The solution to equation (1.95) is rather lengthy and will not be given here (see e.g. Raimes, 1963). The analysis shows that the wave packet is propagated forward with a constant group velocity $\hbar k_0/m_e$ (in accordance with equation (1.79)). The wave profile, however, progressively flattens out, its half-width increasing in proportion to the time elapsed. The spread corresponds exactly to the uncertainty in the position of the electron to be expected at any instant due to the uncertainty in the electron velocity.

1.8 The Schrödinger wave equation

The de Broglie relationship implies that an electron can be represented by a wave equation. Mathematically this will be a differential equation whose solution will yield wave-functions appropriate to the conditions defined in specific situations. A wave-function in the general case will be a function of both position and time so that we write Ψ to mean

$$\Psi = \Psi(x, y, z, t). \qquad \ldots(1.97)$$

However, in many situations of interest, one is dealing with stationary wave solutions either in the form of travelling waves or standing waves, and it will then be necessary only to deal with the space dependence ψ of the function in which case we write

$$\psi = \psi(x, y, z). \qquad (1.98)$$

The prime requirement for a wave equation to represent electron waves is that it be compatible with the dispersion relation

$$E = \hbar\omega = \frac{\hbar^2 k^2}{2m_e} + V \qquad \ldots(1.77)$$

previously derived. In order to arrive at the form of the wave equation let us assume a harmonic wave *solution* representing a stationary state in one dimension. Such a solution will be

$$\Psi = [A\exp(ikx) + B\exp(-ikx)]\exp(-i\omega t), \qquad \ldots(1.52)$$

i.e. a separable solution as used previously. Differentiating this equation yields

$$\frac{\partial \Psi}{\partial t} = -i\omega\Psi$$

and $\qquad \ldots(1.99)$

$$\frac{\partial^2 \Psi}{\partial x^2} = -k^2\Psi.$$

We now need only compare these two results with the dispersion relation (1.77) and by inspection we find that a compatible wave equation is given by

$$-\frac{\hbar}{i}\frac{\partial \Psi}{\partial t} = -\frac{\hbar^2}{2m_e}\frac{\partial^2 \Psi}{\partial x^2} + V\Psi$$

or $\qquad \ldots(1.100)$

$$\frac{\partial^2 \Psi}{\partial x^2} + \frac{2m_e i}{\hbar}\frac{\partial \Psi}{\partial t} - \frac{2m_e}{\hbar^2}V\Psi = 0.$$

This is the time-dependent form of the Schrödinger wave equation and for three dimensions may be written as

$$\nabla^2\Psi + \frac{2m_e i}{\hbar}\frac{\partial \Psi}{\partial t} - \frac{2m_e}{\hbar^2}V\Psi = 0. \qquad \ldots(1.101)$$

This may be compared with equation (1.60) in Section 1.5.2.

Two points need to be made at this stage. First, the equation remains valid when V is a function of position rather than a constant as might be implied from the above. Secondly, the equation is *not* valid for the relativistic region of velocities. This is inherent in the use of the dispersion relation of equation (1.77) rather than a relativistic dispersion relation. The latter leads to a different equation termed the Klein–Gordon wave equation.

As stated previously, in most situations of interest time-independent solutions are required and a simpler form of the wave equation is then available. Writing the separable solution in the form

$$\Psi = \psi \exp(-i\omega t) \qquad \ldots(1.102)$$

we have on differentiation

$$\frac{\partial \Psi}{\partial t} = -i\omega\psi \exp(-i\omega t)$$

and $\qquad \ldots(1.103)$

$$\frac{\partial^2 \Psi}{\partial x^2} = \frac{d^2\psi}{dx^2}\exp(-i\omega t).$$

Substituting in equation (1.100) a derived equation in terms of ψ instead of Ψ is obtained as

$$\frac{d^2\psi}{dx^2} + \frac{2m_e}{\hbar^2}(E-V)\psi = 0, \qquad \ldots(1.104)$$

where $E = \hbar\omega$ (equation (1.76)) represents the total electron energy. The corresponding solution for three dimensions is

$$\nabla^2\psi + \frac{2m_e}{\hbar^2}(E-V)\psi = 0. \qquad \ldots(1.105)$$

This is the time-independent wave equation. Solutions satisfying this equation are termed eigenfunctions and the corresponding values of E are the eigenvalues.

The development of the wave equations given here should not be regarded as formal proofs. The intention has been to show that the procedure is logical and self-consistent. The wave equation may be arrived at from other different postulates, e.g. use of the Hamiltonian operator or indeed the wave equation may be postulated and used as the starting point for development of the theory.

1.8.1 *Interpretation and application of the Schrödinger wave equation*
In looking for solutions of the Schrödinger wave equation in specific problems the potential energy V as a function of position will define the type of solution which will be obtained. Thus in the study of the hydrogen atom, electron wave-functions which are solutions of the three-dimensional stationary wave equation are required. These will represent standing waves and the potential energy function specified is that giving the variation within the Coulomb field surrounding the nucleus. On the other hand, in studying the behaviour of electrons in metals in the so-called free electron model the potential energy V is assumed to have a constant zero value inside the metal and a potential barrier defines the boundaries of the model of the solid. Here again solutions of the time-independent wave equation are sought and we will find that both standing wave and travelling wave solutions are valid.

For any problem it is necessary to determine which are valid or *allowed* solutions. In order to be able to specify the possible limitations on Ψ (or ψ) we need to consider the significance of the quantities involved. Although the wave equation is in fact solved in terms of Ψ there is no measurable physical quantity directly associated with Ψ. Indeed, Ψ itself may be real or complex. However, if Ψ^* is put as the complex conjugate of Ψ then the product $\Psi\Psi^*$ is always real.†

It is *postulated* in the theory that the probability $P(x, t)$ of finding the electron within the range of values x to $x + \delta x$ is given by

$$P(x, t) = \Psi\Psi^* \, dx. \qquad \ldots(1.106)$$

For three dimensions the probability is given in terms of an elementary volume dV so that we have

$$P = \Psi\Psi^* \, dV \qquad \ldots(1.107)$$

The quantity $\Psi\Psi^*$ may therefore be interpreted as a measure of the electron density at any point. This result may be compared with that in electromagnetic theory where the light intensity, or photon density, is proportional to the square of the amplitude of the electric vector.

Let us now consider some of the implications of condition (1.106). If Ψ is to be an acceptable solution to the wave equation then

$$\int_{-\infty}^{\infty} \Psi\Psi^* \, dx = 1 \qquad \ldots(1.108)$$

since the sum of the probabilities of finding the electron at all the separate points must be a certainty (i.e. $P = 1$). This is termed the normalising condition. A corresponding condition holds for waves in three dimensions. If the wave-function we are concerned with is a solution of the time-independent wave equation it must be of the form

$$\Psi = \psi \exp(-i\omega t). \qquad \ldots(1.102)$$

† if $\Psi = a + ib$
then $\Psi^* = a - ib$
and $\Psi\Psi^* = a^2 - b^2$

Since the complex conjugate of $\exp(-i\omega t)$ is $\exp(i\omega t)$† then we must have

$$\Psi\Psi^* = \psi\psi^*. \qquad \ldots(1.109)$$

For a standing wave ψ will be real so that

$$\psi\psi^* = \psi^2, \qquad \ldots(1.110)$$

whereas for a travelling wave ψ is complex and we use the condition (1.109).

A wave-function satisfying equation (1.108) is termed a normalised wave-function. Acceptable solutions of the wave equation can always be subject to normalisation by suitable choice of normalising constant determined by the use of condition (1.108). Examples of the use of the condition will be given in subsequent chapters.

The conditions governing the acceptability of solutions for Ψ (or ψ) to the wave equation can now be specified and are implicit in the meaning assigned to $\Psi\Psi^*$ (and $\psi\psi^*$).

(a) The wave-function must be capable of normalisation.

(b) Ψ (and ψ) must be finite, continuous and single valued.

In addition, since in any real system an abrupt change in V is not possible there is an additional requirement following directly from the wave equation that a discontinuous change in $\partial\psi/\partial x$ is not acceptable. However, in theoretical models abrupt changes in V are often specified and in those cases a discontinuous change in $\partial\psi/\partial x$ can be obtained. The use of boundary and acceptability conditions can be followed in succeeding chapters when specific applications of the wave equation are dealt with.

One further property of eigenfunctions may be dealt with rather briefly, namely that of orthogonality. If Ψ_1 and Ψ_2 are both normalised solutions of the wave equation having different eigenvalues E_1 and E_2 it may be shown that the condition

$$\int_{-\infty}^{\infty} \Psi_1 \Psi_2^* \, dV = \int_{-\infty}^{\infty} \Psi_1^* \Psi_2 \, dV = 0 \qquad \ldots(1.111)$$

applies. By analogy with the result that the scalar product **a**.**b** of two vectors **a** and **b** is zero when the two vectors are orthogonal, the orthogonality of two eigenfunctions means that the two are completely independent of each other. Neither has any component belonging to the other. Thus if Ψ_1 and Ψ_2 represent two atomic orbitals of the same atom this condition means that we can treat them separately. The orthogonality and normalisation conditions can be combined in the form

$$\int \Psi_i^* \Psi_j \, dV = \delta_{ij}, \qquad \ldots(1.112)$$

† $y = \exp(-i\omega t) = \cos\omega t - i\sin\omega t$

$\therefore\ y^* = \cos\omega t + i\sin\omega t$

$= \exp(i\omega t)$

where δ_{ij} is the mathematical symbol termed the Kronecker delta. This has the value unity when $i = j$ and is zero when $i \neq j$. A set of functions satisfying equation (1.112) are then said to be an orthonormal set.

References and further reading

Braddick, H. J. J. (1965) *Vibrations, Waves and Diffraction*, McGraw-Hill.
Coulson, C. A. (1965) *Waves*, Oliver and Boyd.
Crawford, F. S. (1968) *Waves*, Berkeley Physics Course, vol. 3, McGraw-Hill.
French, A. P. (1971) *Vibrations and Waves*, W. W. Norton.
Holden, A. (1971) *Stationary States*, Oxford University Press.
Pohl, H. A. (1967) *Quantum Mechanics for Science and Engineering*, Prentice-Hall.
Raimes, S. (1963) *The Wave Mechanics of Electrons in Metals*, North Holland.
Richards, J. P. G. and Williams, R. P. (1972) *Waves*, Penguin Books.
Smith, R. A. (1963) *Wave Mechanics of Crystalline Solids*, Chapman and Hall.
Speigel, M. R. (1959) *Vector Analysis*, Schaum.
Sproull, R. L. (1964) *Modern Physics*, Wiley.

Problems

1.1 A mass of 0.02 kg is performing SHM of total energy 9 J. Determine the equation of motion of the mass in the form

$$x = a\cos(\omega t + \varepsilon)$$

given that its initial velocity is $-15\ \mathrm{m\,s^{-1}}$ and that its acceleration is $25 \times 10^{-2}\ \mathrm{m\,s^{-2}}$ when its displacement is -9×10^{-2} m.

1.2 A particle is performing SHM. Given that initially its displacement is 0.04 m, its velocity is $-0.06\ \mathrm{m\,s^{-1}}$ and its acceleration is $-0.09\ \mathrm{m\,s^{-2}}$ determine its equation of motion.

At what times will the particle have a velocity of $+0.06\ \mathrm{m\,s^{-1}}$?

1.3 The possible modes of vibration of the air in a pipe depend on whether one end or both ends of the pipe are open. Consider the form of the standing wave equation appropriate to each case and determine for both the allowed values of the propagation constant k in terms of the pipe length. Two such pipes are emitting their third harmonic and these are of the same frequency. Determine the ratio of the lengths of the two pipes.

1.4 Verify by differentiation and substitution that both

$$\Psi = f(x - vt)$$

and

$$\Psi = f(x + vt)$$

are solutions to the wave equation

$$\frac{\partial^2 \Psi}{\partial x^2} = \frac{1}{v^2}\frac{\partial^2 \Psi}{\partial t^2}.$$

What are the limitations to the use of this equation?

1.5 Show that

$$\Psi = f(x+vt)$$

represents a wave travelling without change of form in the negative direction of x with velocity v.

1.6 Determine for the dispersion relationship

$$k = (\omega^2 - \alpha^2)^{\frac{1}{2}}$$

the phase and group velocities of the wave in terms of ω. Show also that these velocities are reciprocals of each other.

1.7 The phase velocity v for waves on deep water obeys the relationship

$$v^2 = \frac{g\lambda}{2\pi} + \frac{2\pi T}{\rho\lambda},$$

where T and ρ are the surface tension and density, respectively. Determine a relationship for the group velocity. Show also that the group velocity is 50 per cent greater than the phase velocity for waves dominated by surface tension and 50 per cent smaller for gravity dominated waves.

Verify also that there exists a minimum phase velocity given by $(4gT/\rho)^{\frac{1}{4}}$. Is it significant that this occurs when the phase and group velocities are equal?

1.8 Electromagnetic waves propagated along a waveguide of square cross-section ($a \times a$) satisfy the dispersion relationship

$$k = \left(\frac{\omega^2}{c^2} - \frac{\pi^2}{a^2}\right)^{\frac{1}{2}}.$$

Derive an expression for the phase velocity v and group velocity u of the wave and show that for the condition

$$\omega > \frac{c\pi}{a}$$

v is greater than c and u is less than c.

Suggest the significance of the condition

$$\omega < \frac{c\pi}{a}.$$

1.9 Electromagnetic waves in the ionosphere follow the dispersion relationship

$$\omega^2 = \omega_p^2 + c^2k^2,$$

where ω_p is the plasma oscillation frequency. Assuming a separable solution for Ψ in the form

$$\Psi = \psi(x)\exp(-i\omega t)$$

show that the waves obey the Klein–Gordon wave equation

$$\frac{\partial^2\Psi}{\partial t^2} = -\omega_p^2\Psi + c^2\frac{\partial^2\Psi}{\partial x^2}.$$

1.10 Verify that for the general wave equation

$$\nabla^2\Psi + g^2(q)\Psi = a\frac{\partial\Psi}{\partial t} + b\frac{\partial^2\Psi}{\partial t^2}$$

the dispersion relationship

$$k = \pm(g^2 + ia\omega + b\omega^2)^{\frac{1}{2}}.$$

is obeyed.

1.11 A metal emits photoelectrons of maximum energies 1.2×10^{-19} J and 7.5×10^{-19} J when light of wavelengths 3.7×10^{-7} m and 1.7×10^{-7} m, respectively, is incident upon it. Calculate the value of Planck's constant and the work function of the metal.

What is the longest wavelength radiation which will excite photoelectrons from the metal?

1.12 The metal sodium has a work function of 2.3 eV. What range of wavelengths in the visible spectrum will excite photoelectrons from the metal?

1.13 Compare, for a wavelength of 10^{-10} m the energies of an electron, a neutron and an X-ray photon.

1.14 Compare the energies of photons emitted from an X-ray tube with a copper target with those from a tungsten lamp and a medium-wave radio transmitter.

1.15 Assuming a luminous efficiency of 10 per cent calculate the number of photons of light emitted by a 100 watt lamp in 1 second.

1.16 An α-particle is the nucleus of a helium atom. If it has energy 5 MeV what is its wavelength?

1.17 Show that the superposition of two infinite trains of sine waves having propagation constants k and $k+\Delta k$ yields a modulated wave train in the form of a series of wave packets each of length $2\pi/k$. By use of the de Broglie postulate show that if one such packet represents a particle the Heisenberg uncertainty principle is obeyed.

2. Atomic structure

2.1 Introduction

In order to develop a theory of the solid state it is of interest first of all to consider the electronic structure of isolated atoms of the elements. In the subsequent chapters we will then discuss how atoms are built up to form crystalline solids, and consider the nature of the binding forces and the electronic structure of solids.

Many elements show striking similarity in both physical and chemical properties to other corresponding elements. Mendeleev (1869) showed that when all the elements were arranged in increasing order of relative atomic mass, then, with a few exceptions, a periodic repetition of elements with similar properties was obtained. This was termed the Periodic Table of the elements, a modern form of which is shown in Table 2.1. The periodicity revealed in the table gives strong evidence for the existence of an internal atomic structure.

The present day accepted model of the isolated atom is that of a central nucleus consisting of uncharged neutrons and positively charged protons surrounded by a cloud of negative electrons, the atom as a whole being electrically neutral. We shall not be concerned with the structure and properties of the nucleus. Early this century Rutherford had established that the atom consisted of a positively charged nucleus of dimensions $\approx 10^{-14}$ m surrounded by electrons occupying a zone $\approx 10^{-10}$ m around the nucleus.

2.2 The Bohr theory

The Bohr theory of the hydrogen atom was a development of the Rutherford model and was based on the application of quantum theory to the description of atomic structure. Although this has since then been superseded by the wave-mechanical approach some of the essential features are retained in later theories and it forms a convenient background and introduction to the more advanced theories of atomic structure.

In terms of classical theory, an electron moving in an orbit around the nucleus would radiate energy continuously since the electron would possess the usual

Table 2.1 The Periodic Table

Period	Sub-Shell	I A	II A	III B	IV B	V B	VI B	VII B	0	III A	IV A	V A	VI A	VII A	VIII			I B	II B
1	1s	1 H							2 He										
2	2s	3 Li	4 Be																
	2p			5 B	6 C	7 N	8 O	9 F	10 Ne										
3	3s	11 Na	12 Mg																
	3p			13 Al	14 Si	15 P	16 S	17 Cl	18 Ar										
4	4s	19 K	20 Ca																
	3d									21 Sc	22 Ti	23 V	24 Cr	25 Mn	26 Fe	27 Co	28 Ni	29 Cu	30 Zn
	4p			31 Ga	32 Ge	33 As	34 Se	35 Br	36 Kr										
5	5s	37 Rb	38 Sr																
	4d									39 Y	40 Zr	41 Nb	42 Mo	43 Tc	44 Ru	45 Rh	46 Pd	47 Ag	48 Cd
	5p			49 In	50 Sn	51 Sb	52 Te	53 I	54 Xe										
6	6s	55 Cs	56 Ba																
	4f																		
	5d									71 Lu	72 Hf	73 Ta	74 W	75 Re	76 Os	77 Ir	78 Pt	79 Au	80 Hg
	6p			81 Tl	82 Pb	83 Bi	84 Po	85 At	86 Rn										
7	7s	87 Fr	88 Ra																
	5f																		

Sub-Shell														
4f	57 La	58 Ce	59 Pr	60 Nd	61 Pm	62 Sm	63 Eu	64 Gd	65 Tb	66 Dy	67 Ho	68 Er	69 Tm	70 Yb
5f	89 Ac	90 Th	91 Pa	92 U	93 Np	94 Pu	95 Am	96 Cm	97 Bk	98 Cf	99 Es	100 Fm	101 Md	102 No	103 Lw

Lanthanides

Main Group Elements **The Transition Elements** **Actinides**

radial acceleration of a particle describing a circular path. As a consequence of this loss of energy the electron would spiral towards the nucleus. A classical model is therefore unable to account for a stable atomic configuration of electrons describing stable orbits around a nucleus. Neither is the associated prediction of the emission of a continuous spectrum of radiation in accord with the experimental observation that emission of optical radiation from free atoms occurs in the form of a number of series of sharp spectral lines of definite frequencies.

In order to resolve this difficulty, Bohr *postulated* that electrons revolved around the nucleus only in certain specific orbits. In such an orbit, termed a stationary state, the electron would not emit radiation. When, however, an electron transfers from one orbit of total energy E_2 to another orbit of lower energy E_1 the atom *emits* a quantum $h\nu$ of radiation of frequency ν where

$$E_2 - E_1 = h\nu. \qquad \ldots(2.1)$$

An electron makes a transition from orbit of energy E_1 to that of energy E_2 when energy $h\nu$ is *absorbed* by the atom. This is illustrated in Fig. 2.1.

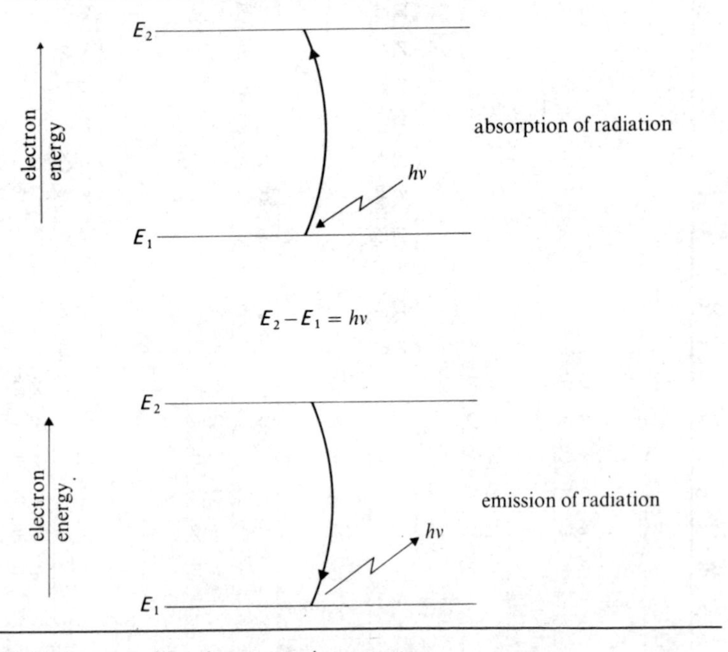

Fig. 2.1 Electron transitions between stationary states.

In specifying the energy of an electron orbit the total energy given by the sum of its potential and kinetic energies is required. Nevertheless, in any physical process, such as for example in the emission process described by equation (2.1), only *changes* in energy are of significance. As discussed previously there is no true zero of energy. As in many other problems the potential energy of an electron at infinite distance from the nucleus is set as zero. The charges on the electron and the hydrogen nucleus are respectively $-e$ and $+e$ so that there is an attractive force between them. Consequently, the potential energy of an electron in an orbit is a negative quantity.

The total energy of an electron orbit may now be calculated. The charge on the nucleus will be put as Ze so that extension of the theory to atoms of higher *atomic number* Z can be included. At a distance x from the nucleus the Coulomb attractive force F on the electron will be given by

$$F = \frac{Ze^2}{4\pi\varepsilon_0 x^2}, \qquad \ldots(2.2)$$

where ε_0 is the permittivity of a vacuum and has a value of $8.854 \times 10^{-12}\ \mathrm{F\,m^{-1}}$. The potential energy of an electron orbit of radius r is then obtained as

$$V = \int_{\infty}^{r} \frac{Ze^2}{4\pi\varepsilon_0 x^2}\,\mathrm{d}x = -\frac{Ze^2}{4\pi\varepsilon_0 r}. \qquad \ldots(2.3)$$

In developing the theory, Bohr assumed that the electron obeys the laws of classical mechanics so that the Coulomb attraction between the electron and the nucleus is exactly balanced by the centrifugal force, i.e.

$$\frac{Ze^2}{4\pi\varepsilon_0 r^2} = \frac{m_e u^2}{r}, \qquad \ldots(2.4)$$

where u is the linear velocity of an electron of mass m_e in this orbit. The kinetic energy K of the electron is therefore

$$K = \tfrac{1}{2} m_e u^2 = \frac{Ze^2}{8\pi\varepsilon_0 r} \qquad \ldots(2.5)$$

and the total energy E corresponding to this orbit is

$$E = K + V = -\frac{Ze^2}{8\pi\varepsilon_0 r}. \qquad \ldots(2.6)$$

Bohr postulated that only certain specific electron orbits were allowed so that in accordance with equation (2.6) only certain values of E are possible, i.e. the values of E given by equation (2.6) are *quantised*. The equation does not, however, tell us *which* particular orbits are allowed. The quantisation condition may most simply be stated in the form that the stationary states correspond to orbits for which the angular momentum L is given by

$$L = m_e u r = \frac{nh}{2\pi} \qquad (n = 1, 2, 3, \ldots). \qquad \ldots(2.7)$$

n is termed the ***principal quantum number.***

It may be noted that the quantisation condition may be deduced from the *Bohr correspondence principle* according to which the predictions of quantum theory should, in the limit of large quantum numbers, i.e. large energies, agree with those of classical mechanics. Further insight into the significance of the quantisation condition may be obtained by consideration of the wave behaviour of the electron (see Section 2.3).

The radii of the allowed orbits are obtained from equations (2.4) and (2.7), which yield

$$r = \frac{n^2 h^2 \varepsilon_0}{\pi Z m_e e^2} \qquad (n = 1, 2, 3, \ldots). \qquad \ldots(2.8)$$

Substituting the appropriate numerical values for the constants and putting $Z = 1$ the allowed radii for hydrogen become

$$\begin{aligned} r &= n^2 \times 5.3 \times 10^{-11}\ \text{m} \\ &= 0.053 n^2\ \text{nm}. \end{aligned} \qquad \ldots(2.9)$$

The radius of the orbit of minimum energy is given by $n = 1$ and is called the Bohr radius. It is given by

$$a_0 = \frac{h^2 \varepsilon_0}{\pi m_e e^2} = 0.053\ \text{nm} \qquad \ldots(2.10)$$

and is used as the unit of length in the atomic unit scale.

The corresponding values of the total energy E of the stationary states are obtained by substituting the values of r from equation (2.8) in equation (2.6) and this yields

$$E = -\frac{m_e Z^2 e^4}{8 n^2 h^2 \varepsilon_0^2} \qquad (n = 1, 2, 3, \ldots). \qquad \ldots(2.11)$$

The quantisation of orbital angular momentum has thus led to the quantisation of the total energy of the electron. This is illustrated in Fig. 2.2 for the hydrogen atom.

It is customary in solid state theory to represent electron energies in terms of the electron volt instead of in joules. The electron volt is the energy gained by an electron in moving through a potential difference of 1 volt. Putting the electronic charge $|e|$ as 1.6×10^{-19} C we have that 1 eV $= 1.6 \times 10^{-19}$ J so that equation (2.11) becomes

$$E = -\frac{13.6}{n^2}\ \text{eV}. \qquad \ldots(2.12)$$

The orbit of minimum energy (i.e. the one with the largest negative energy) corresponds to $n = 1$ and the electron will always occupy this orbit in the normal unexcited state of the atom. It is termed the ground state. We also note that the alternative atomic unit of energy, termed the rydberg, is defined by equation (2.11) by putting $Z = 1$ and $n = 1$. Its value is therefore 13.6 eV.

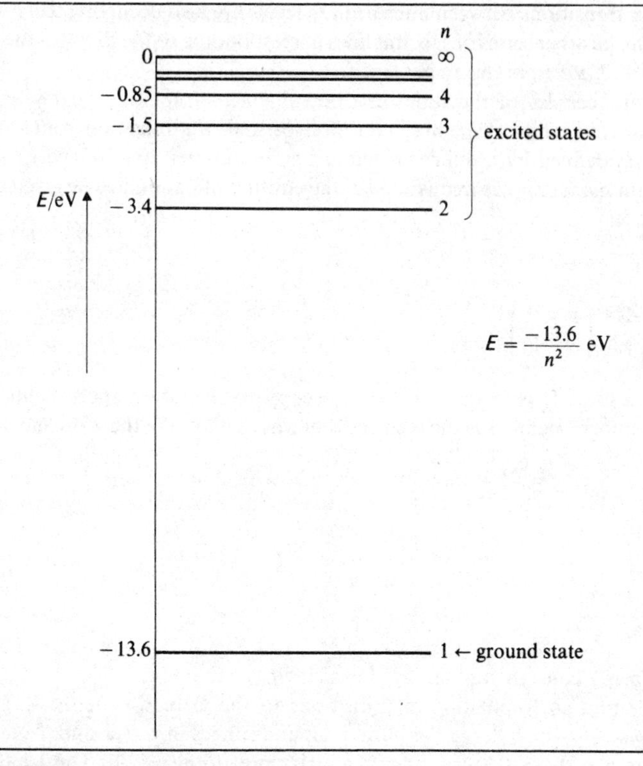

Fig. 2.2 Energy level diagram for the hydrogen atom. The electron energy is zero for an electron at infinite distance from the nucleus.

2.2.1 *Spectral lines*

An important achievement of the Bohr theory was its ability to explain the main features of the hydrogen spectrum. This consists of a number of very sharp line spectra arranged in well-defined groups or series, extending from the ultraviolet through the visible into the infrared region. They are observed when an electric discharge is passed through hydrogen at low pressure and are termed emission spectra. The energy of the discharge is sufficient to eject electrons from many of the atoms in the gas, i.e. the atoms become ionised, and will subsequently tend to revert to the stable ground state of minimum energy. This may be accomplished, for example, by a direct electron transition from outside the atom, i.e. from a state corresponding to $n = \infty$ to the ground state for which $n = 1$, with the emission of a photon of appropriate frequency. Alternatively, there may occur a number of intermediate transitions to states of successively lower energy with the final transition to the ground state occurring from states having a value of n lying anywhere between 2 and ∞. A *series* of spectral lines is thus obtained corresponding to the different ways of filling the ground state level. Since electron transitions between intermediate levels are also occurring there will, in addition, be other series of spectral lines corresponding to the filling of the states for $n = 2, 3$, etc. from the higher levels.

The frequencies of the series can be calculated from the theory by using equations (2.1) and (2.11). Thus, if the electron makes a transition from a state of energy E_i defined by a quantum number n_i to another state of energy E_f and quantum number n_f the frequency ν of the emitted radiation will be given by

$$\nu = \frac{E_i - E_f}{h} = \frac{m_e Z^2 e^4}{8h^3 \varepsilon_0^2}\left(\frac{1}{n_f^2} - \frac{1}{n_i^2}\right) \qquad \ldots(2.13)$$

where $n_i > n_f$. It is practice in spectroscopy to identify a spectral line by a wavenumber $\bar{\nu}$ defined as the reciprocal of wavelength. We therefore have

$$\bar{\nu} = R_\infty Z^2\left(\frac{1}{n_f^2} - \frac{1}{n_i^2}\right), \qquad \ldots(2.14)$$

where

$$R_\infty = \frac{m_e e^4}{8ch^3 \varepsilon_0^2} \qquad \ldots(2.15)$$

is Rydberg's constant and has a value of 1.0968×10^7 m^{-1}.

The series corresponding to transitions to the ground state ($n_f = 1$) is the Lyman series which lies in the ultraviolet and ranges in wavelength from 121.6 nm (for $n_i = 2$) to the series limit at 91.2 nm (for $n_i = \infty$). The transitions occurring to $n_f = 2$ from higher levels yield the Balmer series of lines lying in the visible spectrum, whilst the transitions to $n_f = 3$, 4 and 5 yield respectively the Paschen, Brackett and Pfund series in the infrared. The transitions yielding three series of emission spectral lines are illustrated in Fig. 2.3.

The Bohr theory also explains the *absorption* spectrum of hydrogen. This consists of a series of absorption lines corresponding in wavelength to the Lyman emission spectrum, and may be observed by passing a continuous optical spectrum through the gas. Normally the atoms of the gas are in the ground state corresponding to $n_i = 1$ and selective absorption of radiation of the specific frequencies corresponding to electron transitions to $n_f = 2, 3$ etc. occurs.

2.2.2 *Discussion of the Bohr theory*

At the time the Bohr theory was introduced only the Balmer and Paschen emission spectra had been discovered. In addition to giving the wavenumbers of all the lines in these spectra, the theory predicted the occurrence of other series of emission spectra. These were later discovered by Lyman, Brackett and Pfund. For very accurate comparison of the theory with experiment, account must be taken of the fact that the mass of the nucleus is not infinitely great compared with

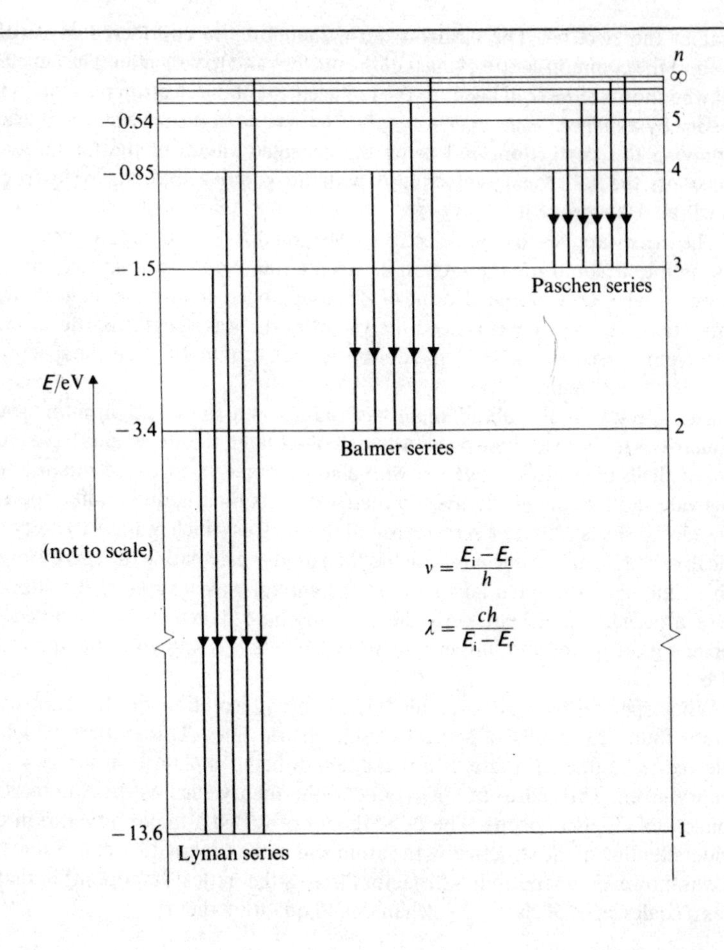

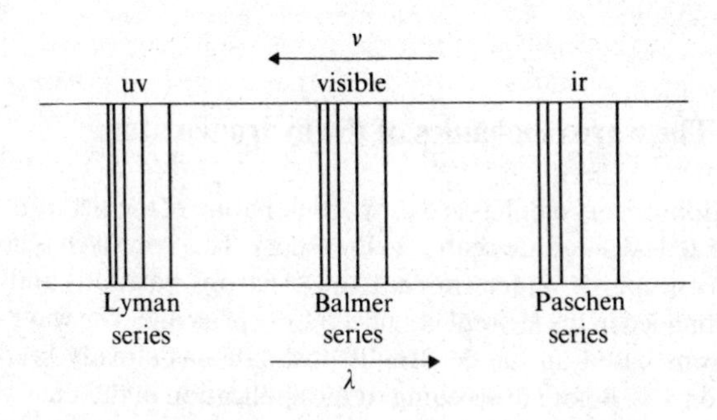

Fig. 2.3 Emission spectral lines of hydrogen.

that of the electron. The electron must, therefore, be considered to revolve around the common centre of mass of the nucleus and the electron. The net effect may be shown to be equivalent to the replacement of the electron mass m_e in the theory by a *reduced mass* $m_e(1+m_e/M)^{-1}$, where M is the mass of the nucleus. Applying this correction, and using the accepted values of the fundamental constants, the Bohr theory values agree with the spectroscopic data for hydrogen to within 0.003 per cent.

The theory applies to a one-electron atom and does not provide any means of taking into account the interaction between a number of orbiting electrons. If, however, a suitably modified value of Z is used, together with the reduced mass correction, the theory yields accurate values of the wavelengths of the spectral lines from ionised helium He^+ and doubly ionised lithium Li^{2+}, i.e. the spectra of one-electron systems.

The spectra of the alkali metals, sodium, potassium and lithium, show similarities to those for hydrogen. As we shall see later, these elements have inner closed shells of electrons together with a single outer valence electron and it is this valence electron which gives to these elements their hydrogen-like spectra. The closed shells provide a screen around the nucleus which reduces to unity the effective charge on the nucleus, which is then used in calculating the energy levels. This simplified approach fails to yield the correct wavelengths of the spectral lines although agreement with the spectroscopic data can be obtained by replacing the quantum number n by $(n-\sigma)$, where σ is known as the quantum defect.

When spectral lines are examined with a high-resolution spectroscope each line is found to consist of several closely spaced lines. This feature, which is referred to as fine structure, is not capable of being explained in terms of the theory given. This led to modifications of the theory such as the Sommerfeld concept of elliptical orbits. The Bohr theory provided a major advance in the understanding of the structure of the atom and in the explanation of line spectra. It was, however, not entirely satisfactory from a theoretical standpoint in that it was a coalescence of classical mechanics and quantum theory.

2.3 The wave mechanics of the hydrogen atom

The Bohr theory emphasised the particle nature of the electron, and orbits which are exactly defined are central to the theory. This approach is not consistent with the concept of indeterminacy of electron position and momentum as propounded in the Heisenberg uncertainty principle. The wave representation of electrons based on the de Broglie postulate has already been introduced (see Section 1.8). Before proceeding to the application of the theory to the hydrogen atom we will briefly show how the de Broglie concept can provide an interpretation of the angular momentum postulate of the Bohr theory. The de Broglie relationship

$$\lambda = \frac{h}{p} = \frac{h}{m_e u}$$

allows the linear momentum in an orbit to be related to the wavelength of the electron. Substituting the linear momentum in the quantisation condition (2.7) it follows that

$$2\pi r = n\lambda \qquad (n = 1, 2, 3 \ldots) \qquad \ldots(2.16)$$

so that the allowed orbits are therefore those which are able to accommodate an exact integral number of wavelengths. This is then the necessary condition for a standing wave to be set up as is illustrated in Fig. 2.4.

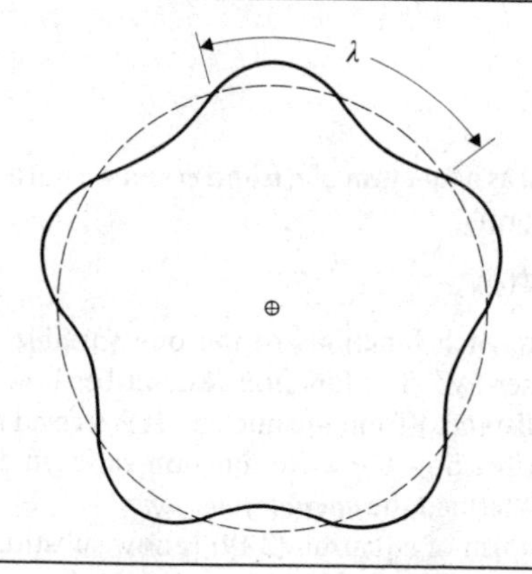

Fig. 2.4 Standing waves in an electron orbit. Five wavelengths fit exactly into this orbit.

Let us now consider the solution of the Schrödinger wave equation for the hydrogen atom. We are interested in the behaviour of an electron bound to the nucleus by the electrostatic Coulombic attraction of the nucleus, i.e. in a central field of force. The solution will allow the electron density surrounding the nucleus to be determined, and clearly solutions of the stationary wave equation

$$\nabla^2\psi + \frac{2m_e}{\hbar^2}(E - V)\psi = 0 \qquad \ldots(1.05)$$

are looked for. A number of different types of solution will in fact be derived.

The potential energy V of the electron in the field of the nucleus is given by

$$V = -\frac{Ze^2}{4\pi\varepsilon_0 r}, \qquad \ldots(2.3)$$

where the zero of potential energy corresponds to infinite separation of the electron and the nucleus. Substituting for V in equation (1.105) we have

$$\nabla^2\psi + \frac{2m_e}{\hbar^2}\left(E + \frac{Ze^2}{4\pi\varepsilon_0 r}\right)\psi = 0 \qquad \ldots(2.17)$$

the solutions of which will describe the possible states of the hydrogen atom. It will also yield for $Z = 2$ the states of the single ionised helium atom or, for $Z = 3$, those of the doubly ionsed lithium atom.

In order to solve equation (2.17) it is necessary to transform it from Cartesian to spherical polar coordinates in accordance with the method outlined in Section 1.4. The transformation yields

$$\frac{1}{r^2}\frac{\partial}{\partial r}\left(r^2\frac{\partial\psi}{\partial r}\right) + \frac{1}{r^2 \sin\theta}\frac{\partial}{\partial\theta}\left(\sin\theta\frac{\partial\psi}{\partial\theta}\right) + \frac{1}{r^2\sin^2\theta}\frac{\partial^2\psi}{\partial\phi^2} + \frac{2m_e}{\hbar^2}\left(E + \frac{Ze^2}{4\pi\varepsilon_0 r}\right)\psi = 0. \qquad \ldots(2.18)$$

ψ is therefore expressed as a function of r, θ and ϕ and we must look for separable solutions for ψ in the form

$$\psi(r, \theta, \phi) = R(r)\cdot\Theta(\theta)\cdot\Phi(\phi), \qquad \ldots(2.19)$$

where R, Θ and Φ are each functions of the one variable indicated but are independent of the other two. The function R describes how the *wave-function* varies with the radial distance r from the nucleus. It is termed the *radial function.* The product $\Theta\Phi$ describes how the wave-function varies in different directions from the nucleus and is termed the *angular function.*

The solution in the form of equation (2.19) is now substituted into equation (2.18) and on multiplying with $r^2 \sin^2\theta/R\Theta\Phi$ and rearranging we obtain

$$\frac{\sin^2\theta}{R}\frac{\partial}{\partial r}\left(r^2\frac{\partial R}{\partial r}\right) + \frac{\sin\theta}{\Theta}\frac{\partial}{\partial\theta}\left(\sin\theta\frac{\partial\Theta}{\partial\theta}\right) + \frac{2m_e}{\hbar^2}r^2\sin^2\theta\left(E + \frac{Ze^2}{4\pi\varepsilon_0 r}\right) = -\frac{1}{\Phi}\frac{\partial^2\Phi}{\partial\phi^2}. \qquad \ldots(2.20)$$

Since on the right-hand side Φ is a function of ϕ only (and not of θ and r), whilst the left-hand side of the equation is independent of ϕ, the only allowed solutions of equation (2.20) are these obtained by equating both sides to a constant. Putting this constant, termed a separation constant, as m^2 we have it in the first instance that

$$\frac{\partial^2\Phi}{\partial\phi^2} = -m^2\phi. \qquad \ldots(2.21)$$

m is termed the *magnetic quantum number*.

To obtain the corresponding equations for R and for Θ the left-hand side of equation (2.20) is now set equal to m^2 so that after rearranging we have

$$\frac{1}{R}\frac{\partial}{\partial r}\left(r^2\frac{\partial R}{\partial r}\right)+\frac{2m_e}{\hbar^2}\left(Er^2+\frac{Ze^2r}{4\pi\varepsilon_0}\right)=\frac{m^2}{\sin^2\theta}-\frac{1}{\Theta}\frac{1}{\sin\theta}\frac{\partial}{\partial\theta}\left(\sin\theta\frac{\partial\Theta}{\partial\theta}\right). \qquad \ldots(2.22)$$

Noting again that in this case the left-hand side is a function only of r and the right-hand side only of θ we may introduce a further separation constant λ. It may then be readily verified that the two differential equations in Θ and R are, respectively,

$$\left(\frac{\partial^2}{\partial\theta^2}+\cot\theta\frac{\partial}{\partial\theta}+\lambda-\frac{m^2}{\sin^2\theta}\right)\Theta=0 \qquad \ldots(2.23)$$

and

$$\left[\frac{\partial^2}{\partial r^2}+\frac{2}{r}\frac{\partial}{\partial r}+\frac{2m_e}{\hbar^2}\left(E+\frac{Ze^2}{4\pi\varepsilon_0 r}\right)-\frac{\lambda}{r^2}\right]R=0. \qquad \ldots(2.24)$$

Summarising, the separable solutions to the wave equation (2.20) are obtained by solving the three equations (2.21), (2.23) and (2.24). Subject to specific conditions these solutions may be combined to yield the complete wave-functions. These are termed atomic orbitals.

2.3.1 *The angular function $\Theta\Phi$*

In order to study the significance of the various possible solutions it is convenient to deal initially with the angular dependence of the wave equation and to consider first the solution to equation (2.21). These are of the form

$$\Phi = A\exp(im\phi) \qquad (m=0,\pm1,\pm2,\ldots), \qquad \ldots(2.25)$$

where A is a constant evaluated separately for each value of m by normalisation of Φ. The integral values of m arise from the requirement that $\Phi\Phi^*$ and Φ be single-valued for each given value of m. Geometrically the points (r,θ,ϕ) and $(r,\theta,\phi+2\pi)$ are equivalent and a non-integral value of m would lead to $m\phi$ and $m(\phi+2\pi)$ being different angles. The values of Φ for different m given by equation (2.25) are included in Table 2.2. These have been normalised in accordance with the condition

$$\int_0^{2\pi}\Phi\Phi^*\,\mathrm{d}\phi=1. \qquad \ldots(2.26)$$

It is clear that instead of the exponential functions the corresponding real trigonometrical functions could have been utilised. In order, however, to obtain the complete angular dependence the solutions for Φ need to be combined with those for Θ. We shall shortly see that suitable combinations of such composite functions, involving the imaginary exponential functions, can lead to real functions which can be represented diagrammatically.

The method used for solving equation (2.23) for Θ is rather complex and the details of solution will not be given here. The acceptable solutions are found to be fairly simple trigonometrical functions and are termed associated Legendre functions of $\cos\theta$ denoted by P_l^m $(\cos\theta)$. Finite single-valued functions are obtained only when

$$\lambda = l(l+1) \qquad (l = 0, 1, 2, \ldots) \quad \ldots(2.27a)$$

and in addition

$$l \geqslant |m|. \qquad \ldots(2.27b)$$

It follows that there are $(2l+1)$ possible values of m for any given value of l and, in accordance with equation (2.23), the same number of solutions for Θ. θ is the azimuthal angle and l is termed the *azimuthal* or *orbital quantum number*. The solutions for Θ given in Table 2.2 are normalised and it is readily verified that they are solutions to equation (2.23) for the different values of l and the corresponding values of λ.

The complete angular dependence of the wave-function is given by the product $\Theta\Phi$. These are termed spherical harmonics and are denoted by Y_{lm}. We therefore have

$$Y_{lm} = A_{lm} P_l^m(\cos\theta)\exp(im\phi), \qquad \ldots(2.28)$$

where the constant A_{lm} is determined by normalising the wave-function. The normalisation condition for the complete wave-function is

$$\int_V \psi\psi^* \, \mathrm{d}V = 1. \qquad \ldots(2.29)$$

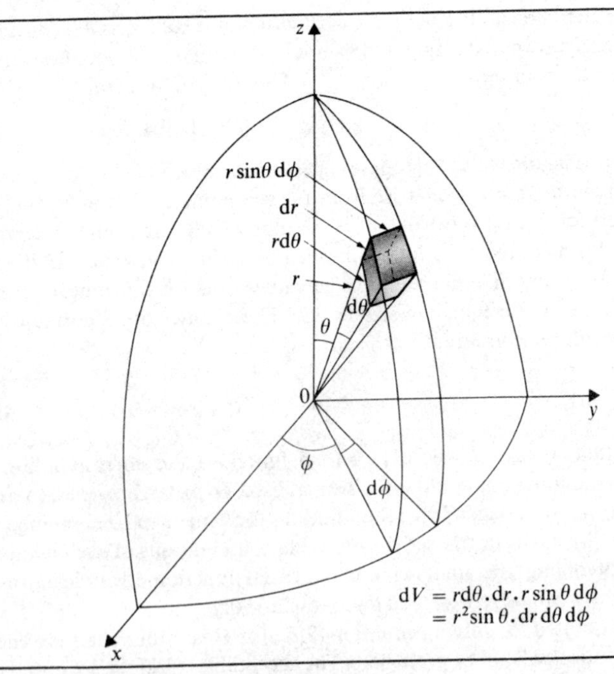

Fig. 2.5 Volume element in spherical polar coordinates.

The volume element $\mathrm{d}V$ in spherical polar coordinates is, in accordance with Fig. 2.5, given by

$$\mathrm{d}V = r^2 \sin\theta\, \mathrm{d}r\, \mathrm{d}\theta\, \mathrm{d}\phi \qquad \ldots(2.30)$$

so that the normalisation condition can be written

$$\int_0^{\infty} r^2 RR^*\, \mathrm{d}r \int_0^{\pi} \sin\theta\, \Theta\Theta^*\, \mathrm{d}\theta \int_0^{2\pi} \Phi\Phi^*\, \mathrm{d}\phi = 1. \qquad \ldots(2.31)$$

Each integral in this expression may separately be set equal to unity and A_{lm} determined for each pair of values of l and m.

Table 2.2 Normalised Θ, Φ and spherical harmonics Y_{lm}

Orbital	l	m	Θ	Φ	Y_{lm}	
s	0	0	$\frac{1}{\sqrt{2}}$	$\frac{1}{\sqrt{2\pi}}$	$\frac{1}{2\sqrt{\pi}}$	Y_{00}
p	1	0	$\frac{\sqrt{6}}{2}\cos\theta$	$\frac{1}{\sqrt{2\pi}}$	$\frac{1}{2}\sqrt{\frac{3}{\pi}}\cos\theta$	Y_{10}
	1	± 1	$\frac{1}{2}\sqrt{3}\sin\theta$	$\frac{1}{\sqrt{2\pi}}\mathrm{e}^{\pm \mathrm{i}\phi}$	$\frac{1}{2}\sqrt{\frac{3}{2\pi}}\sin\theta\,\mathrm{e}^{\pm \mathrm{i}\phi}$	$Y_{1\pm1}$
d	2	0	$\frac{\sqrt{10}}{4}(3\cos^2\theta - 1)$	$\frac{1}{\sqrt{2\pi}}$	$\frac{1}{4}\sqrt{\frac{5}{\pi}}(3\cos^2\theta - 1)$	Y_{20}
	2	± 1	$\frac{\sqrt{15}}{2}\sin\theta\cos\theta$	$\frac{1}{\sqrt{2\pi}}\mathrm{e}^{\pm \mathrm{i}\phi}$	$\frac{1}{4}\sqrt{\frac{15}{2\pi}}\sin 2\theta\,\mathrm{e}^{\pm \mathrm{i}\phi}$	$Y_{2\pm1}$
	2	± 2	$\frac{\sqrt{15}}{4}\sin^2\theta$	$\frac{1}{\sqrt{2\pi}}\mathrm{e}^{\pm 2\mathrm{i}\phi}$	$\frac{1}{4}\sqrt{\frac{15}{2\pi}}\sin^2\theta\,\mathrm{e}^{\pm 2\mathrm{i}\phi}$	$Y_{2\pm2}$

The spherical harmonics are listed in Table 2.2 and, with the exception of those for $m = 0$, involve complex functions. In dealing with the angular dependence it is clearly of advantage to be able to deal with real functions and to represent the functions diagrammatically. To achieve this, appropriate *linear* combinations are taken of spherical harmonics of the same value of l and these are listed in Table 2.3. It should be noted that the number of *independent* solutions is not increased by the formation of these combinations. The designation of the orbitals follows early practice in spectroscopy and is as follows:

value of l	: 0	1	2	3	4
orbital designation:	s	p	d	f	g

Table 2.3 Normalised angular dependence functions

Orbital	Spherical Harmonics	Angular Dependence Functions (Polar form)	Angular Dependence Functions (Cartesian form)
s	Y_{00}	$\frac{1}{2\sqrt{\pi}}$	$\frac{1}{2\sqrt{\pi}}$
p_z	Y_{10}	$\frac{1}{2}\sqrt{\frac{3}{\pi}}\cos\theta$	$\frac{1}{2}\sqrt{\frac{3}{\pi}}\cdot\frac{z}{r}$
p_x	$\frac{1}{\sqrt{2}}(Y_{11}+Y_{1-1})$	$\frac{1}{2}\sqrt{\frac{3}{\pi}}\sin\theta\cos\phi$	$\frac{1}{2}\sqrt{\frac{3}{\pi}}\frac{x}{r}$
p_y	$\frac{i}{\sqrt{2}}(Y_{1-1}-Y_{11})$	$\frac{1}{2}\sqrt{\frac{3}{\pi}}\sin\theta\sin\phi$	$\frac{1}{2}\sqrt{\frac{3}{\pi}}\frac{y}{r}$
d_{z^2}	Y_{20}	$\frac{1}{4}\sqrt{\frac{5}{\pi}}(3\cos^2\theta-1)$	$\frac{1}{4}\sqrt{\frac{5}{\pi}}\left(\frac{3z^2-r^2}{r^2}\right)$
d_{zx}	$\frac{1}{\sqrt{2}}(Y_{21}+Y_{2-1})$	$\frac{1}{4}\sqrt{\frac{15}{\pi}}\sin 2\theta\cos\phi)$	$\frac{1}{4}\sqrt{\frac{15}{\pi}}\frac{2xz}{r^2}$
d_{zy}	$\frac{i}{\sqrt{2}}(Y_{2-1}-Y_{21})$	$\frac{1}{4}\sqrt{\frac{15}{\pi}}\sin 2\theta\sin\phi$	$\frac{1}{4}\sqrt{\frac{15}{\pi}}\frac{2yz}{r^2}$
$d_{x^2-y^2}$	$\frac{1}{\sqrt{2}}(Y_{22}+Y_{2-2})$	$\frac{1}{4}\sqrt{\frac{15}{\pi}}\sin^2\theta(\cos^2\phi-\sin^2\phi)$	$\frac{1}{4}\sqrt{\frac{15}{\pi}}\left(\frac{x^2-y^2}{r^2}\right)$
d_{xy}	$\frac{i}{\sqrt{2}}(Y_{2-2}-Y_{22})$	$\frac{1}{4}\sqrt{\frac{15}{\pi}}\sin^2\theta\sin 2\phi$	$\frac{1}{4}\sqrt{\frac{15}{\pi}}\frac{2xy}{r^2}$

Subscripts, such as for example d_{xz}, refer to the correspondence between the spherical polar and Cartesian designation of each orbital.

Polar diagrams are used to represent the shapes of the different orbitals and those for the s, p and d orbitals are illustrated in Figs. 2.6, 2.7 and 2.8, respectively. These are simply a plot of the magnitude of the angular dependence function of an orbital for each direction defined by θ and ϕ. The value of the function may be positive or negative and in the figures the sign of each surface is indicated accordingly. For s orbitals the angular function is a constant equal to $1/2(\pi)^{\frac{1}{2}}$ so that the orbital is a sphere, i.e. $\Theta\Phi$ for s orbitals is independent of both θ and ϕ.

The angular function for the p orbitals ($l = 1$) consists of two spheres touching at the origin and the orientation of the axis joining their centres is determined by the value of m. This axis lies along the z direction for $m = 0$ and the orbital is referred to as p_z. The p_x and p_y orbitals are obtained by taking appropriate linear combinations of the two spherical harmonics for $m = \pm 1$. It should be noted that in each case the orbital is represented by one positive and one negative sphere. The three orbitals are oriented in mutually perpendicular directions, are independent and have the same energy. Other p orbitals could, alternatively, have been constructed from the spherical harmonics listed in Table 2.2. In each case, however, only a set of three independent orbitals can be so formed.

There are a total of five independent d orbitals. With the exception of the d_{z^2} orbital their form can best be visualised by noting that each complete lobe is three-dimensional and is formed by the rotation of a lobe about its axis of symmetry. Four of the orbitals are thus similar and differ only in their orientation. From symmetry considerations one might have expected to obtain $d_{y^2-z^2}$ and $d_{z^2-x^2}$ orbitals. These two may, however, be combined to give the d_{z^2} orbital giving the total of five independent orbitals. The d_{z^2} orbital consists of two lobes together with a central ring, the whole orbital being formed by rotation about the z-axis.

The interpretation of the angular dependence functions needs to be approached carefully. Referring to Section 1.8.1 it is noted that it is $\psi\psi^*$ which yields the electron density. This is the physically observable quantity and its angular dependence is related to Y_{lm}^2 rather than Y_{lm}. The shapes of both functions are similar, although not identical and, in contrast to ψ and Y_{lm} both $\psi\psi^*$ and Y_{lm}^2 must always be positive. In practice the angular dependences illustrated are used to represent the shapes of orbitals.

The sign of the angular dependence function, and thus of ψ, is important in the

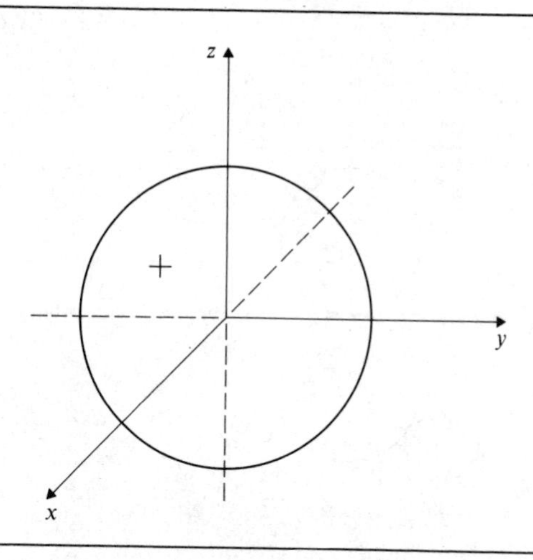

Fig. 2.6 Polar diagram for an s orbital. The orbital has the same sign at all points.

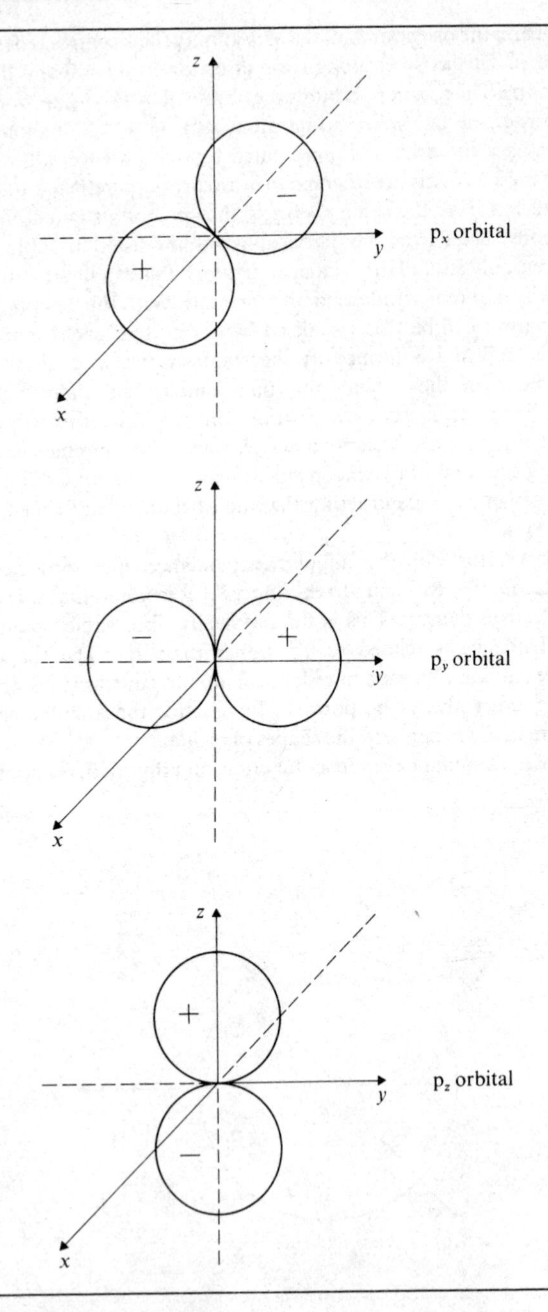

Fig. 2.7 Polar diagrams for p orbitals.

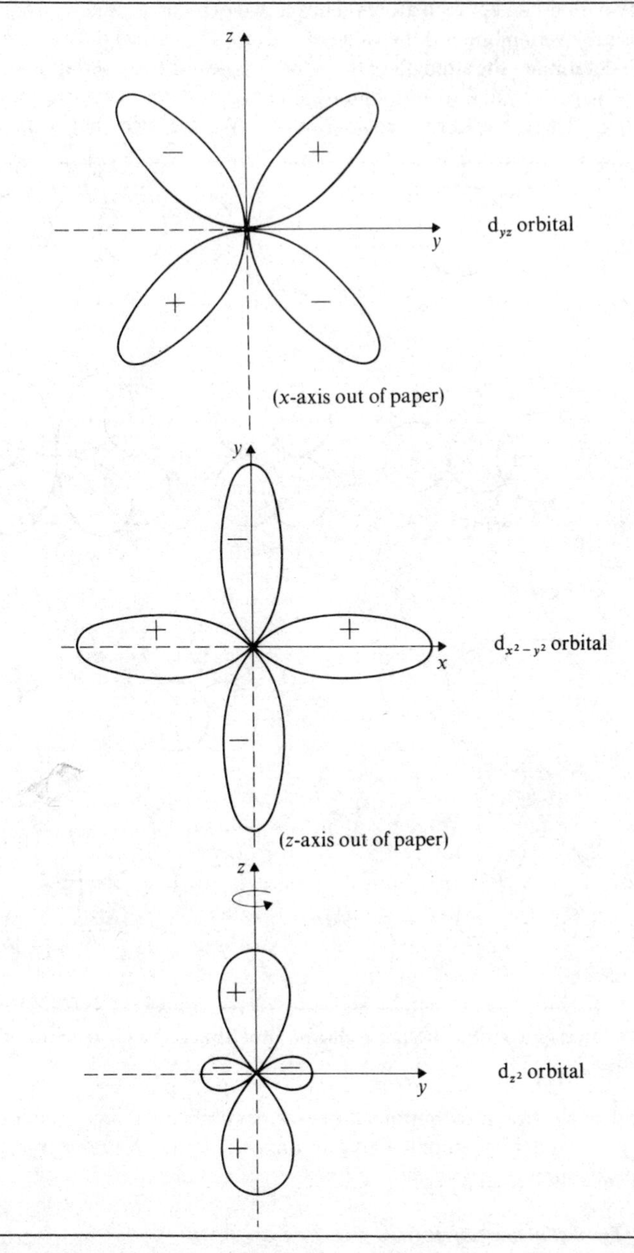

Fig. 2.8 The lobes of the d_{yz} and $d_{x^2-y^2}$ orbitals are three-dimensional surfaces obtained by rotation of each lobe about its axis. The form of the d_{z^2} orbital is obtained by rotating the figure about the z-axis. The d_{xz} and d_{xy} orbitals (not illustrated) are similar in form to the d_{yz} orbital.

theory of bonding between atoms. Thus if two orbitals ψ_A and ψ_B from different atoms are overlapping it is the value of $\int_v \psi_A \psi_B \, dV$, termed the overlap integral, which determines the strength of the bond. Large positive overlap gives a strong bond, a negative value an antibonding situation and a zero value a non-bonding situation. These are illustrated in Fig. 2.9. We see also that a directionally

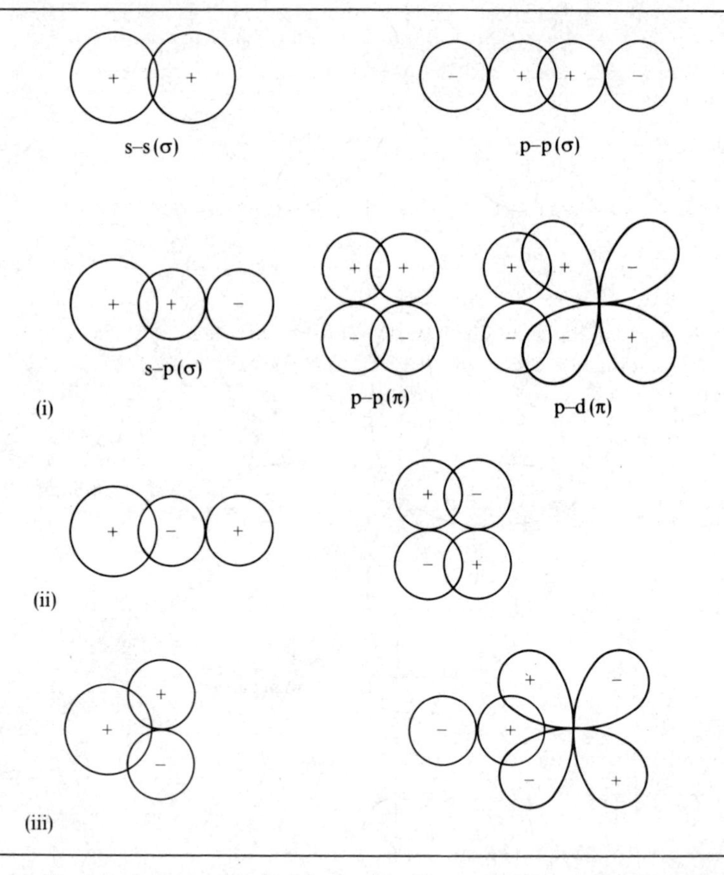

Fig. 2.9 Overlap of atomic orbitals giving positive (i), negative (ii) and zero (iii) overlap integrals.

dependent electron distribution allows the possibility of directed bonds and the form of the different orbitals plays an important part in the interpretation of chemical bonding.

2.3.2 *The radial function R*

The form of the solution of equation (2.24) for the radial function R will depend on λ and therefore on the azimuthal quantum number l. The simplest solution is the one obtained by setting $l = 0$, i.e. an s orbital. For all s orbitals the radial function is required to be a solution of

$$\left[\frac{\partial^2}{\partial r^2}+\frac{2}{r}\frac{\partial}{\partial r}+\frac{2m_e}{\hbar^2}\left(E+\frac{Ze^2}{4\pi\varepsilon_0 r}\right)\right]R = 0. \qquad \ldots(2.32)$$

Mathematically, both positive and negative exponential functions of r satisfy this equation. However, positive exponential solutions give $R \to \infty$ as $r + \infty$ and these are therefore not acceptable solutions. We therefore put as one possible solution

$$R(r) = C\exp\left(-\frac{Zr}{a_0}\right), \qquad \ldots(2.33)$$

where C and a_0 are constants. This will be a solution provided that

$$\frac{Z^2}{a_0^2}-\frac{2Z}{a_0}\frac{1}{r}+\frac{2m_e}{\hbar^2}E+\frac{m_e Ze^2}{2\hbar^2\pi\varepsilon_0}\frac{1}{r} = 0. \qquad \ldots(2.34)$$

In order that the solution may be valid for all values of r the sum of the coefficients of terms in $1/r$ and also of the constant terms must independently be zero. This yields

$$a_0 = \frac{4\pi\hbar^2\varepsilon_0}{m_e e^2} \qquad \ldots(2.35)$$

which as we have seen in Section 2.2 is the first Bohr radius. Additionally we have

$$E = -\frac{m_e Z^2 e^4}{32\pi^2\varepsilon_0^2\hbar^2} \qquad \ldots(2.36)$$

which corresponds to the energy given by equation (2.11) with quantum number $n = 1$.

The solution assumed for R in equation (2.33) is that for $n = 1$ and $l = 0$ and is designated $R_{1\,0}$. It is termed the 1s radial function. The constant C is determined by normalising equation (2.33) in accordance with condition (2.31). A value of $2(Z/a_0)^{\frac{3}{2}}$ is obtained so that

$$R_{1\,0} = 2\left(\frac{Z}{a_0}\right)^{\frac{3}{2}}\exp\left(-\frac{Zr}{a_0}\right). \qquad \ldots(2.37)$$

The 1s radial function for the hydrogen atom is included in Fig. 2.10. Other different solutions for radial functions are obtained for 2s, 3s, ... states and also for p, d and f states. In each case the solution consists of the product of a simple polynomial in r multiplied by a decaying exponential in r. These solutions are obtained only provided that

$$n \geqslant l+1, \qquad \ldots(2.38)$$

where n is the principal quantum number. The normalised radial functions for different values of n and l are given in Table 2.4. and are illustrated in Fig. 2.10.

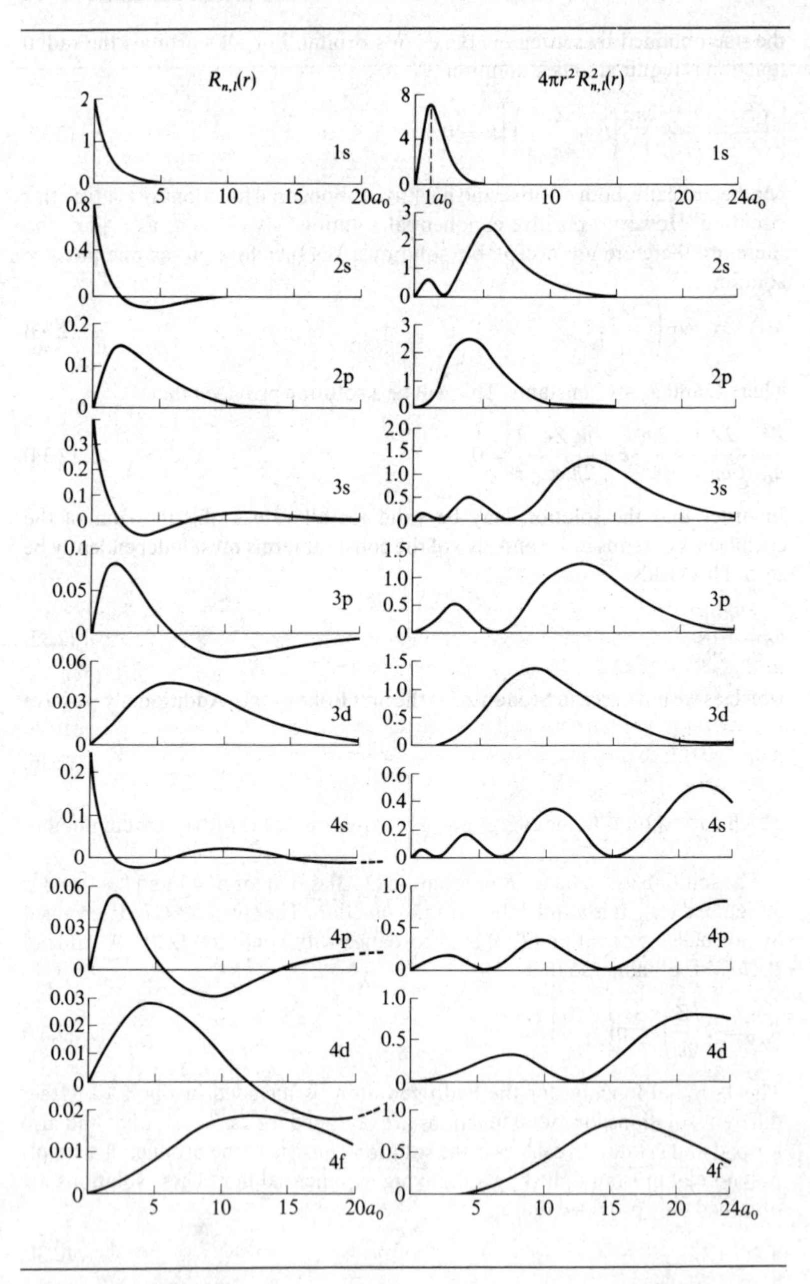

Fig. 2.10 Radial functions for a hydrogen atom. Note that the values of $P(r)$ for the vertical scale have been multiplied by 4π in this representation.

Table 2.4 Normalised radial functions

n	l	R_{nl}	R
1	0	R_{10}	$2\left(\frac{Z}{a_0}\right)^{3/2}\exp\left(-\frac{Zr}{a_0}\right)$
2	0	R_{20}	$\left(\frac{Z}{a_0}\right)^{3/2}\frac{1}{2\sqrt{2}}\left(2-\frac{Zr}{a_0}\right)\exp\left(-\frac{Zr}{2a_0}\right)$
2	1	R_{21}	$\left(\frac{Z}{a_0}\right)^{3/2}\frac{1}{2\sqrt{6}}\left(\frac{Zr}{a_0}\right)\exp\left(-\frac{Zr}{2a_0}\right)$
3	0	R_{30}	$\left(\frac{Z}{a_0}\right)^{3/2}\frac{2}{81\sqrt{3}}\left[27-18\left(\frac{Zr}{a_0}\right)+2\left(\frac{Zr}{a_0}\right)^2\right]\exp\left(-\frac{Zr}{3a_0}\right)$
3	1	R_{31}	$\left(\frac{Z}{a_0}\right)^{3/2}\frac{4}{81\sqrt{6}}\left[6\left(\frac{Zr}{a_0}\right)-\left(\frac{Zr}{a_0}\right)^2\right]\exp\left(-\frac{Zr}{3a_0}\right)$
3	2	R_{32}	$\left(\frac{Z}{a_0}\right)^{3/2}\frac{4}{81\sqrt{30}}\left[\frac{Zr}{a_0}\right]^2\exp\left(-\frac{Zr}{3a_0}\right)$

The increase in radial extent of the functions with increasing n is very apparent and since each radial function has been normalised this is accompanied by a corresponding decrease in the vertical scale. In accordance with the normalisation requirement all these curves decay to zero away from the nucleus. This decay is exponential and is largest for small values of n so that we have here a broad correspondence with the ideas of the Bohr theory. Radial functions may be shown to have zero values $(n-l-1)$ times between 0 and ∞. These are nodal surfaces over which the electron has zero probability of occupation.

Let us now look at the interpretation of the variation of the radial function with distance from the nucleus. In Section 1.8.1 we saw that the quantity $\psi\psi^*\,\mathrm{d}V$ or $|\psi|^2\,\mathrm{d}V$ gives the probability of finding the electron within the volume $\mathrm{d}V$ and the quantity $|\psi|^2$ was then interpreted as the electron density. It is often referred to as the probability density. For s states the angular dependence factor is a constant so that we can compare $|R|^2$ directly with $|\psi|^2$ and thus find that for these states $|\psi|^2$ is finite at the nucleus (see Fig. 2.11). Only for these spherically symmetrical s states is there a finite electron density at the nucleus. On this basis the electron is represented as a spherical charge cloud of varying density surrounding the nucleus.

A complementary approach is to consider the probability of finding the electron within a radial distance between r and $r+\mathrm{d}r$ from the nucleus. Since each integral in condition (2.31) is separately normalised to unity we can put

$$\int_0^\infty P(r)\,\mathrm{d}r = \int_0^\infty r^2\,RR^*\,\mathrm{d}r = 1. \qquad \ldots(2.39)$$

The function $P(r)$ is termed the radial probability density.

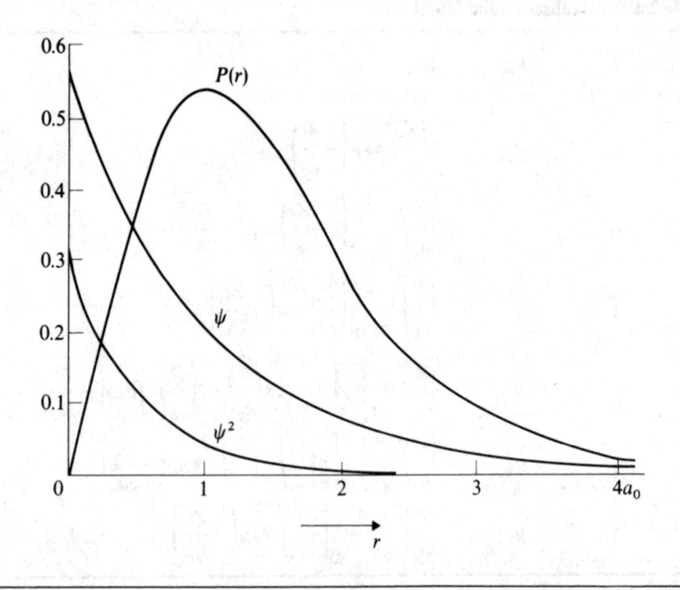

Fig. 2.11 Hydrogen atom ground state (1s) wave-function ψ_{100}, electron density ψ^2_{100} and radial probability density ($P(r)$. The scales are in atomic units.

Radial probability density functions can be determined for each radial function R and like R will depend on both n and l. Thus corresponding to the radial function $R_{1\,0}$ of equation (2.37) there exists the radial probability function

$$P(r) = 4\left(\frac{Z}{a_0}\right)^3 r^2 \exp\left(-\frac{2Zr}{a_0}\right). \qquad \ldots(2.40)$$

This function is shown in Fig. 2.11 for hydrogen and is there compared with the electron density function $|\psi|^2$. The radial probability function $P(r)$ shows a maximum value when r is equal to a_0/Z, i.e. for a hydrogen atom at a value of r equal to the Bohr radius. An exactly defined orbital radius of the Bohr theory has therefore been replaced in the wave-mechanical approach by this concept of a radial probability density.

The radial probability functions for a number of states are included in Fig. 2.10 and the positions of the maxima and the nodal surfaces for the functions may be readily verified. For any given value of principal quantum number n, the orbital of highest angular momentum is that with largest l, i.e. $l = n - 1$. The radial probability functions for such orbitals show a single maximum and this corresponds in position with the Bohr orbit radius for that value of n.

2.4 The complete solution

In accordance with equation (2.19) the complete wave-function is given by the product of the angular and radial functions. The wave-functions depend on n, l and m and are designated in the form ψ_{nlm}. Noting that the angular dependence for an s state is simply $1/2(\pi)^{\frac{1}{2}}$ the ground state wave-function ψ_{100} is obtained directly from equation (2.37) and is given by

$$\psi_{100} = \frac{1}{\sqrt{\pi}}\left(\frac{Z}{a_0}\right)^{\frac{3}{2}} \exp\left(-\frac{Zr}{a_0}\right). \qquad \text{...(2.41)}$$

This function is illustrated in Fig. 2.11 for $Z = 1$ and is the 1s, ground state, for the hydrogen atom. Other wave-functions for values of n up to 3 are listed in Table 2.5. These functions are all normalised and are the *atomic orbitals* for a one-electron atom such as hydrogen, a singly ionised helium atom or a double ionised lithium atom. The energy values of the orbitals depend on n but not on l or m and are given by

$$E = -\frac{m_e Z^2 e^4}{32\pi^2\varepsilon_0^2\hbar^2}\frac{1}{n^2}. \qquad \text{...(2.42)}$$

Consideration of conditions (2.27) and (2.38) shows that for any value of n there are n^2 wave-functions having the same energy. This is termed degeneracy. For non hydrogen-like atoms the degeneracy is less since, as we shall see, the energy of the orbitals depends on l as well as on n.

The shapes of s, p and d orbitals have been discussed in Section 2.3.1. A satisfactory three-dimensional representation for ψ, or for electron density $\psi\psi^*$, is difficult. Contour surfaces can be drawn and an orbital can, for example, be represented by a boundary surface which encloses a volume within which there is a 95 per cent probability of finding the electron. Contours on the xz-plane for a p_z orbital are illustrated in Fig. 2.12.

The extension of the theory to the determination of wave-functions and their eigenvalues of states in many-electron atoms presents many problems. These arise primarily because of the mathematical difficulties encountered in dealing with systems composed of many interacting particles. The simplest system which can be considered is that of the helium atom in which there are three interacting particles, i.e. a nucleus and two electrons. If the ground state energy is calculated on the assumption that the two electrons do not interact the value obtained is about 40 per cent greater than the experimental value, which illustrates that the assumption is not valid. Various methods involving the use of trial functions have, however, been developed and these can yield values in good agreement with experiment.

The problem is basically that in taking into account the effect of one or more electrons on the behaviour of any one particular electron the field of force is no longer a *central* field. As a consequence the form of the function for V is such that the variables r, θ and ϕ are not separable and the method of solution obtained for the hydrogen atom is not appropriate.

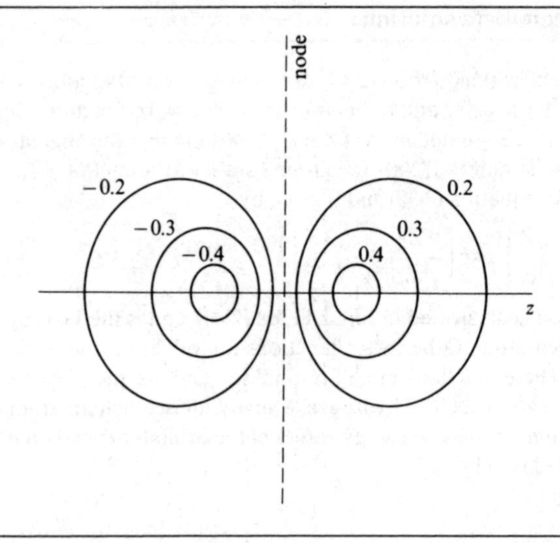

Fig. 2.12 Contours on the xz-plane for a p_z orbital.

The basis of the methods used in overcoming this problem is the one-electron approximation. The wave-function of each electron is calculated on the assumption that it can be regarded as existing in the field of the nucleus together with an additional averaged field representing the effect of all the other electrons. Designating the electron-dependent variable by a subscript i we have for the electron a wave equation

$$\nabla^2\psi_i + \frac{2m_e}{\hbar^2}\left[E_i + \frac{Ze^2}{4\pi\varepsilon_0}\frac{1}{r_i} - V_i(r)\right]\psi_i = 0 \qquad \ldots(2.43)$$

with similar expressions for the other electrons in the atom. The form of the function $V_i(r)$ representing the field due to all the other electrons differs according to the wave-function being examined, but is always chosen to be spherically symmetrical. The error in doing this, even though contributions from p and d type electrons are included, may be shown to be small. The use of a central field, although no longer of coulomic form, does however allow separable solutions to be determined.

In order to arrive at correct value of $V_i(r)$ for each electron the Hartree method of self-consistent fields is used. Briefly, for a system of N electrons, N trial wave-functions are set up each with its appropriate value of $V_i(r)$ obtained as an average due to all the other electrons. Each equation is solved to give improved wave-functions and the solutions are used to calculate corrected potential functions. The process is repeated many times until a self-consistent field is obtained together with a set of wave-functions for the electrons in the atom.

Since in these equations only the form of the potential V as a function of r is

Table 2.5 Normalised atomic orbitals for one electron atoms

n	l	$\|m\|$	**Orbital**	
1	0	0	1s	$\dfrac{1}{\sqrt{\pi}}\left(\dfrac{Z}{a_0}\right)^{\frac{3}{2}}\exp\left(-\dfrac{Zr}{a_0}\right)$
2	0	0	2s	$\dfrac{1}{4\sqrt{2\pi}}\left(\dfrac{Z}{a_0}\right)^{\frac{3}{2}}\left(2-\dfrac{Zr}{a_0}\right)\exp\left(-\dfrac{Zr}{2a_0}\right)$
2	1	0	$2p_z$	$\dfrac{1}{4\sqrt{2\pi}}\left(\dfrac{Z}{a_0}\right)^{\frac{3}{2}}\left(\dfrac{Zr}{a_0}\right)\exp\left(-\dfrac{Zr}{2a_0}\right)\cos\theta$
2	1	1	$2p_x$	$\dfrac{1}{4\sqrt{2\pi}}\left(\dfrac{Z}{a_0}\right)^{\frac{3}{2}}\left(\dfrac{Zr}{a_0}\right)\exp\left(-\dfrac{Zr}{2a_0}\right)\sin\theta\cos\phi$
2	1	1	$2p_y$	$\dfrac{1}{4\sqrt{2\pi}}\left(\dfrac{Z}{a_0}\right)^{\frac{3}{2}}\left(\dfrac{Zr}{a_0}\right)\exp\left(-\dfrac{Zr}{2a_0}\right)\sin\theta\sin\phi$
3	0	0	3s	$\dfrac{1}{81\sqrt{3\pi}}\left(\dfrac{Z}{a_0}\right)^{\frac{3}{2}}\left[27-18\left(\dfrac{Zr}{a_0}\right)+2\left(\dfrac{Zr}{a_0}\right)^2\right]\exp\left(-\dfrac{Zr}{3a_0}\right)$
3	1	0	$3p_z$	$\dfrac{2}{81\sqrt{2\pi}}\left(\dfrac{Z}{a_0}\right)^{\frac{3}{2}}\left[6\left(\dfrac{Zr}{a_0}\right)-\left(\dfrac{Zr}{a_0}\right)^2\right]\exp\left(-\dfrac{Zr}{3a_0}\right)\cos\theta$
3	1	1	$3p_x$	$\dfrac{2}{81\sqrt{2\pi}}\left(\dfrac{Z}{a_0}\right)^{\frac{3}{2}}\left[6\left(\dfrac{Zr}{a_0}\right)-\left(\dfrac{Zr}{a_0}\right)^2\right]\exp\left(-\dfrac{Zr}{3a_0}\right)\sin\theta\cos\phi$
3	1	1	$3p_y$	$\dfrac{2}{81\sqrt{2\pi}}\left(\dfrac{Z}{a_0}\right)^{\frac{3}{2}}\left[6\left(\dfrac{Zr}{a_0}\right)-\left(\dfrac{Zr}{a_0}\right)^2\right]\exp\left(-\dfrac{Zr}{3a_0}\right)\sin\theta\sin\phi$
3	2	0	$3d_{z^2}$	$\dfrac{1}{81\sqrt{6\pi}}\left(\dfrac{Z}{a_0}\right)^{\frac{3}{2}}\left(\dfrac{Zr}{a_0}\right)^2\exp\left(-\dfrac{Zr}{3a_0}\right)(3\cos^2\theta-1)$
3	2	1	$3d_{xz}$	$\dfrac{1}{81\sqrt{2\pi}}\left(\dfrac{Z}{a_0}\right)^{\frac{3}{2}}\left(\dfrac{Zr}{a_0}\right)^2\exp\left(-\dfrac{Zr}{3a_0}\right)\sin 2\theta\cos\phi$
3	2	1	$3d_{yz}$	$\dfrac{1}{81\sqrt{2\pi}}\left(\dfrac{Z}{a_0}\right)^{\frac{3}{2}}\left(\dfrac{Zr}{a_0}\right)^2\exp\left(-\dfrac{Zr}{3a_0}\right)\sin 2\theta\sin\phi$
3	2	2	$3d_{x^2-y^2}$	$\dfrac{1}{81\sqrt{2\pi}}\left(\dfrac{Z}{a_0}\right)^{\frac{3}{2}}\left(\dfrac{Zr}{a_0}\right)^2\exp\left(-\dfrac{Zr}{3a_0}\right)\sin^2\theta\cos 2\phi$
3	2	2	$3d_{xy}$	$\dfrac{1}{81\sqrt{2\pi}}\left(\dfrac{Z}{a_0}\right)^{\frac{3}{2}}\left(\dfrac{Zr}{a_0}\right)^2\exp\left(-\dfrac{Zr}{3a_0}\right)\sin^2\theta\sin 2\phi$

changed the angular functions remain unchanged in form. This is an important result since it justifies the extension of the use of the angular functions representing s, p, d and f states to multi-electron atoms. In contrast to the hydrogen atom, the energies of the orbitals for multi-electron atoms are dependent on l as well as on n, as is illustrated in Fig. 2.13 for sodium.

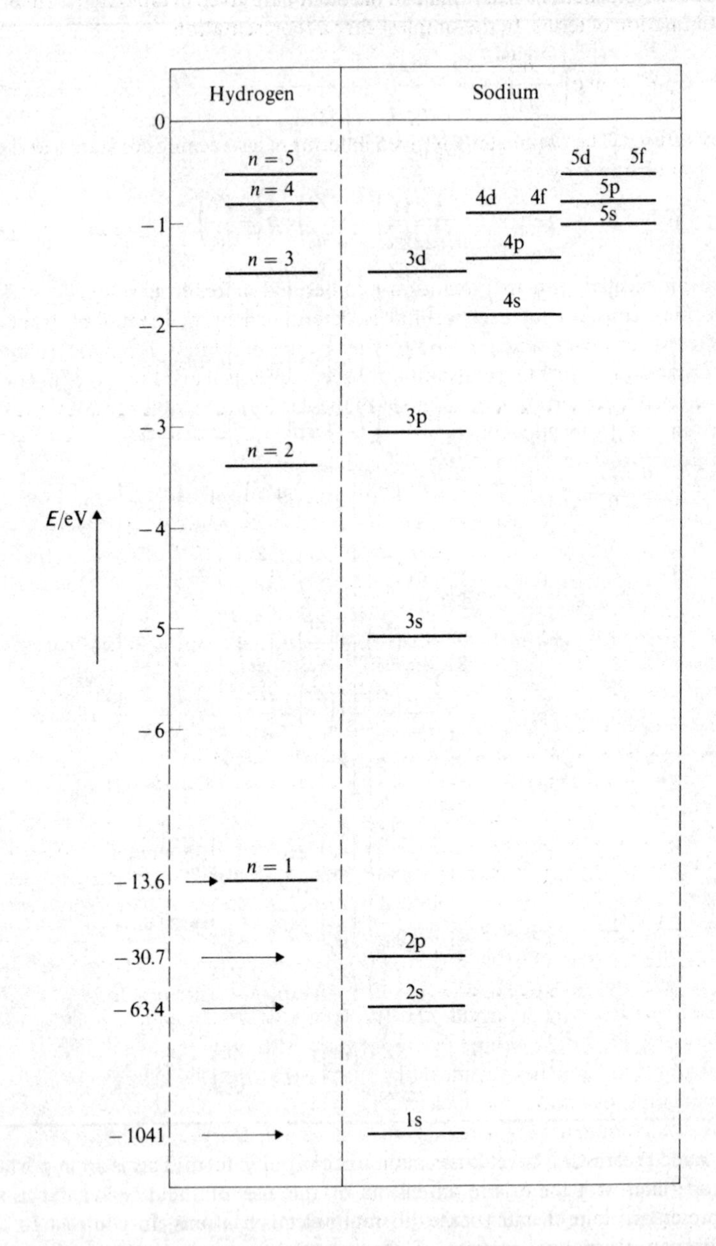

Fig. 2.13 A comparison on the energy levels of hydrogen and sodium.

The radial functions determined in the theory are given in tabulated form or as a summation of terms. In the simplest case a representation

$$R = C_{n^*l} r^{n^*-1} \exp\left(-\frac{\alpha r}{a_0}\right) \qquad \ldots(2.44)$$

may be used. The parameter α is given in terms of a screening constant s and the nuclear charge Z by

$$\alpha = \frac{Z-s}{n^*}. \qquad \ldots(2.45)$$

n^* is an effective principal quantum number and differs from n for $n \geqslant 4$. The screening constant for each orbital is determined by contributions from all electrons occupying similar or lower lying groups of orbitals. Empirical rules for determining the shielding constant have been developed and the wave-functions are termed Slater orbitals. A summary of these rules and a table of Slater orbitals is given by Greenwood (1969).

In order to explain the energy dependence of the different orbitals in multi-electron atoms we need to assume some of the detailed electronic structure of free atoms which will be explained in greater detail in subsequent sections. In multi-electron atoms the electron occupancy of different orbitals follows closely defined rules so that one finds for example that a sodium atom in its ground state will have an electronic configuration $1s^2 2s^2 2p^6 3s^1$ (see Section 2.7). The superscripts indicate the number of electrons occupying the different orbitals. Other excited states are possible, e.g. a 3p or 3d orbital may be occupied instead of the 3s.

We note first that in a multi-electron atom, in accordance with equation (2.44), an increase in nuclear charge Z has the effect of pulling in the shells towards the nucleus. This effect is most marked for the 1s electrons and in Fig. 2.13 we note that the 1s level for sodium lies at -1041 eV compared with -13.6 eV for hydrogen. Other shells will be partly screened from the nucleus by other lower lying shells so that the electronic charge decreases the effective charge on the nucleus. This screening effect modifies the energies of the orbitals of different n values and also leads to a decreasing stability in the order s, p, d, f.

This latter effect may be illustrated for sodium. Figure 2.14 gives the radial probability function $P(r)$ for the 10 core electrons $1s^2 2s^2 2p^6$ and also, separately, for the ground state 3s orbital and the excited states 3p and 3d orbitals. The behaviour of the 3s orbital is important. Although the maximum of the probability function lies outside the core, since the orbital is an s type the electron density must be finite at the nucleus. The orbital is thus penetrating the core and makes maximum use of the region of low potential energy near the nucleus. The 3p and 3d orbitals have lower radial probability in this region, so that in comparison with the 3s orbital will have higher energy and be less stable. A comparison of the energies of the different orbitals has been given in Fig. 2.13.

We may also compare the radial extension of the orbitals in sodium with those of hydrogen. First, the core electrons in sodium are strongly influenced by the

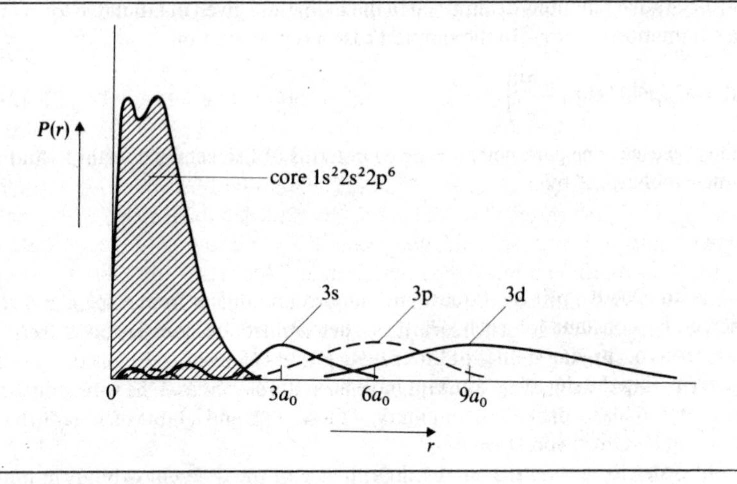

Fig. 2.14 Radial probability functions for sodium orbitals.

increased nuclear charge ($Z = 11$) compared with hydrogen and the radial extension of the 10 core electrons is not greatly different from that for the single 1s electron in hydrogen. The orbital penetration reduces the effective size of the 3s orbital, and to some extent the 3p orbital, so that the maxima lie at $3.5a_0$ and $6a_0$, respectively, compared with $13a_0$ and $12a_0$ for the corresponding orbitals for hydrogen. The 3d orbital is little affected and this exhibits a maximum probability at about $9a_0$ as for hydrogen.

In summary, orbital penetration into the core is important in determining the relative energies of different orbitals in many-electron atoms.

2.5 Angular momentum and electron spin

An aspect of the wave-mechanical theory of atomic orbitals which will be discussed only briefly here is concerned with the explanation of some of the features of spectral lines which were not adequately explained by previous theories. An electron in any given orbital ψ_{nlm} possesses a *total orbital angular momentum* ***L*** of magnitude

$$|L| = [l(l+1)]^{\frac{1}{2}}\hbar. \qquad \ldots(2.46)$$

A corresponding magnetic moment $(e/_2m_e)[l(l+1)]^{\frac{1}{2}}\hbar$ is similarly defined and in an external magnetic field the energy of the atom will be governed by the relative orientations of the moment and field. The angular momentum is a vector quantity and defining the z direction as the direction of an external magnetic field the vector will precess around the z direction at one of a specific number of angles defined by the magnetic quantum number m. The component of angular momentum L_z along the z direction will be quantised according to

$$L_z = m\hbar, \qquad \ldots(2.47)$$

where m_l can take on the $(2l+1)$ values in accordance with condition (2.27b). In the presence of a magnetic field this results in each energy level being split into $(2l+1)$ levels. Electron transitions to or from this split level give rise therefore to a number of closely spaced spectral lines. The allowed transitions are those for which m changes by 0 or ± 1 so that in a magnetic field a single spectral line is split into a triplet of equally spaced components. This is the *normal Zeeman effect*.

In order to account for anomalous Zeeman spectra, i.e. those not explained simply in terms of the quantisation of the orbital angular momentum, and also the occurrence of fine structure it is necessary to introduce the concept of *electron spin*. This was first proposed by Uhlenbeck and Goudsmit. Due to its spin the electron possesses additional intrinsic *spin angular momentum* $\boldsymbol{L}_s$ of magnitude

$$|\boldsymbol{L}_s| = [s(s+1)]^{\frac{1}{2}}\hbar \qquad \ldots(2.48)$$

which relationship is analogous to that for orbital angular momentum. s is the *spin angular momentum* quantum number, and, as explained below, has the single value of $\frac{1}{2}$. The z component of spin momentum L_{sz} is similarly

$$L_{sz} = m_s\hbar, \qquad \ldots(2.49)$$

where m_s has the values $\pm s$, i.e. $\pm\frac{1}{2}$ and is termed the *spin magnetic quantum number*.

Since the electron possesses a permanent spin magnetic moment we should expect it to interact with the magnetic moment due to the orbital angular momentum. This is termed *spin–orbit interaction* and is observed for all states except those for which $l = 0$. The internal magnetic field due to the orbital momentum splits s into $2s+1$ components and since all fine structure states are two in number it follows that $s = \frac{1}{2}$. The orbital and spin moments may be combined to give a total angular momentum $\boldsymbol{M}$ for the electron of magnitude

$$|\boldsymbol{M}| = [j(j+1)]^{\frac{1}{2}}\hbar, \qquad \ldots(2.50)$$

where j is termed the *inner quantum number* given by

$$j = l \pm s = l \pm \tfrac{1}{2}. \qquad \ldots(2.51)$$

Electron states for which $l = 0$ are single but this degeneracy is removed in a magnetic field where again one finds two spin states corresponding to the alignment of the spin magnetic momentum parallel or antiparallel to the applied field. The spin magnetic moment μ of the electron may thus have two possible values $\pm\mu_B$ and these correspond to the two possible values $\pm\frac{1}{2}$ of the spin quantum number. μ_B is the *Bohr magneton*. Radiation of appropriate frequency ν may lead to transitions between the two states provided that

$$\hbar\nu = 2\mu_B B, \qquad \ldots(2.52)$$

where B is the applied magnetic induction.

The frequencies are observed by experiment to lie in the microwave region and the effect is termed electron spin resonance.

The brief outline of the theory given refers to the angular momentum of single electrons. In multi-electron atoms it is the total angular momentum of the group of electrons which is observable and one is particularly interested in the relative energies of various possible electronic configurations of an atom. For a group of electrons their individual angular momenta may be combined vectorially to yield a resultant orbital angular momentum $\boldsymbol{M}_L$ of magnitude

$$|\boldsymbol{M}_L| = [L(L+1)]^{\frac{1}{2}}\hbar. \qquad \ldots(2.53)$$

L is the resultant azimuthal quantum number and, by analogy with the single-electron case, the states are denoted by S, P, D, etc. for $L = 0, 1, 2$, etc. A resultant spin angular momentum $\boldsymbol{M}_S$ of magnitude

$$|\boldsymbol{M}_S| = [S(S+1)]^{\frac{1}{2}}\hbar \qquad \ldots(2.54)$$

is similarly obtained by the vectorial addition of the spin momenta of the individual atoms. S is the resultant spin quantum number.

The resultant total angular momentum of the atom is governed by the value of the resultant inner quantum number J. This depends on the type of coupling between the electrons. The most common is that in which L and S are combined to give J and this is referred to as Russel–Saunders or L–S coupling. The combination of individual j values to give J is more important for the heavier elements and is termed j–j coupling. The different electronic configurations in atoms will be discussed in Section 2.7. We note, however, at this point that one is particularly interested in the relative energies of different states. These are influenced by the values of L, S and J and each state is identified by a so-called term symbol.

2.6 The Pauli exclusion principle

In the preceding sections solutions of the wave equation for a single electron in the field of the nucleus have been derived and an indication given of the extension of the theory to multi-electron atoms. The form of the solutions was found to be governed by three quantum numbers n, l and m and certain limitations were found on the possible values of these quantum numbers. An additional quantum number, that of electron spin m_s has been introduced separately. It may be noted, however, that in relativistic quantum mechanics, developed by Dirac, the electron spin appears as an integral part of the solutions of the wave equation.

There is a very important principle governing the states of electrons in atoms which allows the electron states of free atoms to be explained and classified. This is the Pauli exclusion principle. In its simplest form the principle states that, in any given system, there can at most be only one electron in each quantum state defined by the four quantum numbers n, l, m and m_s. For each orbital wave-function ψ_{nlm} we may therefore have a maximum of two electrons, one corresponding to each state of spin. In a more complete treatment a *total wave-function* is defined and this is written as the product of an orbital, space-dependent function and a corresponding spin function. The latter is usually denoted by α or β according to its state of spin.

The Pauli principle in its more complete form states that the total wave-function, including spin, must be antisymmetric with respect to interchange of electrons. In order to understand this statement we must briefly explore the concept of symmetry and antisymmetry.

Any function which has the property

$$\psi_e(-x) = \psi_e(x) \qquad \ldots(2.55)$$

is termed an even or a symmetric function. If on the other hand

$$\psi_o(-x) = -\psi_o(x) \qquad \ldots(2.56)$$

it is termed an odd or antisymmetric function. The simplest examples are the trigonometric functions $\sin x$ and $\cos x$ which are, respectively, odd and even functions. We shall come across wave-functions of both types.

There is, additionally, another type of symmetry exhibited by wave-functions, namely that of symmetry to interchange of particles. In a multi-electron system the orbital wave-function of the system may be written as a combination of the orbital wave-functions representing the separate electrons. In order that we may look at the symmetry with respect to interchange of particles let us consider two orbitals a and b occupied by two electrons 1 and 2. The wave-function for the system, ignoring inter-electron repulsion, will be based on the product

$$\psi = \psi_{a1}\psi_{b2} = \psi_{a2}\psi_{b1} \qquad \ldots(2.57)$$

This function is itself not a satisfactory solution since it possesses the wrong symmetry. The two linear combinations

$$\psi_S = \frac{1}{\sqrt{2}}(\psi_{a1}\psi_{b2} + \psi_{a2}\psi_{b1})$$

$$\ldots(2.58)$$

$$\psi_{AS} = \frac{1}{\sqrt{2}}(\psi_{a1}\psi_{b2} - \psi_{a2}\psi_{b1})$$

may, however, be used and these conform to the usual normalisation requirement, provided that ψ_a and ψ_b are themselves normalised. If now the two electrons 1 and 2 are interchanged the function ψ_S remains unchanged whereas ψ_{AS} is changed to $-\psi_{AS}$. ψ_S is a symmetric function to the interchange of electrons, whereas ψ_{AS} is antisymmetric. We note also an important additional requirement concerning the two wave-functions in equation (2.58). They must both conform to the fundamental principle of the indistinguishability of electrons. This will be the case for both of these functions since the observable physical quantity $|\psi|^2$ remains unaltered on interchange of electrons.

The Pauli principle refers to the *total* wave-function including the spin function. It may be shown that for two-electron systems there are four possible spin functions. Three of these represent electrons with parallel spins and are symmetric to the interchange of electrons, the other represents antiparallel or paired spins and is antisymmetric to interchange. Since the total wave-function is required to be antisymmetric to exchange of electrons, the symmetrical orbital function ψ_S must be combined with a spin function representing electrons with paired spins and the antisymmetric function ψ_{AS} with a spin function for parallel spins. In particular, if the two electrons are occupying the same orbital (i.e. $\psi_a = \psi_b$) then we only have the single function ψ_S which is symmetrical, and the electron spins must then be paired. This is in accordance with the first statement of the principle.

The restrictions on electron occupancy of wave-functions arising out of the Pauli principle may be regarded as providing a type of interactive force between electrons. This *spin correlation* tends to make electrons of the same spin keep apart and those of opposite spin to come together.

2.7 The electronic structure of free atoms

In order to understand the electronic structure of the elements there are a number of factors which need to be considered. First, the availability of electron states. The restrictions here are a direct consequence of the conditions imposed on the three quantum numbers *n, l* and *m* in the solution of the wave equation and the requirement that the quantum number m_s can take on only the two values $\pm\frac{1}{2}$. Secondly, the mode of occupation of the states. The Pauli principle sets a maximum level of occupation such that no two electrons in an atom can have the same set of four quantum numbers. The number of available electron states for values of *n* up to 5 is given in Table 2.6. In its ground state an atom will

Table 2.6 Number of available energy states in atoms

n	1	2		3			4				5				
l	0	0	1	0	1	2	0	1	2	3	0	1	2	3	4
Type of State	s	s	p	s	p	d	s	p	d	f	s	p	d	f	g
No. of States	2	2	6	2	6	10	2	6	10	14	2	6	10	14	18
Total	2	8		18			32				50				

have an electronic configuration corresponding to the minimum energy. Other excited states of the atom are available in which one or more electrons may occupy electron states or orbitals of higher energy than that giving the minimum energy configuration.

As we have seen in previous sections, factors such as orbital penetration and screening play an important part in determining the relative energies of electron states in an atom. The energy of an orbital is largely determined by the value of the principal quantum number *n* and the effective nuclear charge for each orbital. Except for hydrogen, the energies depend on both *n* and *l*. Consideration of the mode of filling the states in the different elements (Table 2.7) indicates the importance of this dependence on *n* and *l*.

Electrostatic repulsion between electrons in the same or adjacent orbitals is important and is affected by considerations of electron spin as discussed in the previous section. Thus we find that electrons will avoid occupying the same orbital if there are degenerate orbitals available. This is well illustrated in free carbon atoms where all three 2p orbitals have the same energy and the two outer electrons are found to occupy two of these orbitals rather than be paired in any single orbital. As discussed in the previous section, electrostatic repulsion between two electrons is decreased if they have parallel spins which keeps them apart. Electrons in separate orbitals are thus found to have parallel spins; this is the case for example for the electrons in the 2p orbitals of carbon. For degenerate orbitals one therefore finds a maximum number of singly occupied orbitals, i.e. of unpaired electrons. Such electrons will have parallel spins which is termed spin correlation. These effects are examples of Hund's rules. Where degenerate orbitals are not available two electrons may then be occupying the same orbital and must of course have opposite spins. The resultant orbital momentum of a group of electrons and the type of coupling between orbitals and spin momentum is also of importance in determining which electronic configuration is the most stable. This point will be returned to shortly. The ground state electronic configuration of many elements are given in Table 2.7.

The simplest atom is hydrogen and in its ground state the single electron occupies the 1s state ($n = 0$, $l = 0$). The energy required to remove the electron from the atom, termed its ionisation energy, is 13.6 eV and the corresponding ionisation potential is 13.6 V. In an excited state the single electron is less tightly bound, so that taking the usual zero of energy as that of an electron at infinite distance from the nucleus an electron in the 2s state in hydrogen has energy of -3.4 eV.

The two electrons in the helium atom are both in the 1s state so that the electrons must have opposite spins ($\uparrow\downarrow$). The first ionisation energy E_i of 24.6 eV is the highest for any element (see Table 2.7) and this shows that the ground state electronic configuration is extremely stable. As might be expected the second electron is much more tightly bound and the second ionisation potential is 54.4.V. In the absence of repulsion between the two electrons both ionisation potentials might be expected to have this value so that the repulsive coulombic interaction between the electrons contributes about 30 eV to the energy of the atom. The absence of further states for $n = 1$, combined with a substantial excitation energy to raise electrons to states with $n = 2$, explains the chemical inertness and the low electrical conductivity of helium.

The first excited state of helium is the $1s^1 2s^1$ state, and in accordance with the

Table 2.7 The ground state electronic configuration of selected elements

Z	Element		E_i/eV	1s	2s	2p	3s	3p	3d	4s	4p	4d	4f	5s	5p	5d	5f	6s	6p	6d	6f	7s
1	Hydrogen	H	13.6	1																		
2	Helium	He	24.6	2																		
3	Lithium	Li	5.4	2	1																	
4	Beryllium	Be	9.3	2	2																	
5	Boron	B	8.3	2	2	1																
6	Carbon	C	11.3	2	2	2																
7	Nitrogen	N	14.5	2	2	3																
8	Oxygen	O	13.6	2	2	4																
9	Fluorine	F	17.4	2	2	5																
10	Neon	Ne	21.6	2	2	6																
11	Sodium	Na	5.1	2	2	6	1															
12	Magnesium	Mg	7.6	2	2	6	2															
13	Aluminium	Al	6.0	2	2	6	2	1														
14	Silicon	Si	8.2	2	2	6	2	2														
15	Phosphorus	P	10.5	2	2	6	2	3														
16	Sulphur	S	10.4	2	2	6	2	4														
17	Chlorine	Cl	13.0	2	2	6	2	5														
18	Argon	Ar	15.8	2	2	6	2	6														
19	Potassium	K	4.3	2	2	6	2	6		1												
20	Calcium	Ca	6.1	2	2	6	2	6		2												
21	Scandium	Sc	6.5	2	2	6	2	6	1	2												
22	Titanium	Ti	6.8	2	2	6	2	6	2	2												
23	Vanadium	V	6.7	2	2	6	2	6	3	2												
24	Chromium	Cr	6.7	2	2	6	2	6	5	1												
25	Manganese	Mn	7.5	2	2	6	2	6	5	2												
26	Iron	Fe	7.5	2	2	6	2	6	6	2												
27	Cobalt	Co	7.5	2	2	6	2	6	7	2												
28	Nickel	Ni	7.7	2	2	6	2	6	8	2												
29	Copper	Cu	7.8	2	2	6	2	6	10	1												
30	Zinc	Zn	9.4	2	2	6	2	6	10	2												

Table 2.7 *(continued)*

Z	Element		E_i/eV	1s	2s	2p	3s	3p	3d	4s	4p	4d	4f	5s	5p	5d	5f	6s	6p	6d	6f	7s
31	Gallium	Ga	6.0	2	2	6	2	6	10	2	1											
32	Germanium	Ge	7.9	2	2	6	2	6	10	2	2											
33	Arsenic	As	9.8	2	2	6	2	6	10	2	3											
34	Selenium	Se	9.7	2	2	6	2	6	10	2	4											
35	Bromine	Br	11.8	2	2	6	2	6	10	2	5											
36	Krypton	Kr	14.0	2	2	6	2	6	10	2	6											
37	Rubidium	Rb	4.1	2	2	6	2	6	10	2	6			1								
38	Strontium	Sr	5.7	2	2	6	2	6	10	2	6			2								
39	Yttrium	Y	6.4	2	2	6	2	6	10	2	6	1		2								
47	Silver	Ag	7.6	2	2	6	2	6	10	2	6	10		1								
48	Cadmium	Cd	9.0	2	2	6	2	6	10	2	6	10		2								
49	Indium	In	5.8	2	2	6	2	6	10	2	6	10		2	1							
50	Tin	Sn	7.4	2	2	6	2	6	10	2	6	10		2	2							
51	Antimony	Sb	8.6	2	2	6	2	6	10	2	6	10		2	3							
54	Xenon	Xe	14.3	2	2	6	2	6	10	2	6	10		2	6							
55	Cesium	Cs	3.9	2	2	6	2	5	10	2	6	10		2	6			1				
56	Barium	Ba	5.2	2	2	6	2	6	10	2	6	10		2	6			2				
74	Tungsten	W	8.0	2	2	6	2	6	10	2	6	10	14	2	6	4		2				
79	Gold	Au	9.2	2	2	6	2	6	10	2	6	10	14	2	6	10		1				
80	Mercury	Hg	10.5	2	2	6	2	6	10	2	6	10	14	2	6	10		2				
81	Thallium	Tl	6.1	2	2	6	2	6	10	2	6	10	14	2	6	10		2	1			
82	Lead	Pb	7.5	2	2	6	2	6	10	2	6	10	14	2	6	10		2	2			
86	Radon	Rn	10.8	2	2	6	2	6	10	2	6	10	14	2	6	10		2	6			
92	Uranium	U	4.0	2	2	6	2	6	10	2	6	10	14	2	6	10	3	2	6	1		2

previous discussion the space function for this state will be antisymmetrical so as to favour a charge distribution in which the electrons are well separated. The Pauli requirement for the total wave-function to be antisymmetric is met by the electrons having parallel spins. A convenient method of representing the electron occupancy of different states in the elements is given in Table 2.8.

Element	Electronic configuration	Term Symbol	Representation			
			1s	2s	2p	3s
H	$1s^1$	$^2S_{\frac{1}{2}}$	↑			
He	$1s^2$	$^1S_0$	↑↓			
*He	$1s^1\,2s^1$	$^3S_1$	↑	↑		
Li	$1s^2\,2s^1$	$^2S_{\frac{1}{2}}$	↑↓	↑		
C	$1s^2\,2s^2\,2p^2$	$^3P_0$	↑↓	↑↓	↑ ↑ □	
*C	$1s^2\,2s^1\,2p^3$	$^5S_2$	↑↓	↑	↑ ↑ ↑	
F	$1s^2\,2s^2\,2p^5$	$^2P_{\frac{3}{2}}$	↑↓	↑↓	↑↓ ↑↓ ↑	
Ne	$1s^2\,2s^2\,2p^6$	$^1S_0$	↑↓	↑↓	↑↓ ↑↓ ↑↓	
Na	$1s^2\,2s^2\,2p^6\,3s^1$	$^2S_{\frac{1}{2}}$	↑↓	↑↓	↑↓ ↑↓ ↑↓	↑
*Na	$1s^2\,2s^2\,2p^5\,3s^2$	$^2P_{\frac{1}{2}}$	↑↓	↑↓	↑↓ ↑↓ ↑	↑↓

An analogous situation to that in helium occurs with the filling of the $n = 2$ level. This occurs at the element neon ($Z = 10$) which has two electrons in the 1s orbital, two in the 2s, and six in the 2p orbitals. The electronic configuration is designated $1s^2 2s^2 2p^6$. Here again with neon we find a high ionisation potential and an element which is an inert gas. The configuration of filled s and p sub-levels giving a total of eight electrons is particularly stable and gives a spherically symmetrical electron density. Similarly, the noble gases argon, krypton, zenon and radon correspond to the filling of the 3p, 4p, 5p and 6p sub-levels, respectively.

The elements which perhaps most closely resemble hydrogen in their ground state electronic structure are the alkali metals. These elements have one more electron than the noble gas which adjoins them in the Periodic Table. We thus have the configurations shown in Table 2.9, i.e. the valence electron in these metals overlies the noble gas core. The similarity in this configuration to that of hydrogen leads to corresponding similarities in their optical spectra, which is particularly noticeable for transitions to the f states. In all the alkali metals the valence electron is occupying an s state. Energetically this is the most favoured configuration and it arises because of the greater penetration of the s orbitals into the noble gas core compared with that of the p and d orbitals. The difference is,

Table 2.9 A comparison of the ground static electronic configurations of three alkali metals with those for the noble gases which adjoin them in the Periodic Table.

He	$1s^2$	Li	$1s^22s^1$
Ne	$1s^22s^22p^6$	Na	$1s^22s^22p^63s^1$
Ar	$1s^22s^22p^63s^23p^6$	K	$1s^22s^22p^63s^23p^64s^1$

however, not very large as can be seen from Fig. 2.13, which gives the energy level diagram for sodium. This result is due principally to the shielding effect of the core electrons which produces a shallow potential energy variation in the outer part of the atom. For the same reason we find that first ionisation potentials are lower for the alkali metals than for all other elements, e.g. only 4.3 V for potassium. Removal of a second electron must be from the noble gas core, which is energetically much more difficult. One finds therefore for potassium a second ionisation potential of 52.4 V. The looseness of binding of the valence electron leads to a typical metallic type of binding and to a high electrical conductivity for the solid elements.

In contrast to the alkali metals the halogen group of elements—fluorine, chlorine, bromine and iodine—have one electron less in their p sub-levels than is necessary to complete the stable noble gas configuration and form bonds with other atoms which allow these levels to be filled, e.g. ionic bonds with the alkali metals and covalent bonds with carbon. The elements are strongly electronegative in character.

The variation in electron energy in outer electron shells becomes progressively shallower with increasing n. This arises both because of the shielding effect of the core electrons and also because of the $1/n^2$ dependence of the energy on principal quantum number. As a consequence one finds that, for moderately heavy atoms ($Z = 20$–50), the lower angular momentum 4s state, which makes use of the lower potential energy core region, possesses lower energy than the higher angular momentum 3d state. This filling of the 4s states before the 3d leads to the transition elements ($Z = 21$–30) having many similarities in their chemical and physical properties. Thus their metallic and electrical properties are governed by the 4s electrons whereas the magnetic properties of the solids are a result of the unfilled 3d orbitals. Another example of the near degeneracy of orbitals is that shown by the rare earths or lanthanides ($Z = 57$–71). These elements have incompletely filled 4f and 5d orbitals of nearly equal energy.

The Group IV elements are of considerable interest in solid state theory. These are the elements carbon, silicon, germanium, tin and lead which exhibit a gradual transition in behaviour from diamond, which is an insulator, through the semiconductors to the metallic element lead. Taking as an example carbon, we find that the free atom configuration is that of 2 paired electrons in both of the 1s and 2s states, together with single electrons having parallel spins in each of two 2p orbitals. When bonded with other elements, however, or in the allotrope, solid diamond, carbon is quadrivalent and it exhibits a tetrahedral type of bonding. There is thus considerable difference in the electronic configuration between the free atom and the solid. This is common to many solids and is a topic which will be returned to in later chapters.

Excited states of the elements exist and correspond to other configurations of occupancy of orbitals. In agreement with the previous discussion these will have higher energies than the ground state.

It is useful here to refer back to Section 2.5 from which we may recall that each different mode of electron occupancy can be allocated a term symbol which depends on the values of the total quantum numbers *L*, *S* and *J*. For Russell–Saunders coupling this has the form $^{2S+1}L_J$ where the superior prefix and the subscript are given as numbers and *L* as the appropriate orbital momentum symbol. $2S+1$ is termed the spin multiplicity of the state. This allows states of different energies to be codified. In accordance with this nomenclature the ground state configuration of hydrogen ($1s^1$) and helium ($1s^2$) have term symbols $^2S_{\frac{1}{2}}$ and $^1S_0$, respectively, whilst the $1s^12s^1$ first excited state of helium is $^3S_1$. The term symbols for a few electronic configurations are given in Table 2.8.

The most stable term for the ground state configuration is determined in accordance with Hund's rules. For example for a given *nl* configuration the term of highest multiplicity is the most stable and where two terms have the same multiplicity that having the highest *L* is the most stable. The facility with which an atom may be promoted to different excited states is important in determining which valency it exhibits in chemical reactions. An important example is that of carbon. In its ground state carbon has the electronic configuration $1s^22s^22p^2$ and a term symbol $^3P_0$. A number of excited states are possible for the element, e.g. for the same *nl* configuration we may have states such as $^3P_2$ with the two electrons still with parallel spins, or, alternatively, paired electrons in a 2p orbital giving terms such as $^1D_2$ or $^1S_0$. Excitation of one electron from the 1s level to fill the remaining p orbital can yield unpaired electrons in each of four equivalent orbitals. The energy required is less than that released on the formation of, for example, two extra C–H bonds and explains the quadrivalent behaviour of carbon already noted.

References and further reading

Barrett, J. (1970) *Atomic and Molecular Structure*, Wiley.
Cumper, C. W. N. (1966) *Wave Mechanics for Chemists*, Heinemann.
Eisberg, R. M. (1967) *Fundamentals of Modern Physics*, Wiley.
Greenwood, N. N. (1969) *Principles of Atomic Orbitals*, Royal Institute of Chemistry.
Holden, A. (1971) *The Nature of Atoms*, Oxford University Press.
Holden, A. (1971) *Stationary States*, Oxford University Press.
Leighton, R. B. (1959) *Principles of Modern Physics*, McGraw-Hill.
Linnett, J. W. (1960) *Wave Mechanics and Valency*, Methuen.
Murrell, J. N., Kettle, S. F. A. and Tedder, J. M. (1965) *Valence Theory*, Wiley.
Richtmyer, F. K., Kennard, E. H. and Cooper, J. N. (1969) *Introduction to Modern Physics*, McGraw-Hill.

Sproull, R. L. (1964) *Modern Physics*, Wiley.
Stevens, B. (1962) *Atomic Structure and Valency*, Chapman and Hall.
Weidner, R. T. and **Sells, R. L.** (1968) *Elementary Modern Physics*, Allyn and Bacon.

Problems

2.1 What values does the Bohr theory give for the total energy, kinetic energy and potential energy of an electron in the first Bohr orbit?
What is the corresponding value for the electron velocity?

2.2 Show that the Bohr orbital velocity can be expressed in the form

$$u = \left(\frac{Ze^2}{4\pi\varepsilon_0 m_e r}\right)^{\frac{1}{2}}.$$

Derive the analogous expression for the orbital velocity of the moon given that the gravitational attractive force between the moon and the earth is given by

$$F = \frac{G m_m M}{r^2},$$

where m_m and M are the masses of the moon and earth respectively and G is the gravitational constant.
In what way do the two expressions differ?

2.3 Show that the electron orbital velocity u in the Bohr theory of the hydrogen atom can be expressed in the form

$$\frac{u}{c} = \frac{\alpha}{n}$$

where α is a dimensionless ratio involving only fundamental constants. α is known as the fine structure constant and its value determines the fine splitting of energy levels.

2.4 According to classical theory an oscillating charge will emit radiation of frequency equal to that of the oscillation. Derive an expression for the orbital frequency corresponding to the first Bohr orbit of the hydrogen atom and show that radiation of wavelength ≈ 46.5 nm would be expected on this basis. Consider the effect of the emission of radiation on the energy of the electron and confirm that the electron would be expected to spiral inwards to the nucleus.

2.5 Determine the wavelength of the first two lines and that of the series limit for the Balmer and for the Paschen series of the hydrogen spectrum.

2.6 Calculate the energy levels for values of n up to 6 for the hydrogen atom, the He^+ and the Li^{2+} ions. In each of the three cases examine which transitions between the six levels give rise to emission lines within the visible spectrum.

2.7 What is the total number of hydrogen emission lines obtained as a result of transitions occurring between the first six levels?

2.8 Show that for large n the emission line for transitions between adjacent levels is given by

$$\nu = \frac{2R_\infty c}{n^3}.$$

Express this frequency in terms of the radius of the orbit and show that the expression obtained is identical with the orbital frequency. This is an example of Bohr's correspondence principle.

2.9 By setting

$$\int_0^\infty r^2 RR^* \, dr = 1$$

show that the normalised radial function $R_{1\,0}$ is given by

$$R_{1\,0} = 2\left(\frac{Z}{a_0}\right)^{\frac{3}{2}} exp\left(-\frac{Zr}{a_0}\right).$$

2.10 The normalised wave function

$$\psi_{1\,0\,0} = \frac{1}{\sqrt{\pi}}\left(\frac{Z}{a_0}\right)^{\frac{3}{2}} \exp\left(-\frac{Zr}{a_0}\right)$$

was obtained in the text by separately normalising the radial function $R_{1\,0}$ and the angular function $Y_{0\,0}$. Show that the normalisation constant can alternatively be obtained by use of the condition

$$\int_0^\infty 4\pi r^2 \psi^2 \, dr = 1$$

and explain the significance of this condition.

2.11 Show that the radial probability function for the 1s state of the hydrogen atom exhibits a single maximum at a value of r equal to the Bohr radius a_0.

2.12 For a given value of n, radial functions with the largest value of l will possess the *greatest* angular momentum. Which are these if $n = 2$ and if $n = 3$?

Show that the corresponding radial probability functions each exhibit one maximum and that these maxima correspond to the radii of the orbits for $n = 2$ and $n = 3$ of the Bohr theory.

2.13 Confirm the form of the radial function $R_{2\,0}$ and its corresponding radial probability function by investigating the extreme values of the functions and other significant features of the functions.

2.14 The ions He^+, Li^{2+} and Be^{3+} are hydrogen-like and the removal of the remaining electron in each case corresponds respectively to the second, third and fourth ionisation potentials. Calculate these potentials and explain why the method adopted would fail to yield correct values for the other ionisation potentials for these elements.

2.15 The core electrons in sodium shield the 3s electron from the nucleus. Given that the first ionisation potential of sodium is 5.14 eV calculate the effective nuclear charge.

2.16 The 1s and 2p levels in sodium have energies -1041 eV and -30.7 eV, respectively. The K_α characteristic X-radiation is due to electron transitions from the L to K shell. Calculate the wavelength of K_α radiation for sodium.

2.17 In the previous example the 2p electron could be considered to be located in the electric field of the nucleus but shielded by one 1s electron, i.e. the effective atomic number is $Z-1$. Using this modified value in equation (2.13) confirm the value of the wavelength of emitted K_α radiation for sodium.

Deduce the corresponding value for copper K_α radiation.

2.18 In accordance with Section 1.8.1 two wave-functions of an atom are completely independent of each other and are termed orthogonal. For two wave-functions ψ_1 and ψ_2 in the form given in Table 2.5 this condition can be put in the form

$$\int_V \psi_1 \psi_2 \, dV = 0.$$

Show that this condition holds for
(a) any two of the wave-functions $2p_x$, $2p_y$ and $2p_z$; and
(b) the 1s and 2s wave-functions.
(*Hint:* Use equation (2.30) and separate into three integrals as in equation (2.31)).

2.19 Electron energy and atomic distances can be written in dimensionless form by dividing respectively by the corresponding atomic scale unit namely the rydberg $(= m_e e^4/32\pi^2\varepsilon_0^2\hbar^2)$ and the Bohr radius a_0 $(=4\pi\varepsilon_0\hbar^2/m_e e^2)$.
Show that the modified form of the Schrodinger wave equation

$$\frac{d^2\psi}{dx^2}+(E-V)\psi=0$$

is then valid.

2.20 Show that $(Y_{1\,1}+Y_{1\,-1})/\sqrt{2}$ gives the p_x angular dependence function in the two alternative forms quoted in Table 2.3 and confirm that the function is normalised to unity in accordance with equation (2.31). Sketch the function.

2.21 Consider the form of the d_{z^2} orbital (Table 2.3, Fig. 2.8) and show that the function has maximum positive values of $\sqrt{5}/2\sqrt{\pi}$. What is the significance of the independence of the function of ϕ? Note that the central ring corresponds to negative values of the function whereas the two lobes are positive.
Show that over the two planes $r = \pm\sqrt{3}z$ the function has zero value. These planes are termed nodal planes.

2.22 Values of the first ionisation energies of many elements are given in Table 2.7. Plot these values against atomic number and consider the major features of the curve obtained. The curve for the second ionisation energy is of the same general form but is displaced by one atomic number to the right and has all the values roughly doubled. Why should this be so?

3. The structure of solids

3.1 Introduction

Atoms in solids are arranged in an ordered manner. Such configurations are inherently of lower energy, and thereby more stable than a purely random arrangement of atoms. The outward manifestation of the ordered arrangement is the crystalline form exhibited by many solids which mirrors the regularity of their internal atomic arrangement. Although the crystalline nature of many solids such as metals and semiconductors is not often readily apparent, these materials also consist of crystals normally in the form of a large number of interlocking crystals all differently oriented. They can also be produced in the form of large single crystals. There are a considerable number of different atomic arrangements possible, these being termed crystal structures. Many solids, however, crystallise in one of a few simple structures.

3.2 Space lattices

A single crystal of a solid exhibits a regular arrangement of its constituent elements which is based on a three dimensional and periodically repeating unit pattern. In order to introduce the concept of a crystal structure it is first necessary to define a *space lattice.*

Let us first divide space up into a large number of identical parallelepipeds, a single one of which is defined by three non-coplanar vectors $\boldsymbol{a}$, $\boldsymbol{b}$ and $\boldsymbol{c}$. A translation vector $\boldsymbol{T}$ defined by

$$\boldsymbol{T} = n_1\boldsymbol{a} + n_2\boldsymbol{b} + n_3\boldsymbol{c}, \qquad \ldots(3.1)$$

where n_1, n_2 and n_3 are integers, will now define the points of a space lattice. This is illustrated in Fig. 3.1 for a right-handed system of axes. With appropriate values of the integers the vector $\boldsymbol{T}$ will connect any two points of the space lattice.

In a crystal, each *lattice point* has a certain local arrangement of atoms grouped about it, and in particular it should be noted that each lattice point has *identical* surroundings. In some simple cases only one atom is associated with each lattice point, in others a number of atoms, and in the crystals of complex molecules a very large number. When describing the crystal structure of a

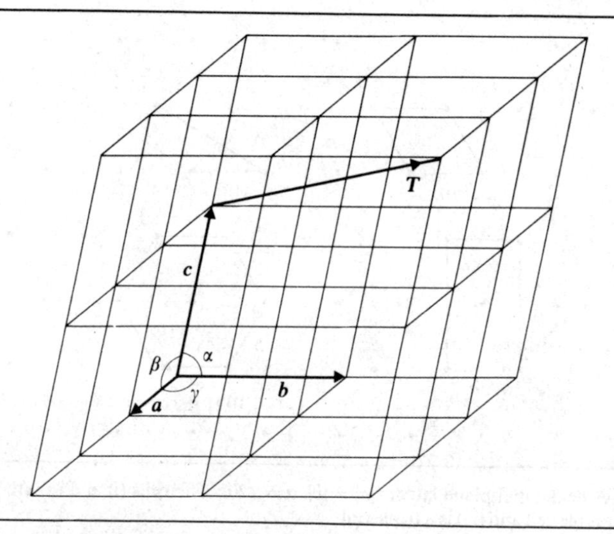

Fig. 3.1 A space lattice. Note that $\boldsymbol{a}$, $\boldsymbol{b}$ and $\boldsymbol{c}$ form a right-handed set of axes. The most general space lattice has $a \neq b \neq c$ and $\alpha \neq \beta \neq \gamma$. One particular lattice vector $\boldsymbol{T}$ is illustrated. This has $\boldsymbol{T} = 3\boldsymbol{a}+2\boldsymbol{b}+\boldsymbol{c}$.

particular solid it is therefore necessary not only to consider the specification of the lattice but also the configuration and symmetry of the *basis* of atoms associated with each lattice point.

The vectors $\boldsymbol{a}$, $\boldsymbol{b}$ and $\boldsymbol{c}$ define a *unit cell* of volume V given by

$$V = \boldsymbol{a} \cdot \boldsymbol{b} \times \boldsymbol{c}. \qquad \ldots(3.2)$$

This result follows directly from the definitions of the products of vectors (see e.g. Speigel, 1959). For a given set of lattice points it is of course possible to partition them in a number of different ways. This is illustrated in Fig. 3.2 for the points of a two dimensional hexagonal lattice defined by $a = b$ and $\alpha = 120°$. Unit cells such as (i) and (ii) contain one lattice point per cell and these are termed *primitive* cells. For certain lattices it is preferable to describe them in terms of a non-primitive or multiple cell which may demonstrate better the inherent symmetry of the particular lattice. For the example chosen this is the triple cell (iv) which shows the hexagonal symmetry of the lattice. A further example is given in Fig. 3.3 which shows the primitive rhombohedral unit cell of the body-centred cubic lattice together with the conventional unit cube which is a double cell.

3.3 Symmetry elements

Lattices are distinguished by their symmetries and it is pertinent to ask how many different *types* of lattices exist. Similar symmetry considerations govern the

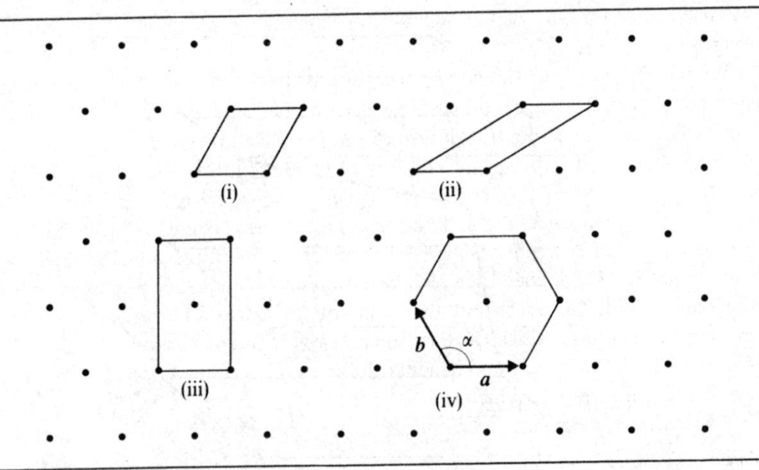

Fig. 3.2 A hexagonal plane lattice, $|a| = |b|$, $\alpha = 120°$. The cells (i) and (ii) are primitive, (iii) is a double cell and (iv) is a triple cell.

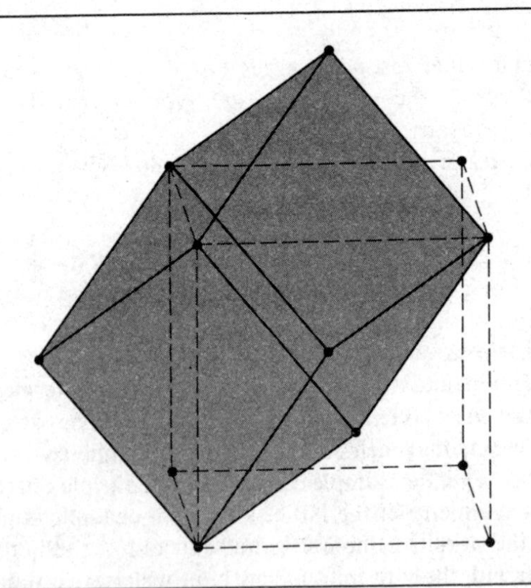

Fig. 3.3 The b.c.c. lattice. The figure shows the conventional unit cube which is a double cell and the rhombohedral unit cell which is primitive.

types of atomic arrangements possible for the basis of atoms located at each lattice point. Finally, in order to be able to specify crystal structures we must be able to specify which bases are compatible with each type of lattice. To proceed, we need therefore to define the various symmetry elements or symmetry operations.

We have already mentioned the necessity for lattices to possess translational symmetry, i.e. lattices are built up of a large number of identical unit cells. Other symmetries not involving translation can be described by symmetry operations which applied at a lattice point will bring the lattice into self-coincidence. This collection of symmetry operations is called a *point group* and will give the symmetry of the lattice about the lattice points. The symmetry of the basis of atoms at each lattice point is also described in terms of the point groups. For each type of lattice defined there will be a number of point group symmetries for the basis which are consistent with that of the lattice. There will also be a minimum symmetry which the base must have for the particular type of lattice. Apart from the point group elements there are also other symmetry elements involving translation to which we shall return later.

The symmetry element of a rotation axis of symmetry is illustrated in Fig. 3.4.

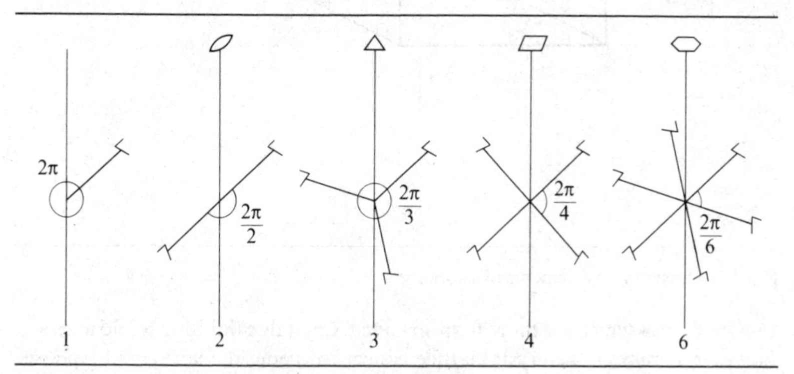

Fig. 3.4 Rotation axes of symmetry. The numbers give the fold of each axes, i.e. the number of repetitions of the rotation angle in one revolution.

Crystals of differing symmetries may thus be brought back into self-coincidence by rotations of either 2π, $2\pi/2$, $2\pi/3$, $2\pi/4$ or $2\pi/6$ and the symmetries concerned are termed to be respectively 1-, 2-, 3-, 4- or 6-fold. We may take a cubic lattice as an example. This possesses 4-fold symmetry along a line parallel to a cube edge, a 3-fold symmetry along the cube body diagonal and a 2-fold symmetry about the direction of a face diagonal (see Fig. 3.5). If the basis consists simply of an atom at each lattice point, as will be the case for many metals, the resulting *structure* will also possess these symmetries. The basis of atoms may, however, have lower symmetry but this must still be consistent with that of the cubic lattice. The *minimum* symmetry for the cubic lattice is that of the four 3-fold cube diagonal symmetry.

It is noted that crystals do not exhibit 5-fold, or any rotational symmetries greater than 6-fold since these are not consistent with the requirement for transitional symmetry; a plane cannot be filled by regular pentagons or by any

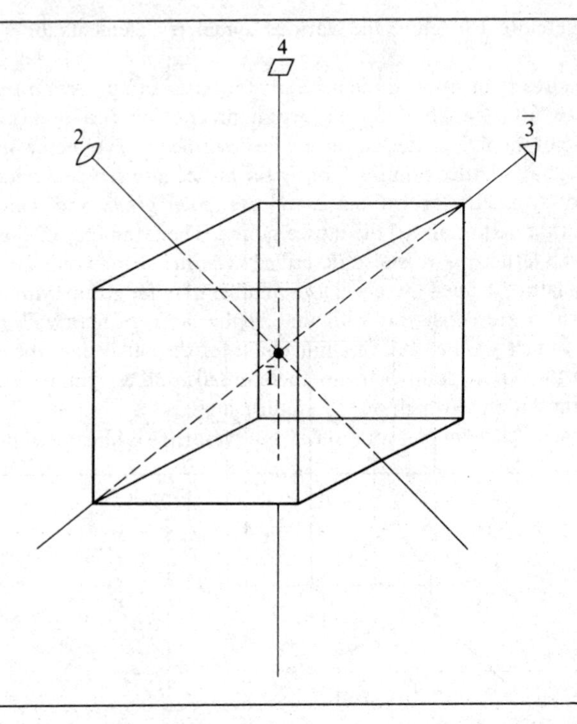

Fig. 3.5 The symmetry elements of a cube.

regular figures which are more than six-sided. Crystals exhibiting 6-fold axes are, however, common, the crystal lattice concerned being the hexagonal type (see Fig. 3.6).

A crystal may possess a *centre of symmetry* or, as it is often termed, *inversion symmetry*. In such a case any point $\boldsymbol{r}$ measured with respect to the centre of symmetry will have a totally equivalent point at the location $-\boldsymbol{r}$. All *lattices* allow the existence of a centre of inversion. The lattice type of lowest symmetry is the triclinic and we find the compound copper sulphate crystallising in this lattice and possessing a centre of symmetry. However, the geometry of the basis of atoms at the lattice points may rule out inversion symmetry in a crystal, e.g. it is absent in sodium tartarate which is also triclinic.

A *plane of symmetry m* is exhibited by a crystal when one half of the crystal is a mirror image of the other half. A cubic lattice exhibits three planes of symmetry parallel to the pairs of opposite faces and, in addition, six diagonal planes of symmetry each of which contains a face diagonal, a body diagonal and a cube edge. As we have already seen, however, it is not necessary for a particular cubic *structure* to possess these symmetries.

There is a fundamental theorem in group theory that one symmetry operation followed by a second is equivalent to a third symmetry operation, e.g. a

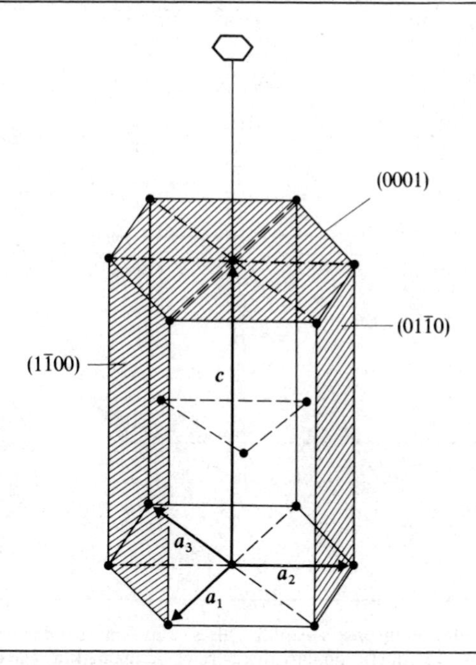

Fig. 3.6 The hexagonal close packed structure showing a 6-fold axis of symmetry. The indices of three planes are given in the hexagonal notation.

combination of 2-fold rotation axes along the *x*- and *y*-axes is equivalent to a 2-fold axis along the *z*-axis. It is sometimes convenient to describe certain symmetry operations in terms of compound operations. Two examples may be cited here. The first is that of rotoreflection which consists of any one of the rotation axes described and this is combined with a mirror plane perpendicular to the axis to give the compound operation. The other example is that of a rotoinversion axis in which a rotation axis is compounded with a centre of inversion. A 2-fold rotation inversion axis (written $\bar{2}$) is illustrated in Fig. 3.7 and is shown to be equivalent to a plane of symmetry *m*. A 1-fold rotation–inversion axis $\bar{1}$ is simply a centre of symmetry, whilst $\bar{3}$ is equivalent to a 3-fold rotation and a centre of symmetry.

The symmetries so far discussed are the so-called point symmetries and these are often apparent from the external form or morphology of a crystal. If, however, we include translations there are two other compound operations which yield other possible symmetries. One such compound operation gives rise to a *glide plane* symmetry. This consists of a mirror reflection followed by a translation of, for example, half the repeat distance in a prominent direction in the crystal. If the gliding direction is a cell edge it is termed an axial glide; if along a cell diagonal it is either a diagonal glide, or, when the repeat distance is one

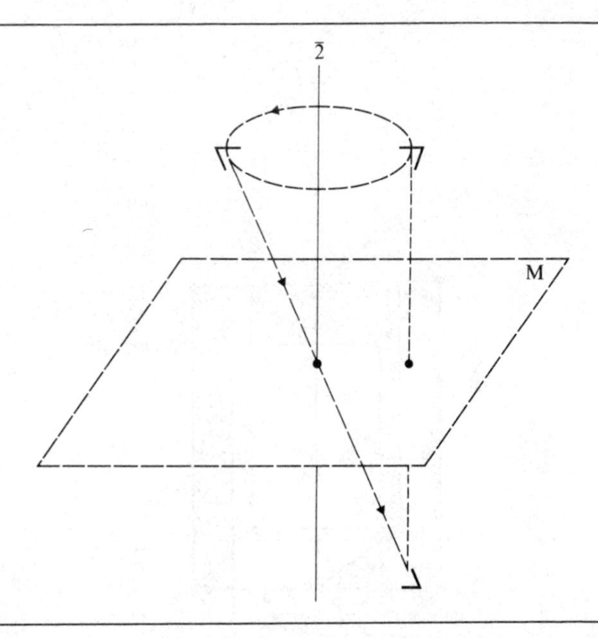

Fig. 3.7 A 2-fold rotation-inversion axis ($\bar{2}$). This is equivalent to a mirror reflection in the plane M. This symmetry operation may therefore be designated m. The operation is also equivalent to a 1-fold rotation together with a mirror reflection designated $\tilde{1}$.

quarter, a diamond glide. A glide plane symmetry is compared with a mirror plane symmetry in Fig. 3.8.

A *screw* axis as the name suggested is a combination of a rotation about an axis together with a translation along the same axis. Both left- and right-handed screw axes are possible and these may be 2-, 3-, 4- or 6-fold. We note that these last two symmetry elements involve very small translations and cannot therefore be inferred from external symmetry. Thus a glide plane cannot by such means be distinguished from a reflection plane or a screw axis from a rotation axis. Their presence may however be inferred from X-ray diffraction patterns which will show extinction of certain reflections when glide planes or screw axes are present.

3.4 Types of lattices and crystal structures

One may now return to the question concerning the number of distinguishable lattices which can exist. The most general plane lattice is the oblique or parallelogram lattice for which $a \neq b$ and $\alpha \neq 90°$. As is clear from Fig. 3.9 the lattice exhibits a 2-fold rotation axis of symmetry about a lattice point and this is

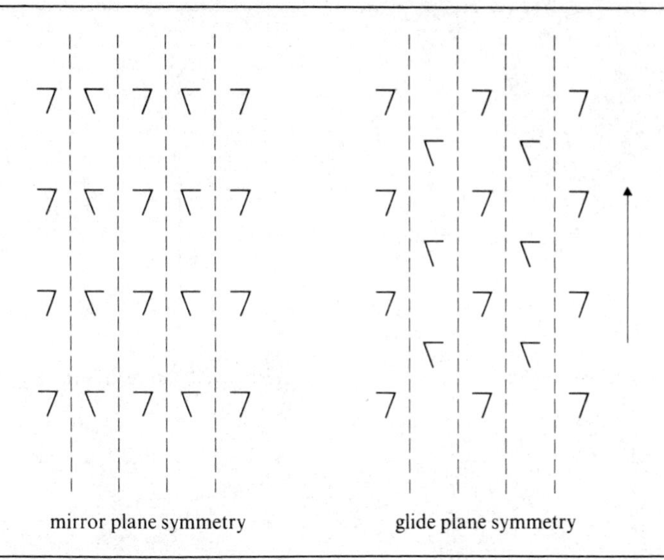

Fig. 3.8 A comparison of a mirror plane and an axial glide plane symmetry. The latter combines a mirror plane symmetry with a translation of $\frac{1}{2}a$. A diagonal glide would have a translation for example of $\frac{1}{2}a+\frac{1}{2}b$.

the minimum symmetry possible for a two-dimensional lattice. If, in addition, a plane of symmetry m is present we obtain the rectangular lattice defined by $a \neq b$, and $\alpha = 90°$ and this is therefore a separate distinguishable lattice. We note that the mirror symmetry together with the 2-fold rotation axis yields another mirror symmetry and the lattice symmetry is designated 2 mm. Additional symmetries are exhibited by square and hexagonal lattices and these together with a centred rectangular lattice make up the total of five possible plane lattices.

In three dimensions the most general lattice is the triclinic for which $a \neq b \neq c$ and $\alpha \neq \beta \neq \gamma$ (see Fig. 3.10). The point symmetry operations described can be used to show that there are a total of 14 types of space lattices which may be distinguished in three dimensions. These are referred to as Bravais lattices. The lattices are divided amongst seven crystal systems according to the values of a, b, c, α, β and γ. This information is summarised in Table 3.1 from which we see that in the cubic system there are three different Bravais lattices. These are the simple cubic, P, body-centred cubic, I, and the face-centred cubic, F. These are illustrated in Fig. 3.10 and we note that the cubes shown represent a primitive cell, a double cell and a quadruple cell. The b.c.c. and f.c.c. lattices may both be represented by primitive rhombohedral cells although the essential cubic symmetry of the lattices is better shown by the non-primitive cubic cells.

In order to arrive at the total number of possible crystal structures we need to combine the different symmetry groups exhibited by the bases with the appropriate Bravais lattices. As indicated in Table 3.1 there are 32 possible point

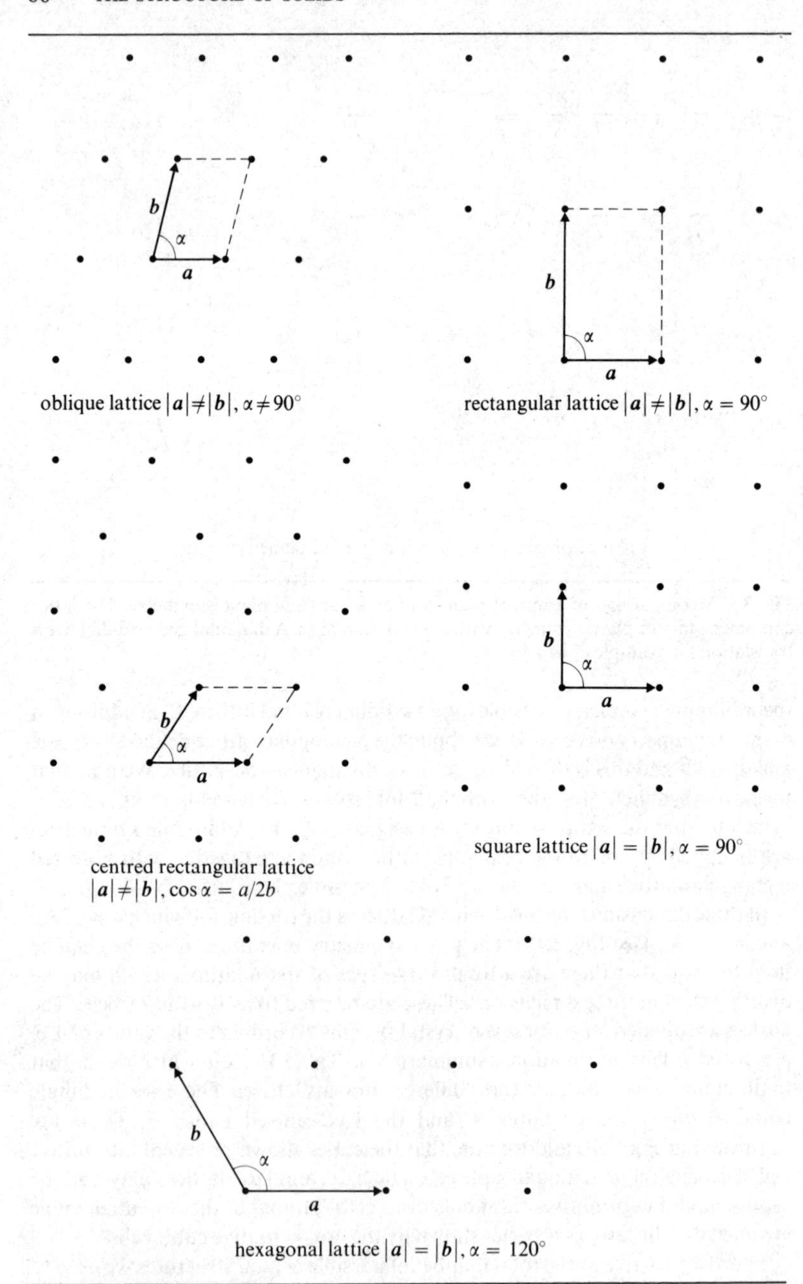

Fig. 3.9 The plane lattices.

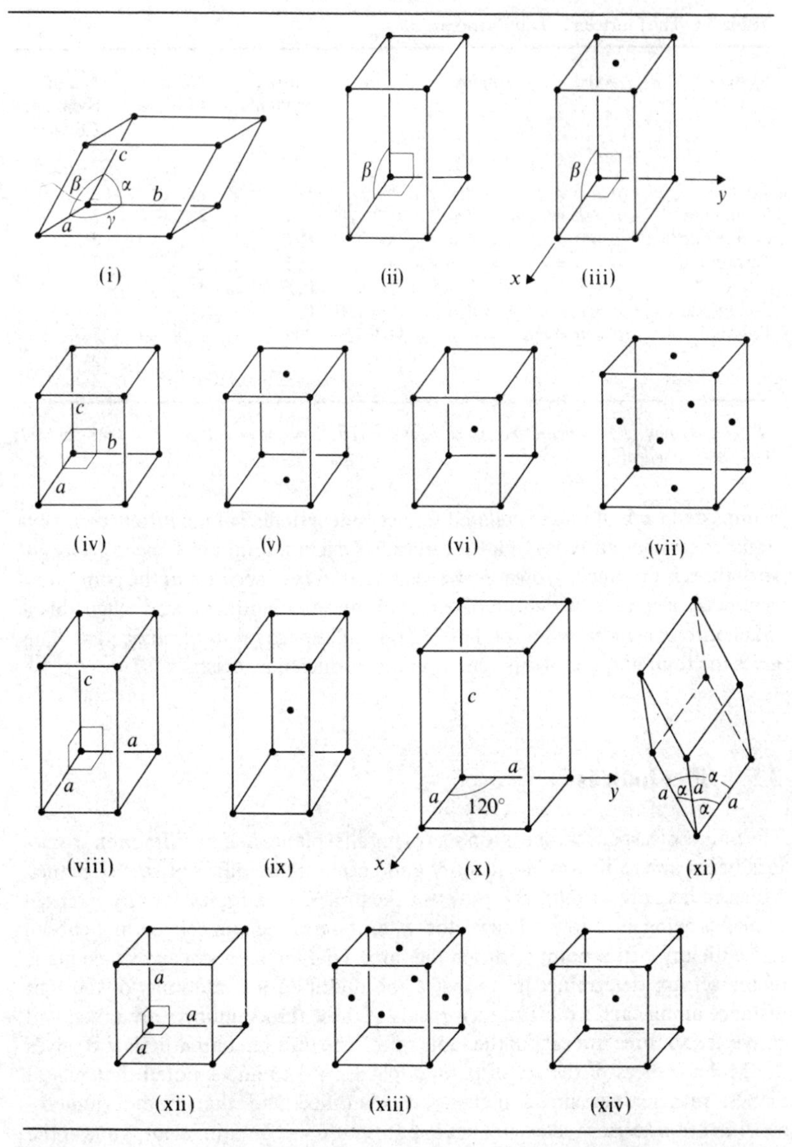

Fig. 3.10 Unit cells of the fourteen Bravais space lattices. (i) Primitive triclinic. (ii) Primitive monoclinic. (iii) Side-centred monoclinic—conventionally the 2-fold axis is taken parallel to *y* and the (001) face is centred (C centred). (iv) Primitive orthorhombic. (v) Side-centred orthorhombic—conventionally centred on (001), i.e. C centred. (vi) Body-centred orthorhombic. (vii) Face-centred orthorhombic. (viii) Primitive tetragonal. (ix) Body-centred tetragonal. (x) Primitive hexagonal. (xi) Primitive rhombohedral (trigonal). (xii) Primitive cubic. (xiii) Face-centered cubic. (xiv) Body-centred cubic.

Table 3.1 The fourteen bravais lattices

System	Axes	Angles	Lattice† Symbols	No. of Lattices	No. of Symmetry Classes
Triclinic	$a \neq b \neq c$	$\alpha \neq \beta \neq \gamma \neq 90°$	P	1	2
Monoclinic	$a \neq b \neq c$	$\alpha = \gamma = 90° \neq \beta$	P, C	2	3
Orthorhombic	$a \neq b \neq c$	$\alpha = \beta = \gamma = 90°$	P, C, F, I	4	3
Tetragonal	$a = b \neq c$	$\alpha = \beta = \gamma = 90°$	P, I	2	7
Cubic	$a = b = c$	$\alpha = \beta = \gamma = 90°$	P, F, I	3	5
Hexagonal	$a = b \neq c$	$\alpha = \beta = 90°, \gamma = 120°$	P	1	7
Trigonal	$a = b = c$	$\alpha = \beta = \gamma \neq 90°$	R	1	5
				14	32

† *Key to symbols:* P = *primitive*; C = *base-centred*; I = *body-centred*; F = *face-centred*; R = *rhombohedral*

groups and each of these is related to a crystal system. Taking into account the number of different types of lattice in each system a total of 71 *space groups* or structures is obtained. However, we also need to take account of the compound symmetry elements, which involve a translation component, and when this is taken into consideration a total of 230 possible space groups is arrived at. This gives the total number of different structures which may exist.

3.5 Miller indices

The points of a space lattice lie on sets of parallel planes and in diffraction studies it is necessary to be able to identify and specify these different sets of planes. Miller indices are used for this purpose, the procedure adopted for any given set of planes being as follows: The vectors ***a***, ***b***, and ***c*** of the unit cell are first chosen and with any lattice point as origin the three axial intercepts of any single plane of the set are determined in terms of the multiples (or fractions) of the unit distance along each axis. The reciprocals of these three numbers are taken and converted to three integers in the same ratio. The result in parenthesis (*hkl*) gives the Miller indices of the set of parallel planes. It should be noted that sets of planes, rather than single planes, are identified and that from symmetry considerations certain sets of planes in a structure will be equivalent. Thus in the cubic system there are six sets of planes parallel to each cube face namely (100), (010), (001), $(\bar{1}00)$, $(0\bar{1}0)$, and $(00\bar{1})$. Such sets of symmetrically equivalent planes are termed a form and are designated by braces, e.g. {100} for the cube face set. The method of determining Miller indices is illustrated in Fig. 3.11 and some prominent planes of the cubic lattice are shown in Fig. 3.12.

A modification of the procedure given is often used when determining the

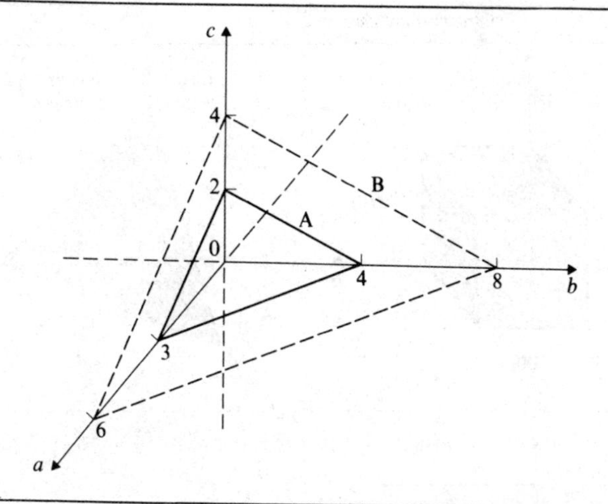

Fig. 3.11 Determination of Miller indices. The plane illustrated makes intercepts of $3\boldsymbol{a}$, $4\boldsymbol{b}$ and $2\boldsymbol{c}$ on the three axes, i.e. in the ratio 3:4::2 in terms of unit distances. The reciprocals are in the ratio $\frac{1}{3}:\frac{1}{4}:\frac{1}{2}$ and yield Miller indices (436). The Miller indices yield a set of parallel planes which are equispaced. One adjacent member is shown dotted.

indices of planes in a hexagonal crystal. The hexagonal axes, to which the planes are referred, are three coplanar a axes at 120° to each other and a fourth c axis at right-angles to them. Referring to Fig. 3.6 the three coplanar vectors $\boldsymbol{a}_1$, $\boldsymbol{a}_2$ and $\boldsymbol{a}_3$ are related by

$$\boldsymbol{a}_1 + \boldsymbol{a}_2 = -\boldsymbol{a}_3$$

so that the indices of a plane are of the form ($hkil$) where $i = -(h+k)$. The index l relates to the intercept on the c axis.

3.5.1 *Atom positions and crystal directions*

The location of any atom or point in the unit cell is specified with respect to the origin 000 of the unit cell in fractions or multiples of the axial lengths a, b, and c. Thus the centre of the unit cell is represented by $\frac{1}{2}\frac{1}{2}\frac{1}{2}$ and a face centre position by for example $\frac{1}{2}\frac{1}{2}0$.

A crystal direction is specified by three integers in square brackets $[hkl]$. When the direction forms a common intersection for a number of different planes it is termed a zone axis. In order to obtain the indices of a direction the multiples of the three unit axial distances defining the direction with respect to the origin are first obtained. These are then reduced to the three smallest integers in the same ratio to give the required set. The indices are in the ratios of the cosines of the angles which the given direction makes with the three axes. In the cubic system the direction $[hkl]$ is always perpendicular to the plane (hkl), but it should be

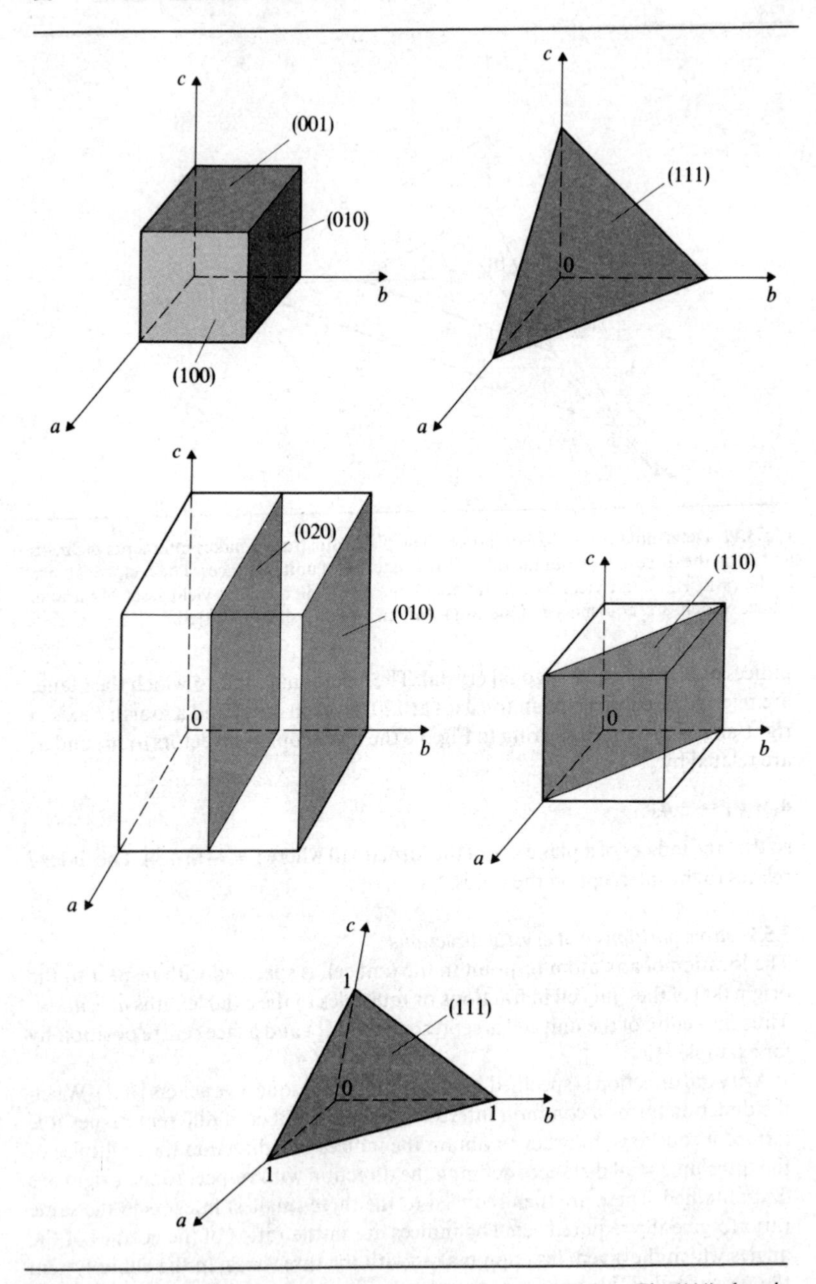

Fig. 3.12 A few of the important planes in the cubic lattice are shown. The (111) plane is also shown for a triclinic lattice.

emphasised that in other crystal systems this will not, in general, be the case. A set of equivalent directions are designated $\langle hkl \rangle$.

3.5.2 *Interplanar spacings*

The determination of interplanar spacings has formed an important aspect of crystallographic studies concerned with the understanding of the structure of matter. The spacings are governed by the indices (hkl) of the planes and the type of crystal lattice. For a cubic lattice of cell edge a, termed the lattice parameter, the first plane from the origin makes intercepts of a/h, a/k, and a/l on the three axes. The perpendicular distance from the origin to this plane gives the required interplanar spacing d_{hkl} and from the geometry will be given by

$$d_{hkl} = \frac{a}{(h^2 + k^2 + l^2)^{\frac{1}{2}}}. \qquad \ldots(3.3)$$

This equation is modified in other crystal systems, and in the most general case of the triclinic system involves all the parameters a, b, c, α, β and γ.

3.6 Diffraction methods

Interatomic spacings in crystals are typically $\approx 10^{-10}$ m and the regular periodic array of atoms on a three-dimensional lattice will diffract waves of the appropriate wavelength in a manner somewhat analogous to the diffraction of light by a grating. The X-ray diffraction method was first proposed by von Laue, and a wide and detailed knowledge of the crystallography of matter has been built up using diffraction techniques. The use of electron and neutron diffraction methods has already been alluded to briefly in Section 1.6.2. We may now review the basis of the use of these three diffraction methods.

3.6.1 *X-ray diffraction*

X-rays are produced when electrons from a cathode are accelerated *in vacuo* to a high potential and then impinge on a metal anode. A small percentage of the total energy of an electron beam is converted into X-rays, the remainder being dissipated as heat in the water-cooled metal anode. For each particular accelerating potential a continuous X-ray spectrum is obtained as illustrated in Fig. 3.13. The energy of an X-ray photon is related to the excitation potential U by

$$h\nu = \frac{hc}{\lambda} \leqslant eU, \qquad \ldots(3.4)$$

where the equality sign corresponds to the threshold wavelength for production of X-rays. For this wavelength the whole of the electron energy is converted to that of the X-ray photon. Most of the electrons, however, have their energies

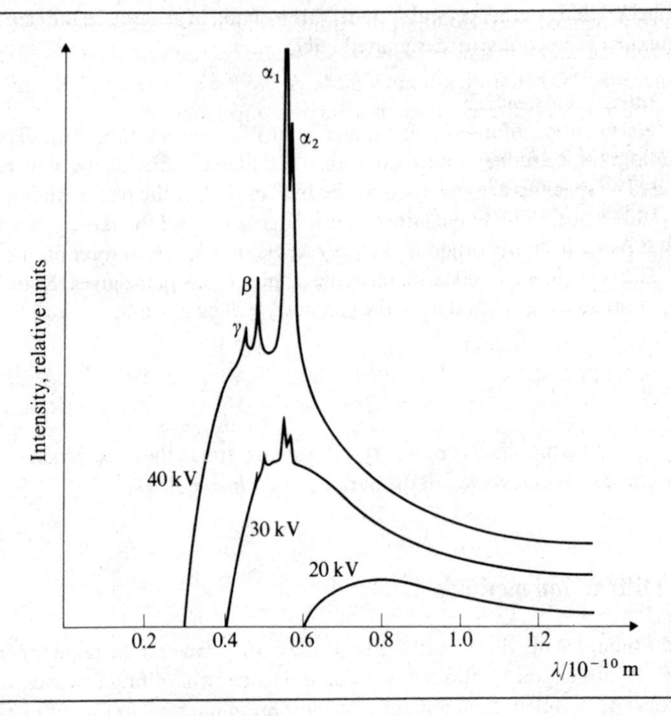

Fig. 3.13 The characteristic X-ray spectrum is superimposed on the continuous spectrum. The intensity of the α_1 line is twice that of the α_2 line. Note also that the separation of the α_1 and α_2 lines is much less than indicated on the diagram.

dissipated in a series of collisions and a spectrum of X-ray wavelengths is obtained. In accordance with equation (3.4) the X-ray wavelengths will be given by

$$\lambda \geqslant \frac{hc}{eU} \qquad \ldots(3.5)$$

so that inserting the values of the physical constants we have

$$\lambda \geqslant \frac{1.243 \times 10^{-6}}{U}\ \text{m}. \qquad \ldots(3.6)$$

The accelerating potentials necessary to produce X-rays of wavelengths comparable to atomic spacings are therefore about 10 KV.

Provided that the accelerating potential is sufficiently large a line spectrum characteristic of the target metal will be produced in addition to the continuous spectrum. The electrons incident on the target under these conditions have sufficient energy to eject electrons from the inner shells and characteristic X-rays of sharply defined wavelengths are emitted when the levels are filled. The important radiations in diffraction work are those corresponding to the filling of the innermost 1s shell from adjacent shells giving the so-called K_{α_1}, K_{α_2} and the K_β lines. For a copper target the characteristic wavelengths are

K_{α_2} 0.154434 nm (strong)
K_{α_1} 0.154050 nm (very strong)
K_β 0.139217 nm (weak)

whereas for molybdenum the wavelength of the K_{α_1} line is 0.0709 nm and for chromium 0.228962 nm. Since metals exhibit sharp absorption edges for X-rays a foil of suitable metal can be used to filter out the K_β line, e.g. a nickel foil will remove copper K_β radiation.

To summarise, there is available for X-ray diffraction studies, in addition to a continuous X-ray spectrum, a wide choice of characteristic K_α lines obtained by use of different target metals. The wavelength spread of each such line is extremely small and each wavelength is known with very high precision.

The theory of diffraction will be treated in Section 3.7. In summary, X-rays are scattered elastically without change of wavelength by the extranuclear electrons of the atoms, the process being comparatively more efficient for heavier elements such as iron than for light elements such as carbon and hydrogen. Coherent, or in phase, scattering of X-rays occurs in directions governed by the wavelength of the radiation and the spacing of the atoms in the crystalline sample. There is, in addition, some inelastic scattering resulting from X-ray photon collisions with loosely bound or free electrons. This is termed the Compton effect and it results in a general background of scattered radiation of longer wavelength.

3.6.2 *Electron diffraction*

In accordance with the de Broglie postulate, electrons of energy E and velocity u will have a wavelength λ given by

$$\lambda = \frac{h}{m_e u} = \frac{h}{(2m_e E)^{\frac{1}{2}}} \qquad \ldots(3.7)$$

and it may be verified from this relationship that electrons of energy 150 eV have a wavelength of 0.1 nm. This should be compared with the photon energy of about 12 KeV for X-rays of the same wavelength. Additionally, the charge on the electrons makes them interact more strongly than X-rays with the atoms of a solid. In diffraction work, therefore, the penetration of electrons is much less than that of X-rays of comparable wavelengths, and for a wavelength of 0.1 nm is only a few interatomic distances. The coherent reflections with electrons of this wavelength are rather diffuse but give useful information about surface structure.

In practice, most electron diffraction work is carried out with much more energetic electrons, e.g. those of energy ≈ 50 KeV which may penetrate many hundreds of interatomic distances into a crystal and for which sharp diffraction reflections are obtained.

The wavelengths of these electrons are very much less (about 20 times) than that of X-rays used in diffraction work. Consequently the reflections obtained are close to the primary beam direction and this necessitates a much larger specimen to detector distance than is usual in X-ray diffraction work. Compared with that of X-rays the depth of penetration in a sample is still small so that electron diffraction is used primarily to study the structure of surfaces and that of thin films.

3.6.3 *Neutron diffraction*

Applying the de Broglie relationship to neutrons we have

$$\lambda = \frac{h}{(2m_n E)^{\frac{1}{2}}} \qquad \ldots(3.8)$$

m_n being the mass of the neutron. Due to the difference in the masses of the neutron and the electron we find that for comparable energies the neutron wavelength is about 40 times less than that of the electron. For a wavelength of 0.1 nm the neutron energy is ≈ 0.08 eV so that *slow* neutrons would be expected to give diffraction effects with solids. In contrast to electrons or X-rays, neutrons are scattered by atomic nuclei rather than by the extranuclear electrons. Light atoms such as hydrogen and carbon scatter neutrons appreciably and neutron diffraction is consequently very useful in the structural studies of organic crystals. In addition there can be interaction between the magnetic moment of the neutron and that of atoms so that neutron diffraction techniques can also be used for the study of the magnetic structure of materials.

Neutrons of suitable energy for diffraction studies are obtained from a reactor by allowing them to reach thermal energy by repeated elastic collisions in a graphite thermal column. Their mean thermal energy is $3kT/2$ or 0.04 eV at 300 K corresponding to neutrons having a mean velocity of 2200 m s^{-1} and a wavelength of 0.14 nm. These neutrons will be diffracted by a crystal and by selecting a strong diffracted beam a monochromatic neutron beam is then available for further diffraction studies.

3.7 The Bragg diffraction law

X-rays incident on a crystalline material are scattered coherently and strong diffracted beams occur in specific directions, these directions being governed by the wavelength of the incident radiation and the nature of the crystalline sample. A relationship for the diffraction condition was first formulated by W. L. Bragg (1913) and a more complete treatment of diffraction was subsequently given by von Laue.

In the Bragg treatment the diffraction of a plane wavefront of monochromatic X-rays incident on a crystal is considered. Since X-rays penetrate into a crystal the interaction of the wavefront with a large number of atomic planes in the crystal is involved. In Fig. 3.14 a plane wavefront is shown incident at glancing

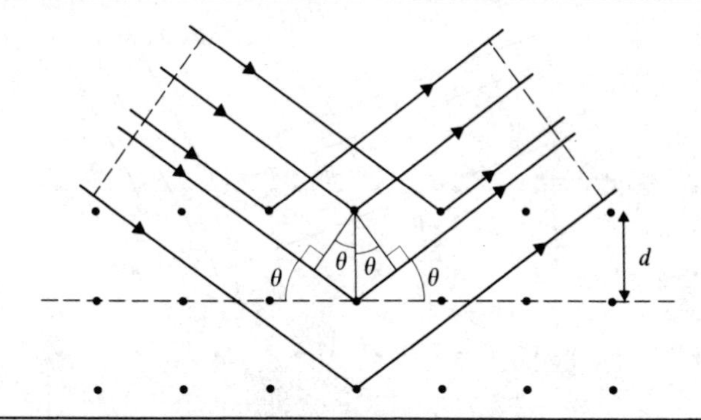

Fig. 3.14 The Bragg law of diffraction. There is in phase scattering of waves from successive planes provided that the path difference of the waves, given by $2d \sin \theta$, is a complete number of wavelengths.

angle θ to a set of planes. The relationship of these planes to the surface of the sample is not in general important. The X-rays are scattered by the atoms on a crystal plane and an in phase 'reflection' of X-rays will occur when there is no path difference between the incident and reflected waves. A similar reflection will occur for scattering from each successive parallel plane of atoms. The intensity of the wave reflected from any given plane is, however, very weak and the necessary condition for a diffracted beam to occur is that reflections from all the planes must be in phase with each other. From the diagram we see that the path difference between waves reflected from adjacent planes is $2d \sin \theta$ so that the condition for constructive interference is

$$2d \sin \theta = n\lambda \qquad (n = 1, 2, 3 \ldots) \qquad \ldots(3.9)$$

where d is the spacing of the set of planes. This is Bragg's law of diffraction.

A beam of monochromatic X-rays incident at angle θ to a set of planes such that it satisfies the condition (3.9) will be reflected. For any other arbitrary angle of incidence to the set of planes the beam will not be diffracted, so that although we speak of X-ray reflections we must not assume that lattice planes act like ordinary mirrors. The condition (3.9) sets a limit, for a given λ and d, to the number of diffracted beams which are obtained. In particular since $\sin \theta \leqslant 1$, we must always have

$$\lambda \leqslant 2d. \qquad \ldots(3.10)$$

In the above derivation the condition for in phase scattering is that corresponding to the reflection of X-rays from a set of planes. We may enquire whether for a given θ and λ other diffracted beams can occur. If the spacing of the atoms *along* the plane is c then the full condition for in phase scattering from that plane is seen from Fig. 3.15 to be

$$c \cos \theta - c \cos \phi = m\lambda \qquad (m = 1, 2, 3 \ldots) \qquad \ldots(3.11)$$

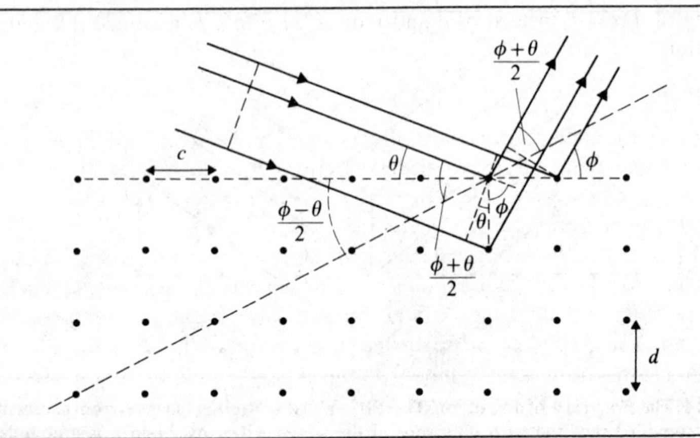

Fig. 3.15 In addition to reflection from a set of planes for which the Bragg condition is obeyed other reflections occur at angle $\phi \neq \theta$. These occur from other prominent crystallographic planes. In the example here the atom positions are given by $l = 1$ and $m = 2$ in the expression $lc + md$.

where ϕ is the angle between the diffracted beam and the plane. In addition there must be constructive interference between waves from two *adjoining* planes for which we have

$$d \sin\theta + d \sin\phi = l\lambda \qquad (l = 1, 2, 3 \ldots). \qquad \ldots(3.12)$$

Combining these two conditions we have

$$\frac{m}{l} \cdot \frac{d}{c} = \frac{\cos\theta - \cos\phi}{\sin\theta + \sin\phi} = \tan[\tfrac{1}{2}(\phi - \theta)]. \qquad \ldots(3.13)$$

If now we construct, as indicated in the figure, a plane lying at an angle of $(\phi - \theta)/2$ to the original set of planes then both the incident and scattered wave will make equal angles (of $[\phi + \theta]/2$) with this plane, i.e. the beam is reflected from it. The significance of condition (3.13) is that it shows the new plane to be a significant crystallographic plane since repeated translations of $lc + md$ from the origin give atom positions on the lattice. We find, therefore, not unexpectedly, that for the given angle of incidence, the additional maxima correspond to those due to reflections from other prominent crystal planes in the lattice.

The Bragg relationship (3.9) shows that, in addition to the first order reflection from a set of planes, second and other higher orders may also be obtained corresponding to the different values of n. It follows that a number of different orientations of the beam relative to a single crystal sample will give reflections from a particular set of planes. A reflection of order n can be regarded as occurring from an imaginary set of planes parallel to the (hkl) planes but having a spacing of $d/n = d'$ instead of d and indices ($nh\,nk\,nl$). A modified diffraction condition

$$2d' \sin\theta = \lambda \qquad \ldots(3.14)$$

is then used for higher order reflections. As an example second order reflections from (100) planes are referred to as 200 reflections.

3.7.1 *Missing reflections*

The lattice parameter a of a cubic crystal, e.g. copper, can be determined by measuring θ for known λ for a number of different sets of reflecting planes. The measurements initially yield d' (in accordance with equation (3.14)) and the lattice parameter a is then determined provided the reflecting planes in equation (3.3) are identified. In determinations of this type it is found that a number of X-ray reflections are missing when either a b.c.c. or a f.c.c. structure is being examined. This is due to the structures having been described in terms of a non-primitive cubic unit cell rather than the primitive rhombohedral cells. Not all the possible index combinations in the non-primitive system describe possible reflections so that in using these combinations certain reflections appear to be missing.

Let us now see why this is so. In the two-dimensional rectangular lattice in Fig. 3.16 the set of planes shown have indices (21) and the rectangular unit cell

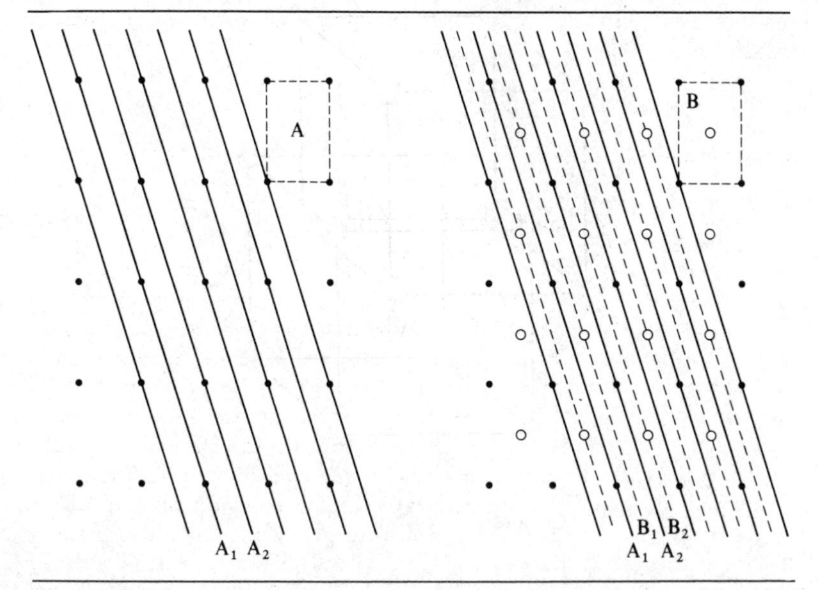

Fig. 3.16 Reflections from a rectangular lattice and a centred rectangular lattice. Atoms marked • belong to the rectangular lattice; the additional atoms ○ give the centred rectangular lattice. Unit cell A of the rectangular lattice is primitive, whilst unit cell B of the centred rectangular lattice is a double cell.

A is primitive. There are no missing reflections and for appropriate θ waves reflected from planes A_1 and A_2 will differ in path by a complete wavelength and will give the 21 reflection. If now atoms are included in the centred positions the rectangular unit cell becomes a double cell and, as indicated, additional planes through the centred atom positions are interposed between the original (21) planes. Waves reflected from the additional planes of atoms such as B_1 will now be exactly out of phase with those from A_1 and A_2 and there will be complete interference. The original reflections are thus extinguished and one would expect to observe only reflections due to planes having half this spacing, i.e. 42 type reflections.

This approach is applicable in three dimensions to the b.c.c. structure, atoms in the body centre positions being responsible for extinguishing, for example, the 100 reflections. If, however, the atoms in these positions are different in type from those at the cube corners the 100 reflections will still be observed and their intensities will be governed by the different scattering powers of the two types of atoms. In such a case, however, it should be noted that the structure is simple cubic, rather than b.c.c., with a basis of atoms 000 and $\frac{1}{2}\frac{1}{2}\frac{1}{2}$. This is the cesium chloride structure illustrated in Fig. 3.17. The unit cell describing the structure is therefore primitive and missing reflections would not be expected.

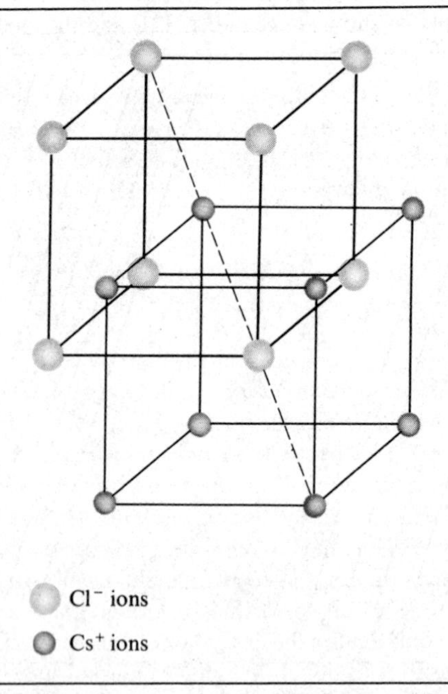

Fig. 3.17 The cesium chloride structure. The lattice is simple cubic with a basis of Cs^+ at 000 and Cl^- at $\frac{1}{2}\frac{1}{2}\frac{1}{2}$. One of the unit cells shown contains the equivalent of one Cs^+ and one Cl^- ion.

In order to determine the extinctions expected due to non-primitive unit cells it is necessary to compare the primitive and non-primitive indices using appropriate transformation relations. This shows that for a b.c.c. lattice reflections are missing for which the sum of $h+k+l$ is odd whilst for f.c.c. reflections are obtained when there are no mixed indices, i.e. h, k and l must all be even or all odd. Extinctions also occur due to the presence of glide planes or screw axes in a structure. The study of extinctions enables the space group symmetry elements of a structure to be determined and are thus of considerable importance in X-ray crystallography.

3.8 The Laue equations

An alternative approach to X-ray diffraction is to regard the phenomenon as one of scattering from the three-dimensional array of atoms in a manner analogous to that of the diffraction of light by a grating.

If an X-ray beam is incident at an angle of α_0 to a row of atoms spaced a apart (as in Fig. 3.18) then the X-rays scattered from all the atoms will be in phase provided that

$$a(\cos\alpha - \cos\alpha_0) = h\lambda \qquad (h = 0, 1, 2, \ldots) \quad \ldots(3.15)$$

where α is the direction of the scattered beam with respect to the row of atoms. This equation defines, with the various values of h, a set of coaxial cones each corresponding to a possible order of diffraction. This treatment may clearly be generalised for the three-dimensional array of atoms on a crystal lattice. The two additional conditions

$$b(\cos\beta - \cos\beta_0) = k\lambda \qquad (k = 0, 1, 2\ldots) \quad \ldots(3.16)$$

$$c(\cos\gamma - \cos\gamma_0) = l\lambda \qquad (l = 0, 1, 2\ldots) \quad \ldots(3.17)$$

then hold, the angles α_0, β_0 and γ_0 defining the direction of the incident beam and α, β, γ that of the diffracted beam. There are three sets of cones, with their axes along the a, b and c directions, defined by these three equations and their intersection will be governed by the direction of the incident beam. The necessary condition for in phase scattering by all the atoms on the lattice to occur is that, for some set of values of h, k, l, the three cones must possess a common *line* of intersection. This will define the direction of the diffracted beam and therefore only for certain specific angles of incidence should one obtain a diffracted beam.

The three equations (3.15) to (3.17) are termed the Laue conditions for diffraction and are equivalent to the Bragg law of diffraction. This equivalence of the Laue conditions to reflection from a plane of atoms may be seen as follows. The conditions imply that an atom adjacent to the origin along the a direction scatters X-rays h complete wavelengths out of phase with the scattering from the atom at the origin. Hence atoms situated at a repeat distance kl along the a axis

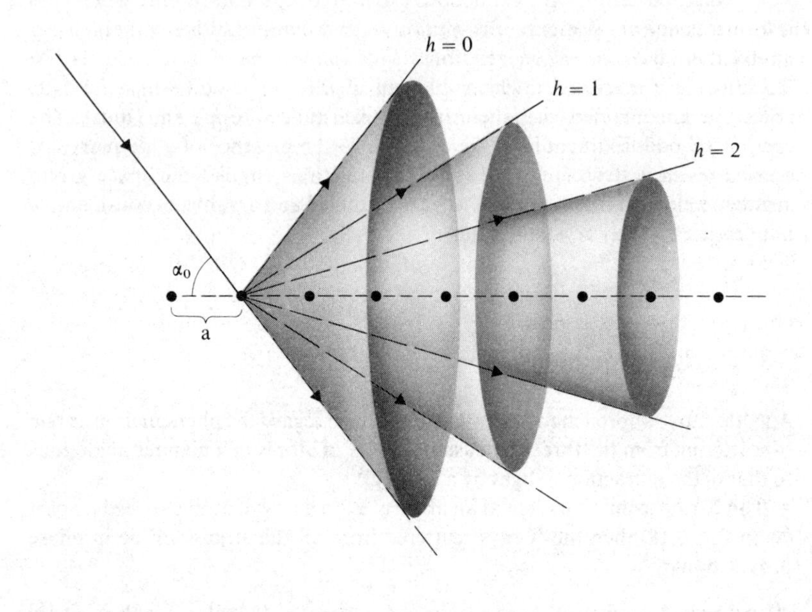

Fig. 3.18 The Laue approach. A beam incident at angle α_0 to a row of equispaced atoms will be scattered to give diffracted beams which lie on the surface of a set of coaxial cones. Each cone corresponds to a particular order of diffraction.

will scatter at *hkl* complete wavelengths out of phase. This is also true for atoms at *hl* repeats along the *b* axis and for *hk* repeats along the *c* axis. Consequently these atoms are scattering in the *same* absolute phase and also in phase with the atoms at the origin. The three repeat distances *kl*, *hl*, and *hk* define planes of index (*hkl*) and the phase relationships imply a reflection of X-rays from the planes, i.e. the Bragg reflection condition obtains.

3.9 The reciprocal lattice

In dealing with specific problems in crystallography one is interested in the orientation of the different sets of planes in a structure and in their spacing. The use of the reciprocal lattice enables a very convenient representation to be made of these quantities and one which considerably simplifies the interpretation of diffraction photographs of crystals.

Two definitions of the reciprocal lattice differing slightly from each other are in current use, one in the field of crystallography and the other in solid state theory.

The former definition will be developed first and it will become clear later how it is modified in solid state theory.

To construct the reciprocal lattice from the corresponding direct lattice the following procedure is followed: From any point chosen as origin a line is drawn corresponding to each set of reflecting planes of length equal to the reciprocal of the spacing of the set of planes and in a direction normal to the planes. The end points of such lines define a lattice of points. By including the imaginary sub-multiple spacing planes corresponding to the second and higher orders of reflection the lattice of points so defined is of infinite extent. This is the reciprocal lattice. As in the direct lattice each point has identical surroundings. Each lattice vector joining two lattice points corresponds to a set of planes; the direction of the vector is normal to the set of planes and its length is equal to the reciprocal of the spacing of the set of planes.

The procedure adopted is shown in Fig. 3.19 for an oblique plane lattice. The set of planes (11) in the direct lattice together with its sub-multiple planes give rise to reflections 11, 22, 33, etc. In the reciprocal lattice the distances OA, OB, OC are, therefore, equal to $1/d_{11}$, $1/d_{22}$, $1/d_{33}$ and form a set of equispaced points. We

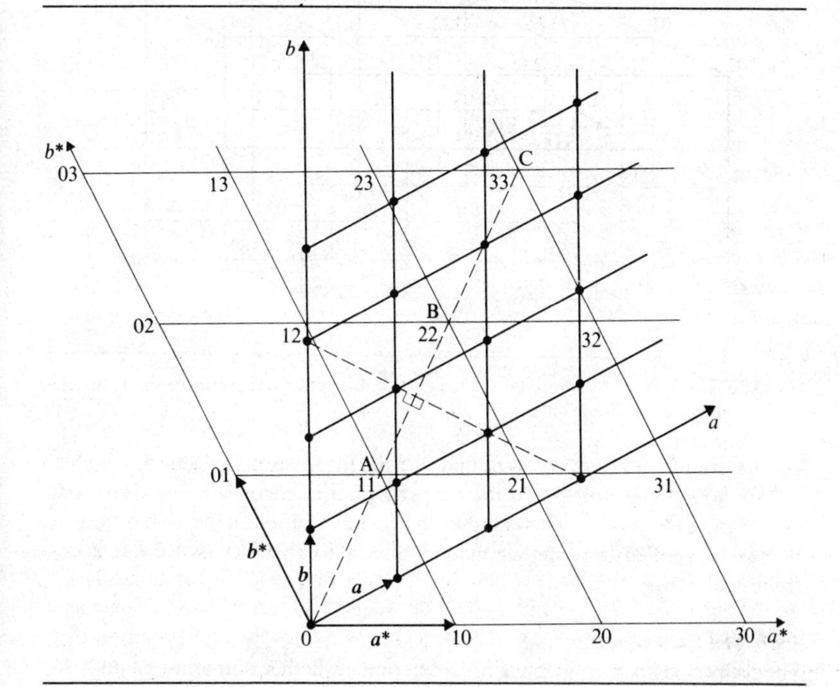

Fig. 3.19 The relationship between the direct and reciprocal lattices illustrated for a 2-dimensional oblique lattice. This is equivalent to the view along the *c*-axis of a monoclinic crystal.

note that the reciprocal lattice points 10, 20, 30 define the a^* axis of the reciprocal lattice. Bearing in mind the method of constructing the reciprocal lattice it is clear that the a^* axis must be normal to the b axis of the direct lattice, and for a three-dimensional lattice to the c^* axis as well. Putting $\boldsymbol{u}$ as a unit vector along the a^* axis we have, for the three-dimensional case, the fundamental translation vector $\boldsymbol{a}^*$ given by

$$\boldsymbol{a}^* = \frac{1}{d_{100}}\boldsymbol{u} \qquad \ldots(3.18)$$

with similar relationships for $\boldsymbol{b}^*$ and $\boldsymbol{c}^*$. It follows that the reciprocal lattice to a direct cubic lattice is itself cubic and correspondences of this type are found in other crystal systems.

The reciprocal lattice to the simple cubic direct lattice can readily be constructed by following the procedures outlined. The reciprocal lattice points corresponding to the different reflections are shown in Fig. 3.20 and clearly form

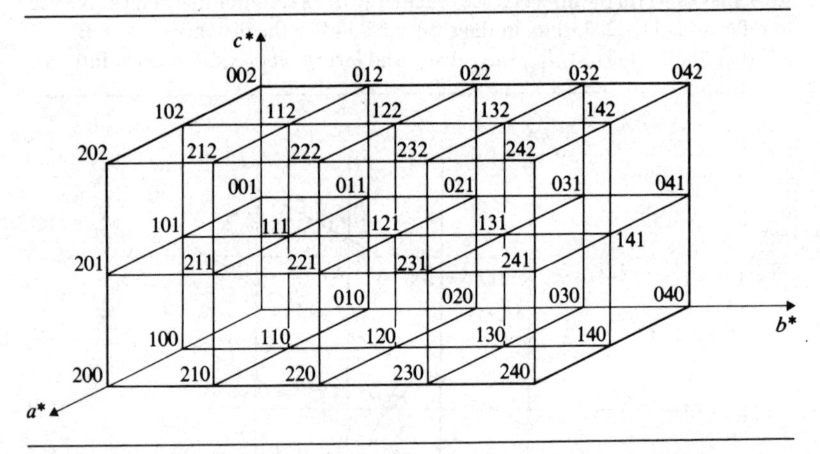

Fig. 3.20 The reflections obtained for a simple cubic lattice. These form the reciprocal lattice which is itself simple cubic.

another simple cubic lattice. By excluding from the figure all reflections for which hkl are not all even or all odd the reciprocal lattice corresponding to the f.c.c. direct lattice is obtained. This is shown in Fig. 3.21 and is seen to be a b.c.c. lattice. It may be verified in a similar manner that a b.c.c. direct lattice has a f.c.c. reciprocal lattice.

3.9.1 *The Ewald construction*

The reciprocal lattice may be used to determine whether diffraction of an X-ray beam of a given wavelength will occur with a particular crystal. The construction in Fig. 3.22 is due to Ewald. A vector $\boldsymbol{AB}$ of length $1/\lambda$ is drawn parallel to the incident beam direction and terminating at any reciprocal lattice point B. If now

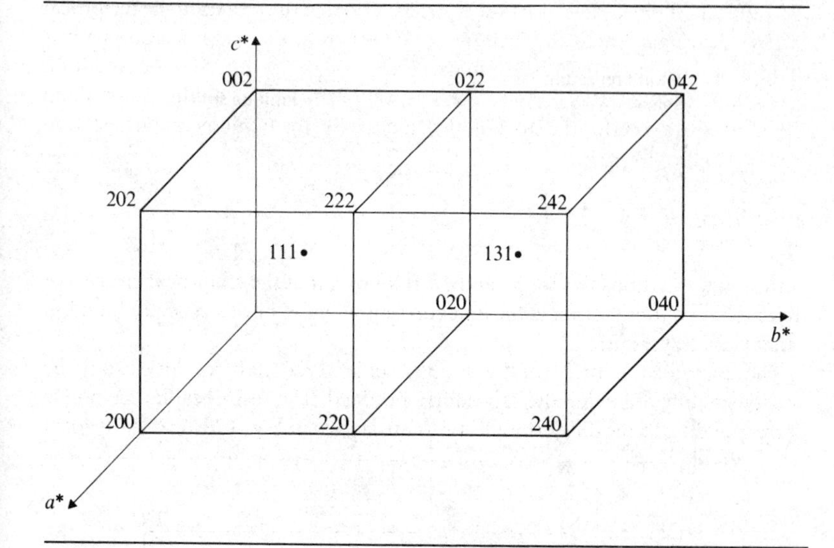

Fig. 3.21 The reciprocal lattice for a direct f.c.c. lattice. This figure should be compared with Fig. 3.20 and it is noted that all reflections have k, k and l all even or all odd. This yields the b.c.c. lattice illustrated.

the sphere centred on A and of radius $1/\lambda$ intersects another lattice point such as C then diffraction of the beam represented by $\boldsymbol{AB}$ must occur. This may be shown as follows:

The vector $\boldsymbol{BC}$ is a vector $\boldsymbol{g}$ of the reciprocal lattice and is of length

$$g = \frac{1}{d}, \qquad \ldots(3.19)$$

where d is the spacing of the corresponding set (or sub-multiple set) of planes in the direct lattice. Since $\boldsymbol{g}$ is normal to the planes, $\boldsymbol{AD}$ must be parallel to them so that BAD is the Bragg angle θ. We thus have

$$d = \frac{1}{g} = \frac{1}{2AB\sin\theta} = \frac{\lambda}{2\sin\theta}. \qquad \ldots(3.20)$$

The Bragg condition is therefore obeyed and diffraction of the beam occurs.

For a random setting of the crystal with respect to the beam it is unlikely that the sphere drawn will intersect the additional lattice point C. However, rotation of the crystal with respect to the beam will, in general, bring planes into the appropriate reflecting positions. This rotation is equivalent to the rotation of the 'reflecting' sphere about the point B, i.e. the wavelength is kept constant but the direction of the beam with respect to the crystal is varied. If, for example, the sphere is rotated about the vertical axis shown, the point C will pass through its

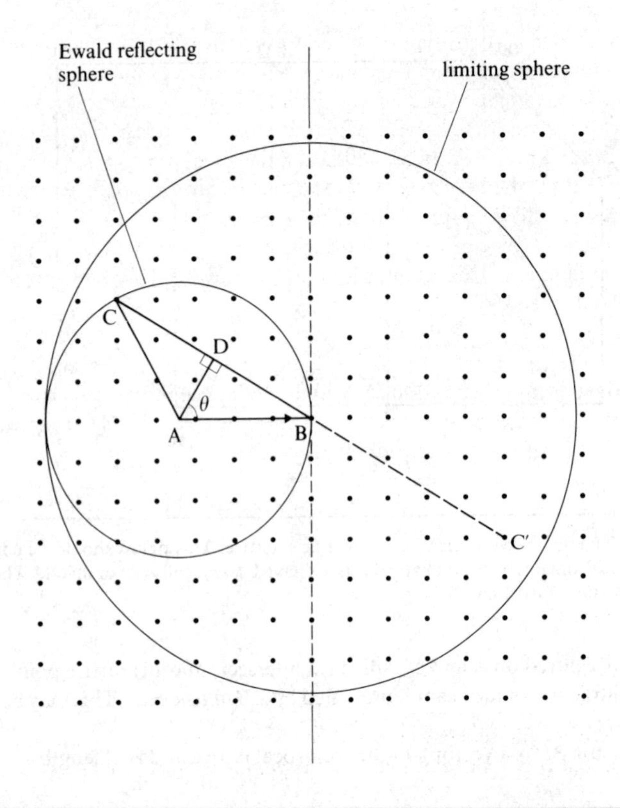

Fig. 3.22 The Ewald construction. The point B is taken as the origin of the reciprocal lattice. $|\boldsymbol{AB}| = 1/\lambda$ and is in the direction of the incident beam. All reciprocal lattice points lying on the surface of the Ewald reflecting sphere give rise to reflections. For a given wavelength λ the corresponding reciprocal lattice points lie within the limiting sphere of radius $2/\lambda$.

surface *twice* and will lead to two symmetrical reflections. Since every lattice possesses inversion symmetry there will exist a reciprocal lattice point C′ symmetrically displaced to C with respect to the origin B and two further reflections will be obtained. Thus for an axis of crystal rotation normal to the beam one obtains both horizontal and vertical symmetry in the X-ray reflections. The reflecting sphere is always contained within the outer 'limiting' sphere of radius $2/\lambda$ and all reciprocal lattice points lying within this outer sphere will give rise to diffracted beams. This gives rise to the condition that $d \geqslant \lambda/2$ in accordance with equation (3.10).

3.10 Wave-vector representation

As stated earlier it is customary in solid state theory to define the reciprocal lattice vectors in a slightly different manner. In this representation the reciprocal lattice vectors are defined as $2\pi/d$ instead of $1/d$ so that the lattice is expanded by a factor of 2π in each direction as compared with the previous representation. An incident X-ray photon of wavelength λ is represented by a vector of magnitude $2\pi/\lambda = k$, i.e. by the wave-vector $\boldsymbol{k}$ and the reciprocal space defined in this manner is termed a wave-vector or $\boldsymbol{k}$-space.

The points of the reciprocal lattice in $\boldsymbol{k}$-space are defined in an exactly analogous manner to that adopted for the direct lattice. Thus a reciprocal lattice vector $\boldsymbol{G}$ is defined by

$$\boldsymbol{G} = h\boldsymbol{a}^* + k\boldsymbol{b}^* + l\boldsymbol{c}^*, \quad \ldots(3.21)$$

where h, k and l are integers and $\boldsymbol{a}^*$, $\boldsymbol{b}^*$ and $\boldsymbol{c}^*$ are the fundamental vectors of the lattice, i.e. they represent the unit cell of the reciprocal lattice. They are related to the translation vectors $\boldsymbol{a}$, $\boldsymbol{b}$ and $\boldsymbol{c}$ of the direct lattice by the three equations

$$\boldsymbol{a}^* = 2\pi\frac{\boldsymbol{b} \times \boldsymbol{c}}{\boldsymbol{a}\cdot\boldsymbol{b} \times \boldsymbol{c}} \quad \ldots(3.22)$$

$$\boldsymbol{b}^* = 2\pi\frac{\boldsymbol{c} \times \boldsymbol{a}}{\boldsymbol{a}\cdot\boldsymbol{b} \times \boldsymbol{c}} \quad \ldots(3.23)$$

$$\boldsymbol{c}^* = 2\pi\frac{\boldsymbol{a} \times \boldsymbol{b}}{\boldsymbol{a}\cdot\boldsymbol{b} \times \boldsymbol{c}}. \quad \ldots(3.24)$$

Let us see how these relationships compare with our previous definition. If we put $\boldsymbol{u}$ as a unit vector perpendiulcar to the plane of $\boldsymbol{b}$ and $\boldsymbol{c}$ we have

$$\boldsymbol{b} \times \boldsymbol{c} = bc \sin \alpha\, \boldsymbol{u}, \quad \ldots(3.25)$$

where α is the angle between $\boldsymbol{b}$ and $\boldsymbol{c}$.

Since in addition

$$\boldsymbol{a}\cdot\boldsymbol{u} = d_{100} \quad \ldots(3.26)$$

equation (3.22) gives for $\boldsymbol{a}^*$

$$\boldsymbol{a}^* = \frac{2\pi}{d_{100}}\boldsymbol{u}. \quad \ldots(3.27)$$

$\boldsymbol{a}^*$ is thus normal to the vectors $\boldsymbol{b}$ and $\boldsymbol{c}$ of the direct lattice and in accordance with our initial premise is 2π times greater than the value found previously in equation (3.18). We note also that in a crystal system in which $\boldsymbol{a}$, $\boldsymbol{b}$ and $\boldsymbol{c}$ are orthogonal $\boldsymbol{a}^*$, $\boldsymbol{b}^*$, and $\boldsymbol{c}^*$ are parallel to $\boldsymbol{a}$, $\boldsymbol{b}$ and $\boldsymbol{c}$, respectively.

The wave-vector representation can be applied to the Bragg treatment of reflection. Referring to Fig. 3.23 an incident wave-vector $\boldsymbol{k}$ is scattered through an angle 2θ when the Bragg condition is satisfied and a different wave-vector $\boldsymbol{k}'$

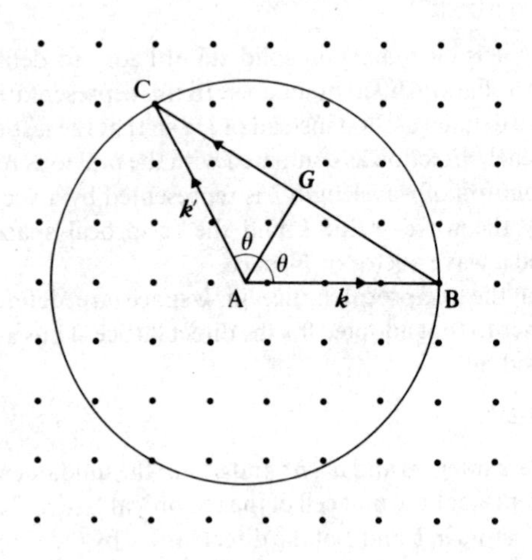

Fig. 3.23 The wave vector representation of Bragg diffraction. Note that ***AB*** representing the incident wave has now magnitude $k = 2\pi/\lambda$ not $1/\lambda$ as in Fig. 3.22.

obtained. The scattering is coherent, i.e. there is no change of wavelength so that

$$k = k' = \frac{2\pi}{\lambda}. \qquad \ldots(3.28)$$

The necessary condition for diffraction to occur is that $\boldsymbol{BC}$ is a reciprocal lattide vector and putting this as $\boldsymbol{G}$ the Bragg condition can be put in the form

$$\boldsymbol{k}' - \boldsymbol{k} = \boldsymbol{G}$$

or $\qquad \ldots(3.29)$

$$\Delta \boldsymbol{k} = \boldsymbol{G}.$$

We note here that since both $\boldsymbol{G}$ and $-\boldsymbol{G}$ are equally valid lattice vectors there is no significance to the sign of $\boldsymbol{G}$ in this equation. An alternative form of this condition is obtained by rearranging and squaring. This gives

$$k'^2 = k^2 + 2\boldsymbol{k}\cdot\boldsymbol{G} + G^2 \qquad \ldots(3.30)$$

and since the scattering is elastic, i.e. with no change in energy or wavelength, we have

$$2\boldsymbol{k}\cdot\boldsymbol{G} + G^2 = 0. \qquad \ldots(3.31)$$

An application of the use of this equation in the theory of Brillouin zones will be given in Section 7.7.

3.11 X-ray diffraction methods

The methods of X-ray diffraction are extensive and a complete coverage of them cannot be attempted here. Many different types of X-ray cameras have been developed and the choice of technique in a particular investigation will largely be determined by the type of sample and the kind of information required. The use of counter diffractometry, instead of film, coupled with computer analysis is of growing importance.

A method which has proved to be of wide applicability is the Debye–Scherrer powder method which is illustrated in Fig. 3.24. The sample normally consists of

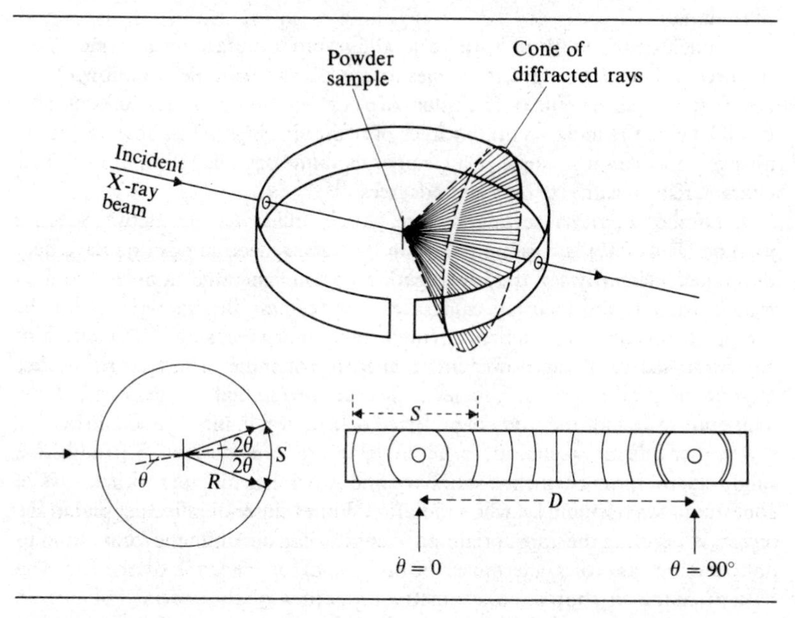

Fig. 3.24 The Debye–Scherrer powder method. For a camera of radius 5.726 cm, the arc D corresponding to a θ difference of 90° is 9 cm, i.e. 1 mm on the film corresponds to a 1° arc.

a fine crystalline powder which may either be coated on a supporting fibre or contained within a very fine quartz tube. A collimated beam of monochromatic X-rays falls normally on the specimen mounted centrally and a recording film in the form of a short cylinder coaxial with the axis of the specimen records the diffracted beams. The specimen is normally rotated although this is not essential. Since the sample is polycrystalline the Bragg condition for a given set of planes can be satisfied by crystallites suitably oriented with respect to the vertical fibre axis. The diffracted beams are, therefore, in the form of cones, coaxial with the incident beam direction, and they will intersect the film in short arcs. For any particular reflection there will be two arcs symmetrically displaced on either side of the central beam direction. Their measured separation S on the film for a given reflection is related to the Bragg angle θ for that reflection according to

$$\theta = \frac{S}{4R}, \qquad \ldots(3.32)$$

where R is the radius of the film cylinder or camera.

The measurements allow the spacing of each set of planes to be determined and, provided that the different reflections can be identified, will allow the parameters of the structure to be determined. The number of reflections recorded on a film and their intensities is governed by the crystal structure of the sample. The lower the crystal symmetry the greater will be the number of separate reflections.

In much of the work in X-ray crystallography accurate measurements are required and lattice parameter values for metals for example are known to an accuracy of 1 part in 50000. The values of θ corresponding to back reflection (i.e. $\theta \rightarrow 90°$) give the most accurate values of plane spacings. This, however, is not always feasible, e.g. some clay minerals and organic compounds have characteristic θ values of only a few degrees.

A number of other techniques are closely allied to the Debye–Scherrer method. Thus flat plate cameras used in the back reflection position have been developed primarily for the investigation of solid metallic samples. Another modification is the focusing camera of the Seeman–Bohlin type where the samples form part of the camera circumference. Sharp focusing of the reflections and increased resolving power are a property of these arrangements. Other modifications allow crystal samples to be examined at high temperature, at low temperature and for the study of preferred orientation in metals and alloys.

A rather different technique is employed in the Laue method. This utilises a single crystal sample which is sationary and which is irradiated with X-rays of continuous wavelength, i.e. white radiation. For each set of reflecting planes the crystal is selecting the appropriate wavelength λ for the diffraction condition to hold and the value of θ determines the position of the reflection on the film. The pattern obtained shows the symmetry properties of the crystal but not its dimensions since a range of wavelengths is contained in the beam. The main use of the Laue method is now in routine orientation of crystal samples although it can also be used for the study of the symmetry of crystals.

The fulfilment of the Bragg condition for diffraction is achieved in the Laue method by the use of white radiation and in the Debye–Scherrer method by the use of polycrystalline samples. In the rotating crystal method the condition is enabled to be satisfied by the rotation of a single crystal specimen. The axis of rotation is a prominent crystallographic axis and as shown in Fig. 3.25 this is normal to a monochromatic X-ray beam. As mentioned previously diffraction spots are obtained corresponding to the points of the reciprocal lattice. Crystal planes parallel to the rotation axis give a series of reflections, on either side of the direct beam, in a horizontal plane. Other sets of planes give reflections in vertically displaced horizontal rows which are termed layer lines. The diffraction pattern forms essentially a map of the reciprocal lattice of the crystal. This feature

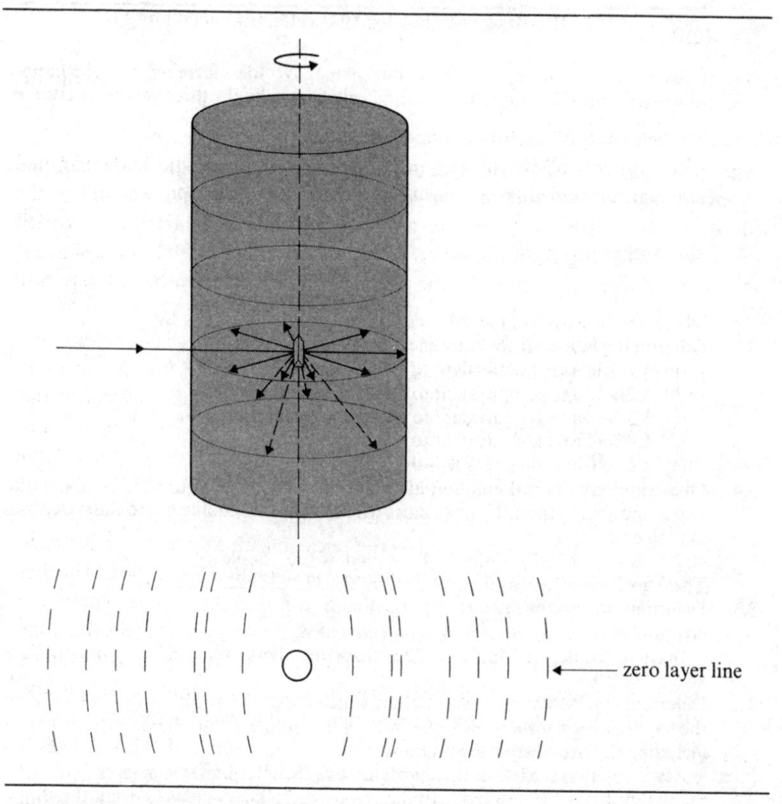

Fig. 3.25 The rotating crystal arrangement.

makes the reciprocal lattice approach a powerful tool in the examination of crystal structures.

References and further reading

Azároff, L. V. (1968) *Elements of X-ray Crystallography*, McGraw-Hill.
Azároff, L. V. (1960) *Introduction to Solids*, McGraw-Hill.
Barrett, C. B. (1953) *Structure of Metals*, McGraw-Hill.
Buerger, M. J. (1949) *X-ray Crystallography*, Wiley.
Hume-Rothery, W. (1950) *The Structure of Metals and Alloys*, Institute of Metals.
Kelly, A. and **Groves, G. W.** (1960) *Crystallography and Crystal Defects*, Longmans.
Peiser, H. S., Rooksby, H. P. and **Wilson, A J. C.** (1955) *X-ray Diffraction by Polycrystalline Materials*, Chapman and Hall.
Speigel, M. R. (1959) *Vector Analysis*, Schaum.
Sproull, R. L. (1964) *Modern Physics*, Wiley.

Problems

3.1 Show if a, b and c are three non-planar vectors forming a right-handed system of axes the volume V of the unit cell defined by the three vectors is given by

$$V = a \cdot b \times c.$$

3.2 Show that for a cubic lattice the direction $[hkl]$ is normal to the plane (hkl). Show also that the interplanar spacing d_{hkl} of the set of planes is given by

$$d_{hkl} = \frac{a}{(h^2 + k^2 + l^2)^{\frac{1}{2}}},$$

where a is the length of the cubic cell edge.

3.3 Confirm the following for rotoreflection axes:
(a) A 1-fold axis is equivalent to a mirror plane m;
(b) A 2-fold axis is equivalent to a centre of symmetry $\bar{1}$;
(c) A 3-fold axis is equivalent to a 6-fold rotoinversion $\bar{6}$;
(d) A 4-fold axis is equivalent to a 4-fold rotoinversion $\bar{4}$;
(e) A 6-fold axis is equivalent to a 3-fold rotoinversion $\bar{3}$.

3.4 One symmetry operation followed by another is equivalent to a third. Show that two 2-fold axes at the following angles to each other are equivalent to the stated axis in each case:
(a) 90°, 2-fold; (b) 60°, 3-fold; (c) 45°, 4-fold; (d) 30°, 6-fold.
These give respectively the point groups 222 322 422 622.

3.5 Determine for a cube the number of separate
(a) 4-fold axes (b) 3-fold axes; and (c) 2-fold axes.
In what way do the 3-fold axes differ from the others? What other symmetries are also present?

3.6 Determine the Miller indices of a plane which makes intercepts of $6a$, $4b$ and $2c$ on the x-, y- and z-axes and give the intercepts of other planes having these indices including the two nearest the origin.
Assuming that the lattice is cubic determine the interplanar spacing.

3.7 Show that the error in the determination of interplanar spacings using the Bragg law of diffraction will be least for back reflection positions, i.e. for $\theta \to 90°$.

3.8 The metal copper crystallises on a f.c.c. lattice with one atom at each lattice point. If the cell edge $a = 0.361$ nm what is the closest interatomic distance in the structure?

3.9 Lithium has a b.c.c. lattice with cell edge $a = 0.351$ nm. What is the closest interatomic distance in the structure?

3.10 Show that for the ideal h.c.p. structure the axial ratio $c/a = 1.633$.
For many crystals there is significant departure from ideality such that c/a differs from the ideal value. Show that the second nearest neighbour distance is then increased to $(a^2/3 + c^2/4)^{\frac{1}{2}}$.
The metal zinc has a h.c.p. structure with $a = 0.266$ nm, $c = 0.494$ nm. Determine the second nearest neighbour distance.

3.11 Silver crystallises on a f.c.c. lattice with one atom at ech lattice point. How many atoms are accommodated in each unit cell?
Given that the unit cell edge $a = 0.409$ nm determine the density of silver (relative atomic mass of silver = 108).

3.12 In an X-ray diffraction experiment using nickel K_α radiation a doublet was registered corresponding to Bragg reflection angles of 75.08° and 75.60°. Given that the 75.08° reflection corresponds to the K_{α_1} line of wavelength 0.16578 nm determine the wavelength of the K_{α_2} line.

3.13 Many ionic solids crystallise in the so-called cesium chloride structure. This possesses a simple cubic space lattice with Cs atoms at 000 and C1 atoms at $\frac{1}{2}\frac{1}{2}\frac{1}{2}$. The unit cell edge = 0.412 nm. Draw a sketch diagram of four adjacent unit cells. Specify the first eight reflections which would be obtained for the structure and determine the corresponding reflection angles for a wavelength of 0.154 nm.

In what way would the diffraction be modified if all the atoms were of the same type?

3.14 The following series of reflection angles θ were measured in a diffraction experiment on a cubic sample using cobalt K_α radiation of wavelength 0.1790 nm: 14.47°, 20.68°, 25.63°, 29.97°, 33.97°, 37.70°, 44.95°, 48.53°. Determine the type of lattice and the unit cell edge.

3.15 Repeat the procedure given in Problem 3.12 for the two following sets of data:
(a) 20.68°, 30.02°, 37.75°, 45.02°, 52.22°, 60.02°, 69.30°;
(b) 17.81°, 20.7°, 30.02°, 35.90°, 37.77°, 44.98°, 50.35°, 52.23°.

3.16 In a lattice parameter determination using copper K_α radiation of wavelength 0.1542 nm the minimum Bragg reflection angle was 14.9°. What is the value of the unit cell edge if the lattice was (a) simple cubic, (b) body-centred cubic and (c) face-centred cubic.

At what angles would the next reflection be obtained for each of these structures?

3.17 The sodium chloride structure is equivalent to two f.c.c. lattices displaced by half a cube diagonal with respect to each other. The lattice is therefore f.c.c. with Na atoms at 000 and Cl atoms at $\frac{1}{2}\frac{1}{2}\frac{1}{2}$. Specify the first eight reflections which would be obtained for a crystal of this structure. Given that the unit cell edge for a sodium chloride crystal is 0.564 nm, determine the reflection angles θ expected with radiation of wavelength 0.1544 nm.

3.18 The density of a sodium chloride crystal is 2.15×10^3 kg m^{-3} and the unit cell edge is 0.564 nm. Using values of relative atomic masses for sodium and chlorine determine the number of formula weights in a unit cell. Sketch the structure and confirm the result obtained for formula weights.

3.19 The interplanar spacing d for a h.c.p. structure is given by

$$d = \left(\frac{4}{3a^2}(h^2 + hk + k^2) + \frac{l^2}{c^2}\right)^{-\frac{1}{2}}$$

in terms of hexagonal indices ($hkil$). If the $21\bar{3}0$ and $20\bar{2}4$ reflections for a zinc crystal correspond to θ values of 71.91° and 79.37°, respectively, determine the values of the two lattice parameters a and c and the axial ratio for the crystal. The data refer to reflections obtained with nickel K_α radiation of wavelength 0.16578 nm. What would be the measured arc length for a $21\bar{3}1$ reflection in a Debye–Scherrer camera or radius 0.1 m?

3.20 Show that a b.c.c. direct lattice has a f.c.c. reciprocal lattice.

4. Crystal bonding

4.1 Introduction

In the last chapter the ways in which atoms come together in symmetrical arrays to form crystals were described. In this chapter we shall principally be concerned with the reasons why the bonding of atoms occurs in the solid state.

It is observed experimentally that any particular element or compound will, at a given temperature and pressure, always crystallise in a particular structure. The dimensions of the crystal lattice under the given physical conditions are also found to be highly reproducible and for many solids the same crystal structure is found to be stable over a wide temperature range. In other solids a transition from one crystal type to another occurs at a particular temperature and pressure. The stable configuration of atoms is the one which leads to the minimum free energy for the solid. That this energy is less than that of the isolated atoms accounts for the formation of crystalline solids. It is apparent that for many solids large forces are required both to extend and to compress them and one thus concludes that there must exist both strongly attractive and repulsive forces between the atoms in a solid. The equilibrium separation observed experimentally corresponds to an exact balance between these forces and it also gives the condition of minimum energy of the system.

The range of possible crystal structures, the observed differences in physical properties, such as compressibility, melting points, electrical and optical properties as well as chemical properties, all point to the existence of different types of bonding mechanisms in solids. The forces present, however, are all electrical in origin and gravitational forces may be shown to be insufficiently large to give any appreciable binding. A preponderance of one type of force gives rise to a corresponding type of bonding and its associated physical and chemical properties, e.g. strong coulombic forces give rise to ionic bonding.

The energy of a crystal can be written as the sum of a number of terms, each term corresponding to a separate mechanism of repulsion or attraction. A simplified relationship for the energy $E(r)$ of a crystal can be written in the form

$$E(r) = -\frac{A}{r^n}+\frac{B}{r^m} \qquad \ldots(4.1)$$

where r is the separation of nearest neighbour atoms and A, B, n and m are

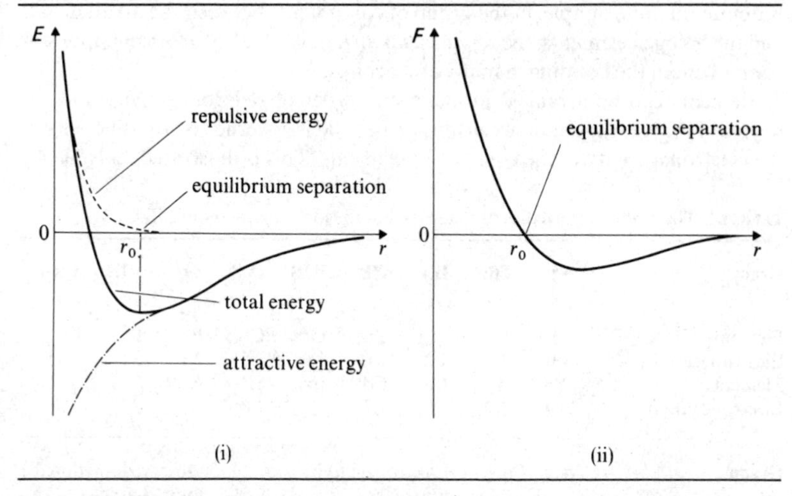

Fig. 4.1 (i) The energy of a solid as a function of the separation of the atoms. The minimum of the curve corresponds to the equilibrium separation $r = r_0$. (ii) The force between atoms in a solid as a function of the separation of the atoms. At the equilibrium separation the repulsive and attractive forces are in balance.

constants characteristic of the crystal. This function is shown in Fig. 4.1 from which it should be noted that the first term corresponds to the attractive energy and the second to the repulsive energy. Only provided that $m > n$ will a minimum be obtained and it follows that the repulsive forces are short range forces. The resultant force F between atoms is given by $\mathrm{d}E/\mathrm{d}r$ and will be zero at the equilibrium atomic separation $r = r_0$.

The bonds between atoms may be classified under four main types, namely, ionic, van der Waals, covalent and metallic, and the different mechanisms responsible for the attractive and repulsive forces will be brought out in the discussion of these types of bonding.

4.2 The ionic bond

The origin of the ionic bond lies in the strong coulombic attraction which exists between two spherical charges of opposite sign and it is the predominant bonding mechanism in solid compounds formed between strongly electropositive and electronegative elements. In contrast to the covalent or homopolar bond, where, as we shall see, there is a sharing of negative electronic charge, the ionic bond is a heteropolar bond in which ions of opposite signs are responsible for the attractive energy. The electron transfer necessary for this type of bonding is dominant, for example, in the group of compounds between the alkali metals and the halogen elements, the so-called alkali halides. It also forms an important component in the bonding in many other solids.

Elements can be arranged in increasing order of electronegativity and the degree of charge transfer between ions in the solid is governed by the difference in the electronegativity of the constituent elements. This is illustrated in Table 4.1

Table 4.1 The electronegativities of elements and the ionic character of solids

Group	**IA**	**IB**	**IIA**	**IIB**	**IIIB**	**IVB**	**VB**	**VIB**	**VIIB**
Element	Li	Cu	Mg	Zn	Ga	C	P	O	F
Electronegativity	1.0	1.9	1.2	1.6	1.6	2.5	2.1	3.5	4.0
Element	Na	Ag	Ca	Cd	In	Si	As	S	Cl
Electronegativity	0.9	1.9	1.0	1.7	1.7	1.8	2.0	2.5	3.0

Group	Crystal	Electronegativity difference	Observed fractional ionic character
IA–VIIB	LiF	3.0	0.92
IIA–VIB	MgO	2.3	0.84
IB–VIIB	CuCl	1.1	0.75
IIB–VIB	ZnS	0.9	0.62
IIIB–VB	GaAs	0.4	0.32
IVB–IVB	SiC	0.7	0.18

for a few selected elements. One thus finds that compounds formed between elements in Groups I and II with those in Groups VI and VII show fairly strong ionic character whilst compounds between elements in the middle of the table are predominantly covalent (see Section 4.4). Solids therefore exhibit a gradation in the degree of charge transfer and this is reflected in the binding mechanism and in the crystal structure.

The reason for the existence of the ionic bond becomes clear when we look at the electronic structure of the atoms of the two types of element involved in the bond. A closed electronic configuration, which as seen previously is the most stable, will occur when an alkali metal atom *loses* a single electron and when a halogen element atom *gains* an electron. In an alkali halide this transference yields a crystal having a lower energy than that of the separate component atoms. There is strong binding between the alkali cations and the halide anions and since the ions are spherically symmetrical the bond between them is non-directional. This is reflected in the types of crystal structures which are exhibited by ionic crystals and which are such that each ion is surrounded by a group of ions of opposite sign.

The two crystal structures favoured by strongly ionic compounds are the cesium chloride and sodium chloride structures. The former structure has already been illustrated (Fig. 3.17) and it consists of a simple cubic lattice with a basis of a Cs^+ ion at 000 and a Cl^- ion at $\frac{1}{2}\frac{1}{2}\frac{1}{2}$. Figure 4.2 shows the sodium

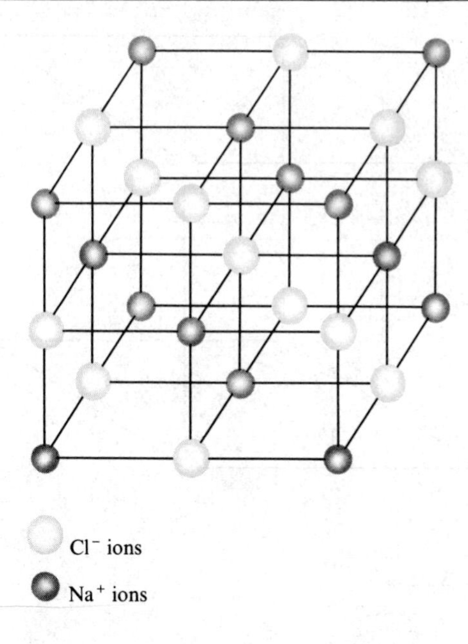

Fig. 4.2 The sodium chloride structure. The lattice is f.c.c. with a basis of Na^+ at 000 and Cl^- at $\frac{1}{2}\frac{1}{2}\frac{1}{2}$. The unit cell shown contains the equivalent of four Na^+ and four Cl^- ions.

chloride structure which has a face-centred cubic lattice with a basis of a Na^+ ion at 000 and a Cl^- ion at $\frac{1}{2}\frac{1}{2}\frac{1}{2}$. The two structures thus have atoms of each type in a unit cell and the compounds are referred to as being of AB type. In both structures an ion of one type is always surrounded by a group of ions of opposite type and the number of nearest neighbours possessed by an atom is termed the coordination number. For the cesium chloride structure it is (8:8) and for the sodium chloride (6:6). Less strongly ionic compounds crystallise either in the cubic zinc-blende structure (Fig. 4.3) or in the closely related hexagonal wurtzite structure. The zinc-blende structure will be discussed later in connection with covalent bonding but we note at this stage that each ion is surrounded by a group of ions of different sign and that the coordination number of the structure is (4:4). The wurtzite structure differs from it only in the arrangement of the second nearest neighbours.

Let us briefly look at some of the reasons underlying the choice of structure favoured by the AB type compounds. The metal ion is formed by losing an electron to expose the noble gas core and taking into account the nuclear charge the size of a cation will in general be less than that of an anion in the compounds. Values of ionic radii can be estimated from experimentally determined

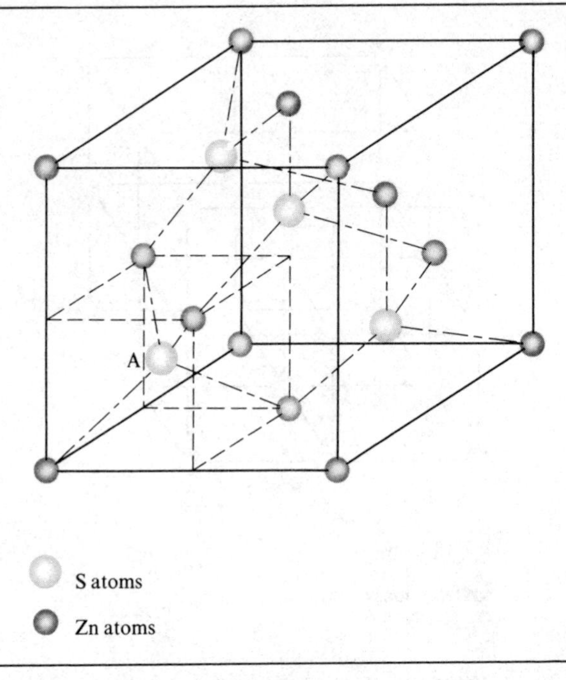

Fig. 4.3 The zinc-blende structure. The lattice is f.c.c. with a basis of Zn at 000 and S at $\frac{1}{4}\frac{1}{4}\frac{1}{4}$. Each atom is tetrahedrally bonded with an atom of different type. Atom A at centre of small cube is bonded to atoms at alternate corners of small cube.

interatomic spacings in ionic crystals and published values are based on the work of Goldschmidt, of Pauling and of Zachariasen. Values of typical ionic radii are given in Table 4.2. The governing factor, determining the preferred

Table 4.2 Ionic radii of selected elements†

Ion	Na^+	K^+	Rb^+	Cs^+	Cu^+	Zn^{2+}	Be^{2+}	F^-	Cl^-	Br^-	S^{2-}
Ionic radius/nm	0.098	0.133	0.148	0.167	0.095	0.083	0.030	0.133	0.181	0.196	0.190

† *With the exception of Cu^+ and Zn^{2+} the values refer to the stable noble gas configuration*

structure for an ionic solid, is the relative size of the two ions. If both ions are of comparable size, the structure of highest coordination number, i.e. the cesium chloride structure, will be preferred since this leads to the greatest interattraction between the ions and a minimum energy for the crystal. The structure may be visualised as consisting of ions of roughly equal sizes all touching each other. If, however, the anion is substantially larger than the cation, as will be the case for many ionic solids, a point will be reached for the cesium chloride structure when contact between ions of opposite sign is prevented. This leads to a fall in the attractive energy and the alternative sodium chloride structure is then preferred. A simple geometrical calculation based on this notion shows that the cesium chloride structure is favoured for ionic ratios of up to 1.37, the sodium chloride structure for ratios in the range 1.4 to 2.4 and the zinc-blende structure for ratios from 2.4 to 4.5. Table 4.3 indicates the general type of agreement

Table 4.3 Ionic radius ratios and observed structures of selected ionic solids

Solid	Ionic radius ratio	Observed structure
CsCl	1.1	Cesium chloride
CsBr	1.2	
NaCl	1.9	Sodium chloride
NaBr	2.0	
KCl	1.4	
RbF	0.9	
ZnS	2.3	Zinc-blende
BeS	6.3	
CuCl	1.9	

obtained and we see that the model serves only as a pointer to the structure which is most likely to be shown by a particular compound.

The energy of a structure compared with that of the isolated components is a measure of the strength of binding. In order to estimate the value of this energy it will be useful first to consider the binding in the isolated sodium chloride ion pair Na^+Cl^-.

The ground state atomic configuration of the isolated sodium and chlorine atoms are as follows

Na $1s^2 2s^2 2p^6 3s^1$

Cl $1s^2 2s^2 2p^6 3s^2 3p^5$

and the Na^+ Cl^- ion pair will have the following closed shells

Na^+ $1s^2 2s^2 2p^6$ (neon core)

Cl^- $1s^2 2s^2 2p^6 3s^2 3p^6$ (argon core)

The energy balance in forming the ion pair may be estimated as follows. First, the energy necessary to form a sodium ion from a neutral sodium atom is the ionisation energy of sodium and by experiment this amounts to 5.1 eV. The resulting ion is the spherically symmetrical neon core. In contrast the chlorine atom on ionisation must gain an electron and it is found experimentally that the Cl^- ion formed is stable. Chlorine has an electron *affinity* of 3.6 eV and this energy is *released* on forming the ion. The additional electron allows the ion to have all the 3p orbitals filled, leading again to the spherically symmetrical noble gas core configuration. Although the formation of the ion must result in an increase in inter-electron repulsion its stability shows that this is more than compensated by the additional coulombic attraction brought about by the presence of the extra electron.

In order, therefore, to form the ion pair expenditure of energy of 5.1 − 3.6 = 1.5 eV is required and on its own the formation of the pair of ions would not give a stable situation. However, the ion pair formed will have a strong mutual attraction and the coulombic attractive energy leads to a strong binding in the molecule. From basic electrostatics theory the attractive energy E will be given by

$$E = -\frac{e^2}{4\pi\varepsilon_0 r_0}, \qquad \ldots(4.2)$$

where r_0 is the experimentally observed separation of the ions in the vapour. The negative sign shows that the energy is one of attraction. For the sodium chloride ion pair $r_0 = 0.25$ nm so that the attractive energy $E = 5.7$ eV. The energy balance in forming the ion pair is illustrated in Fig. 4.4.

	Energy required eV	Energy released eV
$Na \rightarrow Na^+ + e^-$ (Ionisation energy)	5.1	
$e^- + Cl \rightarrow Cl^-$ (Electron affinity)		3.6
Formation of separate Na^+ and Cl^- ions	1.5	
Coulomb energy of $Na^+ Cl^-$ ion pair		5.7
Repulsive energy of $Na^+ Cl^-$ ion pair	0.2	
Formation of $Na^+ Cl^-$ ion pair		
... from separate ions		5.5
... from separate atoms		4.0

Fig. 4.4 Formation of a Na^+Cl^- ion pair. The dissociation energy of the ion pair is 4.0 eV.

Account must also be taken of the repulsive energy and this arises as a result of the overlap of the electron cores. As discussed in Section 2.6 the restriction on electron occupancy described by the Pauli exclusion principle provides a repulsive force between the cores and will limit the minimum separation of the ions. Although the repulsive *force* must equal the coulombic attractive force at the equilibrium separation the energy contribution due to the exclusion principle is quite small being ≈ 0.2 eV. The energy of the sodium chloride ion pair compared with that of the separate ions is therefore 5.5 eV, and compared with the separate atoms 1.5 eV less, i.e. 4.0 eV. This is termed the *dissociation energy* of the ion pair and the value given indicates an ion pair of high stability.

We may next consider the nature of the forces binding atoms together in an ionic solid. In addition to the energy contributions already discussed, account must now be taken of the interaction between all the ions forming the crystal. As we shall see this yields additional binding associated with the cohesive energy of an ionic solid. The latter quantity is defined as the energy per ion pair which is required to separate the solid into neutral atoms.

As pointed out, the structures of the ionic crystals are such that ions of one sign are always surrounded by a symmetrical array of ions of the opposite sign. The sodium chloride structure, in which almost all the alkali halides and many other compounds such as the alkaline earth oxides crystallise, is equivalent to two interpenetrating face-centred cubic arrays displaced from each other by one half of the cube diagonal distance. The anions are associated with one lattice, the cations with the other. Both the cesium chloride and the zinc-blende structures can be described in a similar manner.

The minimum interatomic distance r_0 in the sodium chloride structure occurs between ions of different signs and for sodium chloride is 0.281 nm, i.e. it is somewhat larger than the observed separation in the isolated ion pairs. By reference to Fig. 4.2 it can be seen that each cation is surrounded at this distance by 6 anions, giving a strong attractive force. At a distance $\sqrt{2}r_0$ there are a further 12 cations giving repulsion and with increasing distance alternating sets of anions and cations.

The coulombic electrostatic energy for each ion pair in an ionic crystal can be expressed in the form

$$E = -\frac{\alpha e^2}{4\pi\varepsilon_0 r_0}, \qquad \ldots(4.3)$$

where α is the sum of a series which, for the sodium chloride structure is given by

$$\alpha = \left(6 - \frac{12}{\sqrt{2}} + \frac{8}{\sqrt{3}} - \frac{6}{\sqrt{4}} + \frac{24}{\sqrt{5}} - \cdots\right) \qquad \ldots(4.4)$$

$$= 1.748.$$

α is termed the Madelung constant. Since the environment of an atom differs according to the structure, different values of α are obtained for each structure. The values are 1.763 for the cesium chloride, 1.638 for the zinc-blende and 1.641 for the wurtzite structure. Convergence of series of this type is very poor and special methods, involving the division of the array of ions into electrically neutral groups, have been developed in order to improve convergence.

At the observed interatomic separation of 0.281 nm the attractive energy for a sodium chloride crystal given by equation (4.3) is 8.9 eV per ion pair. As in the case of isolated ion pairs it is also necessary to take account of ion core repulsion in order to arrive at the total energy. We know that this repulsive energy component is strongly dependent on atomic separation and is of importance only for small separations. A repulsive energy term of the form A/r^n may thus be used so that the energy per ion pair relative to infinitely separated ions will be given by

$$E = -\frac{\alpha e^2}{4\pi\varepsilon_0 r} + \frac{A}{r^n}, \qquad \ldots(4.5)$$

where A and n are constants for the solid. The value of E at the equilibrium separation $r = r_0$ is required. At this value the repulsive and attractive forces must balance and the energy then has a minimum value. Putting $\mathrm{d}E/\mathrm{d}r = 0$ at $r = r_0$ we can then eliminate A and obtain

$$E = -\frac{\alpha e^2}{4\pi\varepsilon_0 r_0}\left(1 - \frac{1}{n}\right). \qquad \ldots(4.6)$$

The value of n can be determined from compressibility measurements, a value of $n \approx 9$ being found for a sodium chloride crystal. In accordance with equation (4.6) the value of E is reduced from 8.9 eV to 7.9 eV per ion pair. The cohesive energy referred to neutral atoms is, as in the case of the dissociation energy of the ion pair, 1.5 eV less, i.e. 6.4 eV. This value is in good agreement with experimental

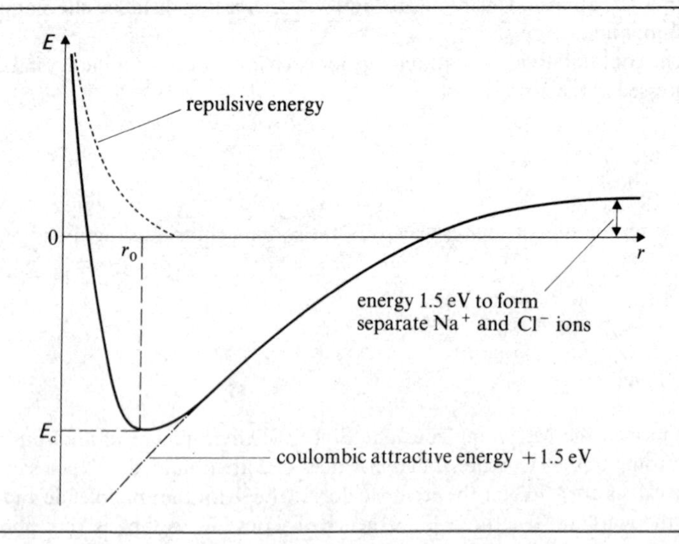

Fig. 4.5 Energy of a Na^+Cl^- lattice referred to neutral atoms. The cohesive energy $E_c \simeq 6.4$ eV.

values. This agreement, however, is to some extent fortuitous since a contribution to bonding arising from van der Waals forces has not been included. This, however, is small and is balanced by a repulsive component due to zero point energy. The energy of the Na^+Cl^- lattice as a function of the separation of the ions is given in Fig. 4.5.

In conclusion it should be emphasised that there is a complete gradation in degree of ionic character from solids which are almost completely ionic, such as sodium chloride and rubidium fluoride, to diamond in which the bonding is purely covalent. The typical ionic compound has very low conductivity, being an insulator at ordinary temperatures, and it will possess the associated properties of hardness and transparency. At a temperature close to its melting point, and also in appropriate solvents, conductivity is exhibited since the ions are then mobile. Of the compounds which are intermediate in character the three compounds PbS, PbSe, and PbTe have the sodium chloride structure, are primarily ionic in character but exhibit some properties associated with covalent compounds, e.g. they are semiconductors. The corresponding beryllium compounds BeS, BeSe and BeTe are much more strongly covalent, although exhibiting some ionic character, and they crystallise in the zinc-blende structure.

4.3 The van der Waals bond

The van der Waals bonding mechanism is present in all solids. However, only in the absence of other bonding, such as is the case in the solid form of the noble gases, does it form the dominant mechanism. The binding energy for these solids is given in Table 4.4 and it is clear that, compared with the ionic bond, it is very weak. The bond is also that occurring between molecules in the solid form of hydrogen, oxygen, nitrogen and the halogens. These are termed molecular

Table 4.4 The cohesive energies of solids bonded by van der Waals forces

Solid		**Cohesive energy**		**Melting point**
		/eV atom^{-1}	/kJ mol^{-1}	/K
Neon	Ne	0.026	2.5	24
Argon	Ar	0.088	8.5	84
Krypton	Kr	0.12	12	117
Xenon	Xe	0.17	16	161
		/ev molecule^{-1}	/kJ mol^{-1}	/K
Hydrogen	H_2	0.01	1.0	14
Oxygen	O_2	0.090	8.6	54
Nitrogen	N_2	0.081	7.8	63
Chlorine	Cl_2	0.32	31	172
Hydrogen Chloride	HCl	0.22	21	158

crystals and virtually all organic crystals are also in this category. The bonding between the atoms in the individual molecules is of the covalent type and is much stronger, e.g. hydrogen molecules have a dissociation energy of 4.75 eV compared with a binding energy of 0.01 eV per molecule for the solid.

Let us consider the bonding as exhibited in the crystals of the noble gases and deal first with the repulsive energy component. The electronic configuration in the noble gases is spherically symmetrical and decreasing the atomic separation gives increased overlap of the closed electron shells of the atoms. As discussed, the restriction on the spatial extent of the electron wave-functions forces electrons into higher energy states since the Pauli exclusion principle prevents multiple occupancy of lower lying states. This is equivalent to an inter-electron repulsive force and a corresponding repulsive energy. Using a power law expression for the repulsive energy term (as in equation (4.1)) the experimental data for the ion core repulsion in the noble gases indicates a dependence proportional to $1/r^{12}$.

In order to account for the binding we need to look for an attractive force between the neutral atoms which make up these crystals. On average the centre of gravity of the electrons in each atom coincides with the nucleus. However, fluctuating displacements of the electrons which occur even at absolute zero, due to zero point energy, give to each atom a dipole moment (defined by the products of the charges and their separation). This is a fluctuating dipole moment and it has a zero time average. The electric field due to the dipole polarises neighbouring atoms so that an adjacent atom will have a component of its fluctuating dipole moment in phase, and correlated, with that of the first atom. From electrostatics theory the electric field on the axis of a dipole of moment p and at a distance r from the dipole is given by $2p/4\pi\varepsilon_0 r^3$. Resulting from the interaction a pair of adjacent atoms will have an interaction energy proportional to $-(p/r^3)^2$. This will give an attractive energy component since the time average of p^2 must be positive.

The energy of the crystal bound by van der Waals forces can therefore be expressed in the form

$$E = -\frac{A}{r^6} + \frac{B}{r^{12}} \qquad \ldots(4.7)$$

which is known as the Lennard–Jones potential. The curve is of the general form of Fig. 4.1 where, however, both repulsive and attractive energies decrease rapidly with increasing r. It follows that the attractive van der Waals forces are proportional to $1/r^7$ and are, therefore, of much shorter range than the ionic forces which are proportional to $1/r^2$.

The atoms concerned in the bonding have spherically symmetrical electron distributions and the bond is non-directional. This favours a close-packed arrangement of atoms and one finds that the noble gases crystallise in the face-centred cubic structure. As can be seen from Table 4.4 the weakness of the bond is reflected in the low melting points of the crystals.

4.4 The covalent bond

Understanding of the covalent bond in solids can perhaps be best approached by looking first at the nature of the bond in hydrogen and then seeing how the concepts developed apply to other molecules and compounds.

4.4.1 *The hydrogen molecule ion*

Perhaps the simplest bond which one may consider is the one electron bond existing in the hydrogen molecule ion H_2^+. This molecule ion consists of two protons having a mean separation of 0.106 nm together with a single electron. The binding energy referred to the isolated components is 16.4 eV, whilst the dissociation energy referred to a separated H atom and a H^+ ion is 2.7 eV, i.e. the molecule ion is the more stable. The difference between the two values is of course the ionisation energy of the hydrogen atom. Three separate energy components may be identified in the system. A repulsive energy which arises from the coulombic repulsion of the two protons, and the kinetic and potential energies of the electron. The energies are all functions of the separation of the two nuclei and their sum will be a minimum, equal in magnitude to the binding energy, at the observed experimental separation in the molecule ion.

As shown in Section 2.4 the ground state 1s wave-function of the hydrogen atom gives an electron density which falls away exponentially from the nucleus. The electrostatic attraction confines the electron to the region close to the nucleus and this may be represented as a *potential well* whose shape is determined by the variation of electron potential energy with position. This model is illustrated in Fig. 4.6. What is the corresponding picture for the hydrogen molecule ion? The single well afforded by the isolated nucleus is now shown in Fig. 4.7 as being replaced by a broader well which gives a region of low potential energy between the two nuclei. The two atomic orbitals for the separate nuclei are, therefore, replaced by *molecular orbitals* which extend over both. The electron is under the influence of both nuclei and this type of situation is common to all molecular systems.

A method known as the Linear Combination of Atomic Orbitals (LCAO) may be used to find the form of the molecular orbitals. Let us designate the 1s atomic orbitals of two nuclei A and B by ψ_A and ψ_B. When the two nuclei are well separated the electron will be associated with only A or B but not with both and the ground state energy of the system will clearly be the same in both cases. The normalised wave-functions, in accordance with equation (2.41) will be for the two cases

$$\psi_A = N \exp\left(-\frac{r_A}{a_0}\right)$$
$$\psi_B = N \exp\left(-\frac{r_B}{a_0}\right), \qquad \ldots(4.8)$$

where r_A and r_B are the distances of the electron from the two nuclei A and B,

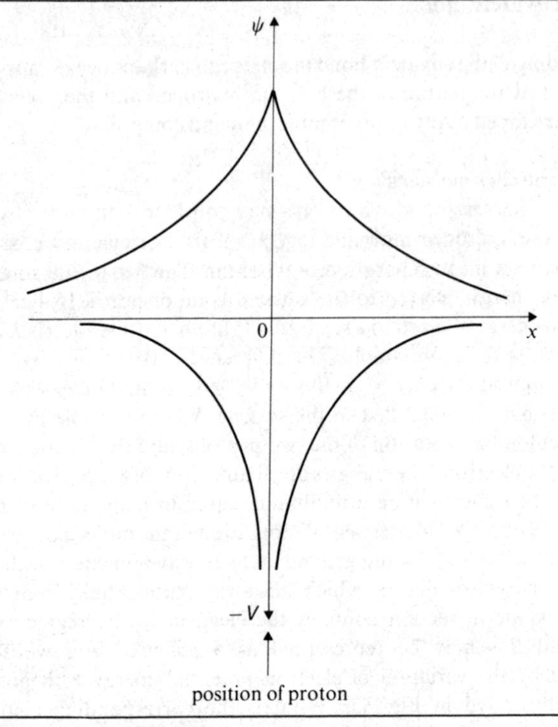

Fig. 4.6 The potential well given by a proton and the corresponding 1s wave-function. Note the two separate vertical scales.

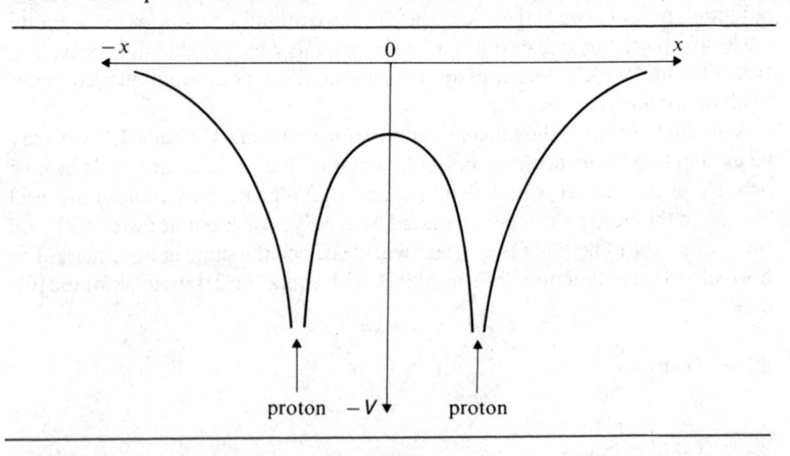

Fig. 4.7 The pair of wells provided by the two protons in a hydrogen molecule. These yield a region of low potential energy between the nuclei. The corresponding wave-functions are given in Fig. 4.8.

respectively. N is the normalising constant, which from symmetry will be the same for both ψ_A and ψ_B. If now the nuclei are brought closer together two linear combinations of ψ_A and ψ_B will be found to give satisfactory molecular orbitals. These are

$$\psi_S = N_S(\psi_A + \psi_B)$$

and

$$\psi_{AS} = N_{AS}(\psi_A - \psi_B), \qquad \ldots(4.9)$$

where N_S and N_{AS} are normalising constants. The symmetry of the system requires that no other constants be introduced. The two functions are shown in Figs. 4.8 and 4.9 and it is clear that ψ_S and ψ_{AS} are, respectively, symmetrical and

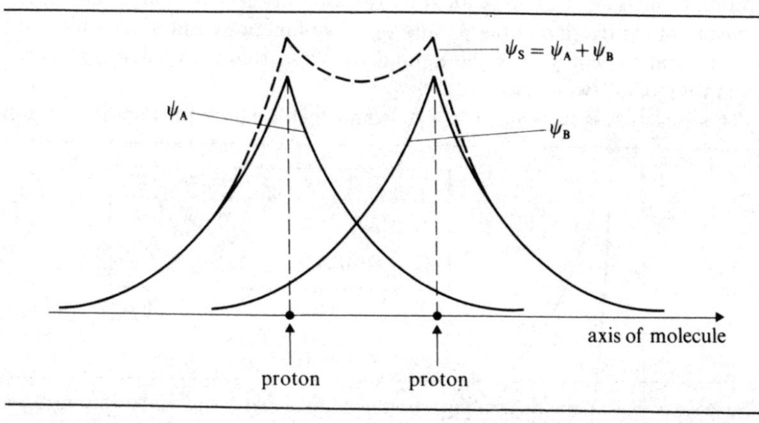

Fig. 4.8 The symmetrical wave-function ψ_S given by the overlap of two 1s atomic orbitals ψ_A and ψ_B.

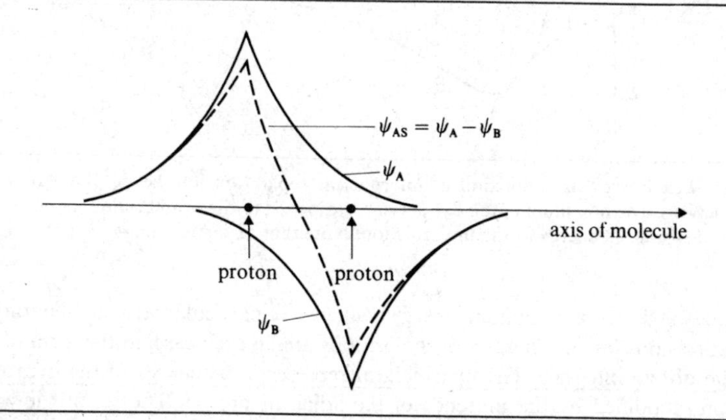

Fig. 4.9 The antisymmetrical wave-function ψ_{AS} given by the overlap of two 1s atomic orbitals ψ_A and ψ_B.

antisymmetrical about the mid point of AB. For the orbitals to be normalised we must have, in accordance with Sections 1.8.1 and 2.3.1,

$$\int_V \psi_S^2 \, dV = 1$$
$$\int_V \psi_{AS}^2 \, dV = 1. \qquad \ldots(4.10)$$

Noting also that ψ_A and ψ_B are themselves normalised, the values of N_S and N_{AS} will be governed by the value of $\int_V \psi_A \psi_B \, dV$. This is termed the overlap integral and its value is strongly dependent on the separation of the two nuclei. It is a measure of the overlap of the ψ_A and ψ_B wave-functions and it determines the variation in the energies of the ψ_S and the ψ_{AS} orbitals as a function of the separation of the two nuclei.

This variation is shown in Fig. 4.10 from which we have the important result

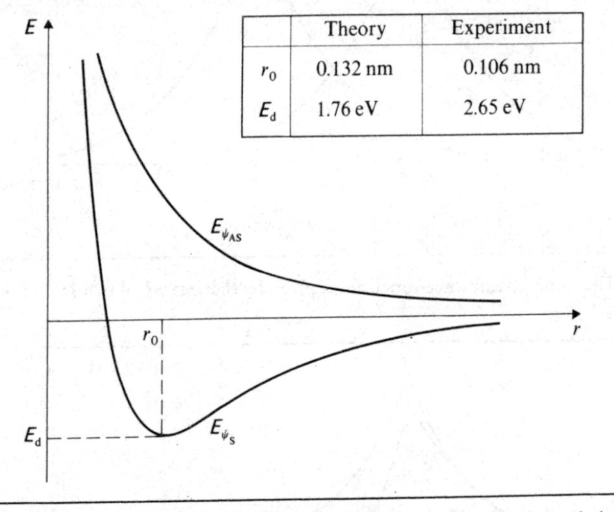

	Theory	Experiment
r_0	0.132 nm	0.106 nm
E_d	1.76 eV	2.65 eV

Fig. 4.10 Energy as a function of internuclear separation for the ψ_S and ψ_{AS} wave-functions. The minimum of the E_{ψ_S} curve corresponds to the equilibrium separation and the dissociation energy of the molecule ion. Note that zero point energy is not included here.

that ψ_S shows a minimum energy value at a particular atomic separation. Expressions for the energies of the orbitals are, in each case, in the form of the sum of two integrals. The first integral represents the energy of the hydrogen atom, modified by the presence of the adjacent proton. The second integral shows how the energy is affected by the overlap electron density.

In order to see why the overlap plays an important part in determining the energies of the orbitals we need to compare the electron densities for the ψ_S and ψ_{AS} orbitals. These are shown in Fig. 4.11 and the marked differences in the

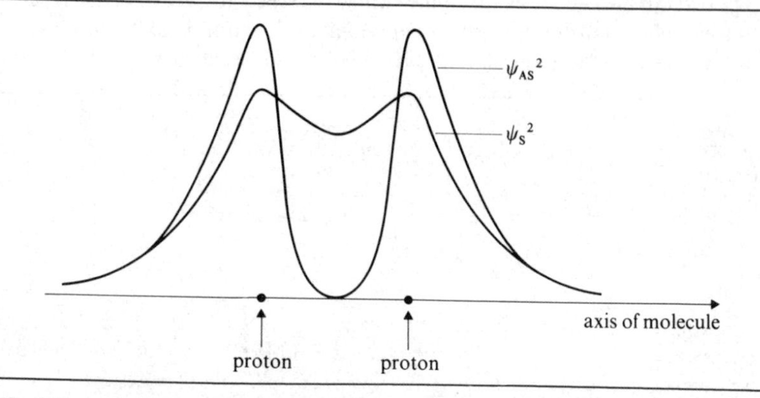

Fig. 4.11 A comparison of the electron densities of the ψ_S and ψ_{AS} wave-functions. The densities are symmetrical about the axis of the molecule.

internuclear region should be noted. The density ψ_S^2 for the symmetric orbital in this region is greater than would be the case for separate non interacting orbitals, for which $\psi_A^2 + \psi_B^2$ is a measure, and this indicates the increased probability for an electron in the ψ_S orbital to lie in this region. Since it is a region of low potential energy this has the effect of reducing the total energy of the system. It therefore leads to a stable bonding situation and the symmetrical ψ_S function is termed a bonding orbital. A parallel interpretation is that the binding of the molecule ion is a result of the attraction between each proton and the electron which, on average, is located between the two protons.

The stable bonding situation for the ion corresponds to the minimum of the energy curve in Fig. 4.10. This is achieved as a balance between the repulsive coulombic energy of the two protons, which increases with decreasing separation of the protons and, on the other hand, the lowering of the potential energy of the electron which increases the attractive energy. The energy of the bonding state can be calculated as a function of the internuclear spacing. As indicated on the figure the value of the dissociation energy differs by as much as 40 per cent from the experimental value. The actual dissociation energy of a bond is also slightly less due to zero point energy. The solution obtained is therefore an approximate one, and indeed exact solution of the dynamics of three interacting charged particles is not possible. The functions ψ_S and ψ_{AS} are essentially trial solutions and much improved agreement with experiment has been found possible by using modified wave-functions and introducing additional adjustable parameters.

As illustrated in Fig. 4.11 the electron density for the ψ_{AS} orbital is low in the region between the nuclei. The energy of this orbital increases continuously with decreasing separation of the nuclei and the absence of a minimum leads to a repulsive energy and an antibonding orbital. The electron in the ground state configuration of the molecule ion is therefore located in the bonding ψ_S orbital. The orbitals are cylindrically symmetrical with respect to the axis of the molecule and are referred to as sigma (σ) orbitals. A further illustration of the bonding and antibonding situations in the hydrogen molecule ion is given in Fig. 4.12.

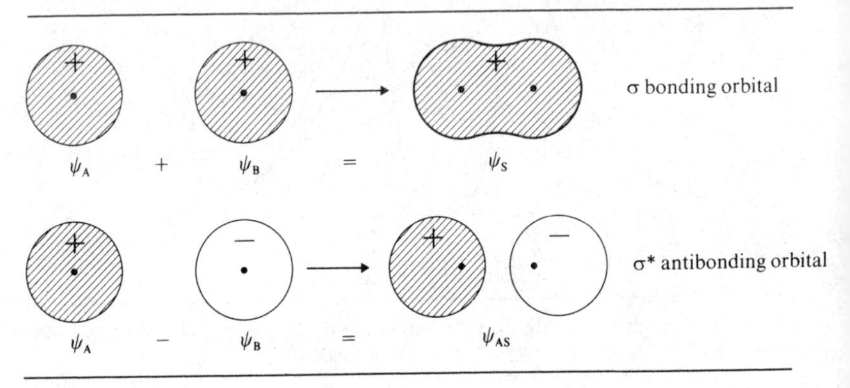

Fig. 4.12 The formation of the σ bonding orbital and the σ^* antibonding orbital for the hydrogen molecule ion. The + and − refer to the signs of each orbital.

4.4.2 *The hydrogen molecule*

We have seen that the method outlined, termed the molecular-orbital approach, can explain the bonding in the hydrogen molecule ion. It may be extended to deal with the bonding in the two electron hydrogen molecule by writing a total wave-function for the molecule as the product of two separate molecular orbitals, one for each of the two electrons. A symmetrical and an antisymmetrical orbital are obtained as for the hydrogen molecule ion. In the ground state the two electrons will both occupy the symmetrical wave-function, which is, of course, allowable in accordance with the Pauli exclusion principle. The symmetrical wave-function has the lower energy, and in accordance with the discussion in Section 2.6 the electrons will have paired spins.

The electrons involved in the bond suffer mutual repulsion and will therefore give an additional repulsive energy component in the system. However, since there are two electrons they will also provide an increased attractive energy component drawing the protons towards the region of high electron density separating them. This is reflected in the decreased bond length of 0.074 nm compared with a value of 0.106 nm for the molecule ion. Overall the occupation of the bonding orbital by two electrons leads to a bond strength which is nearly double that of the molecule ion. The dissociation energy of the hydrogen molecule is 4.5 eV compared with 2.7 eV for the molecule ion.

The two-electron bond in the hydrogen molecule, formed by the overlap of the two 1s orbitals is a σ bond. As we shall see, overlap of other types of orbitals can also give electron pair bonds and they are variously referred to as covalent, homopolar or valence bonds. Electron pair bonds are common and they form, for example, the basic bonding in organic compounds.

We may enquire about the bond strength when a different number of electrons is involved. The strength of the one-electron bond in H_2^+ has been seen to be considerably weaker than the two-electron bond in H_2. A further weakening results if the bond occurs between atoms of different elements. The electron may then tend to locate on one or other of the nuclei, charge concentration between the nuclei is lessened, and the bond becomes weak. Under the same circumstances a two-electron bond remains strong since inter-electron repulsion prevents charge concentration on a single nucleus.

A rough guide to the strength of bonding of a molecule is given by the net excess of the number of electrons in bonding orbitals over those in antibonding orbitals. Taking as an example the bonding in the helium molecule He_2^+. This ion is found experimentally to be stable and its bond strength is roughly comparable to that in H_2^+, i.e. rather greater than half that of the electron pair bond in H_2. In He_2^+ two of the three electrons occupy the bonding orbital whilst the third occupies the antibonding orbital. The observed bond strength is, therefore, consistent with a net excess of one electron in a bonding orbital.

4.4.3 *Covalent bonding in gaseous molecules and solids*

Covalent bonding accounts for the stability at ordinary temperatures of gaseous molecules such as the halogens, nitrogen and oxygen and compounds such as hydrogen sulphide (H_2S), phosphine (PH_3) and ammonia (NH_3). The type of electronic configuration involved can best be explained by looking at one or two specific examples.

The covalent bonding in nitrogen (N_2) is the strongest in any diatomic molecule and it has the very high dissociation energy of 9.8 eV. The nitrogen atom has the ground state electronic configuration $1s^2 2s^2 2p^3$, with the 1s and 2s states both occupied by paired electrons. In accordance with Hund's rules electrons do not share an orbital if other unoccupied orbitals of equal energy are available. Accordingly one finds in the ground state one electron in each of the three equivalent p_x, p_y, and p_z orbitals and the electrons will have their spins parallel. The electronic configuration is illustrated in Fig. 4.13. For the nitrogen molecule there is a total of six electrons in the three symmetric molecular orbitals and the electron spins in each orbital will be paired. In chemical structure this is referred to as a triple bond and is written $N{\equiv}N$. One component of the bond in nitrogen is formed by the overlap of two p_z orbitals and is directed along the axis of the molecule. It is therefore another example of a σ bond. The other two molecular orbitals are formed by linear combinations of p_x and p_y orbitals and give maximum overlaps in directions at right angles to the molecular axis. These are referred to as π bonds and are illustrated in Fig. 4.13.

In the molecule of ammonia the three unpaired electrons of the nitrogen atom partake in molecular orbitals with three electrons from atoms of hydrogen. There are therefore paired electrons in each of three symmetric molecular orbitals and these are spatially oriented at 107° to each other. The reasons for

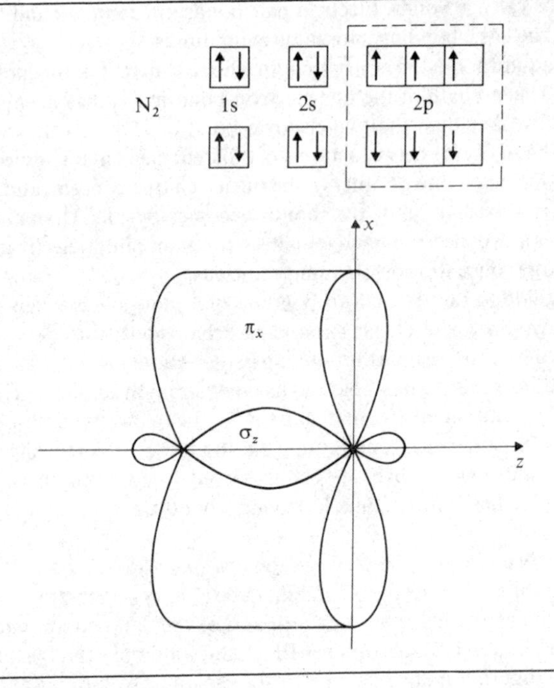

Fig. 4.13 The nitrogen molecule illustrating overlap of orbitals to form σ and π bonds. Note that the square of the angular parts of the wave-functions (Y_{lm}^2) are utilised in this representation. The π_y orbital is not illustrated but is perpendicular to the π_x orbital. Since both orbitals contain electrons the charge cloud is symmetrical about the internuclear axis.

this deviation from the axial orientation of p orbitals are complex. It has been attributed to repulsion between the bonds and to an ionic component in the bond. Significantly, bond angles much closer to 90° are found in the compound PH_3 where there is less difference in the electronegativity characteristics of the two elements.

Solids formed by the very weak van der Waals bonding of covalent molecules or neutral atoms have been discussed already. In some solids a slightly stronger bond can be formed by the electrostatic attraction between a proton and electronegative atoms. It is termed a hydrogen bond. The strongest is that occurring in the ion HF_2^- and in this the proton lies midway between the fluorine atoms. A more common situation is that observed in ice where the hydrogen atom is nearer to one or other of its adjacent oxygen atoms. In the structure each oxygen atom has four hydrogen atoms tetrahedrally disposed, two of which are covalently bonded with it and have a bond length of 0.1 nm, i.e. only slightly greater than in the isolated molecule. The other two bonds are the weaker hydrogen bonds which have a bond length of 0.18 nm.

The covalent bonding of carbon is of interest, not only because of the very wide range of organic compounds, but also because the tetrahedral bonding in diamond itself is closely patterned by other solids. This crystal structure and its associated covalent bonding underlies the physical properties of an important group of insulators and semiconductors.

Isolated carbon atoms in the ground state have the electronic configuration $1s^2 2s^2 2p^2$ and in accordance with Hund's rule the two p electrons will occupy separate p_x and p_y orbitals and will possess parallel spins. It might, therefore, be supposed that carbon would be divalent. However, in almost all its compounds carbon is tetravalent and the simplest stable compound with hydrogen is methane (CH_4). As mentioned at the end of Chapter 2 the excited configuration $1s^2 2s^1 2p^3$ allows carbon to attain a tetravalent state. This is illustrated in Fig. 4.14 for the compound methane. The energy required to promote the excited state is more than counterbalanced by the energy gain in bond formation.

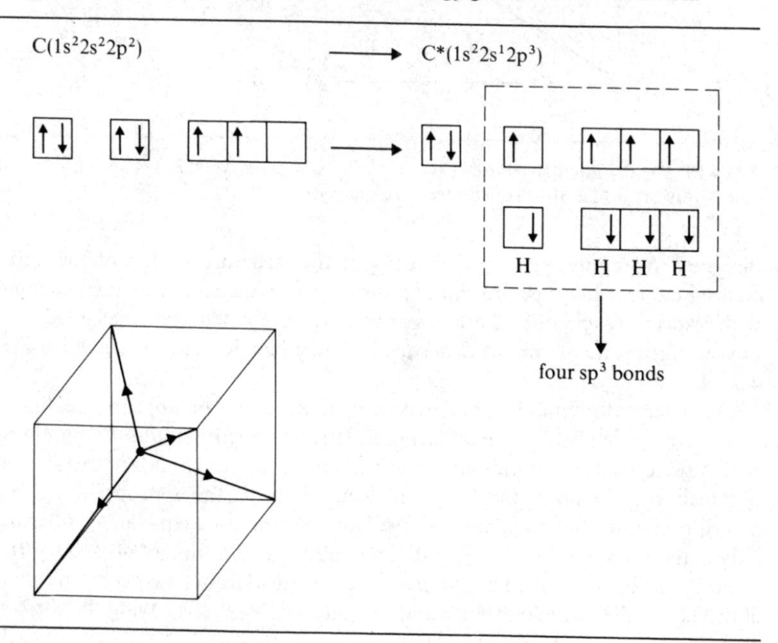

Fig. 4.14 The tetrahedral bonds in methane CH_4. The angle between each pair of bonds is 109.5°.

The four electron-pair bonds give rise to the tetravalent bonding exhibited in the very many organic compounds. Due to inter-bond repulsion they align themselves towards the four corners of a regular tetrahedron giving a bond angle of 109° 28′. This is the arrangement in the diamond crystal structure shown in Fig. 4.15. The diamond lattice is face-centred cubic with a basis of atoms at 000 and $\frac{1}{4}\frac{1}{4}\frac{1}{4}$ and in the structure each atom has four nearest neighbours tetrahedrally

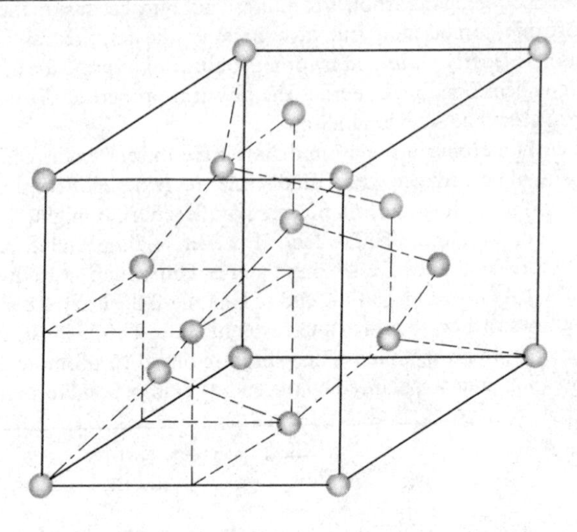

Fig. 4.15 The diamond structure. This should be compared with Fig. 4.3 from which it differs only in that the atoms at all sites are of one type.

disposed. An equivalent representation of the structure is that of two face-centred cubic sub-lattices displaced a distance of a quarter of a cube diagonal with respect to each other. If atoms of two different elements are associated with the two lattices we obtain the closely related zinc-blende structure shown in Fig. 4.3.

The tetrahedral bonds can be described in terms of appropriate linear combinations of s and p wave-functions. This is termed hybridisation and the bonds are termed sp^3 hybrids. Other hybrid bonds such as sp and sp^2 can also be obtained and the bond angles in a molecule indicate the appropriate hybrid description of the bonding. Since all the four sp^3 bonds are equivalent, differing only in their orientation, it is clear that the ratio of s to p content of each hybrid orbital must be the same. The sp^3 orbitals are formed from linear combinations of the 2s and the three 2p orbitals and the four combinations having the correct balance are

$$\begin{aligned} \psi_1 &= \tfrac{1}{2}(s+p_x+p_y+p_z) \\ \psi_2 &= \tfrac{1}{2}(s-p_x-p_y+p_z) \\ \psi_3 &= \tfrac{1}{2}(s-p_x+p_y-p_z) \\ \psi_4 &= \tfrac{1}{2}(s+p_x-p_y-p_z) \end{aligned} \qquad \ldots(4.11)$$

Provided that the component atomic orbitals are normalised these form a set of normalised and orthogonal wave-functions.

The direction of each of these orbitals can be inferred from that of their constituent orbitals, e.g. the ψ_2 hybrid orbital is in the $[-1\ -1\ 1]$ direction. The angular dependence for each orbital may be derived using the angular dependence function for s and p orbitals given in Table 2.2. For the ψ_1 orbital this yields

$$\psi_1 = \frac{1}{2\sqrt{\pi}}\left\{1+3[\cos\theta+\sin\theta(\sin\phi+\cos\phi)]\right\}. \qquad \ldots(4.12)$$

It is readily confirmed that this has a maximum value of $1/\sqrt{\pi}$ which is twice that for an s orbital (see Section 2.3.1.)). Thus if we take the magnitude of an s orbital as 1, that of a p orbital is $\sqrt{3}$ and of an sp^3 orbital is 2.

In covalent bonding the overlap of orbitals is a measure of the bond strength and the energy of the bond is roughly proportional to the product of the magnitude of the orbitals concerned. Taking the energy of an s–s bond as unity, that of an s–p bond is $\sqrt{3}$, of an s–sp^3 bond is 2 and of two sp^3 orbitals is 4.

The types of bonding observed in the three compounds ethane, ethylene and acetylene may serve to illustrate some of these points. In ethane $H_3C–CH_3$ each carbon atom forms the tetrahedral sp^3 hybrid orbitals, three of which give s–sp^3 bonds of the σ type with the hydrogen atoms whilst the fourth gives the sp^3–sp^3 σ bond between the two carbon atoms. This has a bond strength of 2.7 eV and has length 0.154 nm. The bonding in the molecule is illustrated in Fig. 4.16.

Each carbon atom in ethylene $H_2C=CH_2$ provides three sp^2 hybrid orbitals and a single p orbital. Two of the hybrid orbitals provide σ bonds with the two hydrogen atoms whilst the third provides another σ bond between the carbon atoms, i.e. an sp^2–sp^2 bond. The remaining p orbital of each carbon atom gives the additional π bond and the double C=C bond strength is increased to 4.4 eV with a decreased bond length of 0.135 nm.

Finally, in acetylene HC≡CH each carbon atom has two sp hybrid orbitals, which form the σ bonds, and two p orbitals giving p–p bonds of the π type. As illustrated in Fig. 4.16, the triple bond C≡C therefore consists of a σ bond and two π bonds. The bond energy is 5.5 eV and it is of length 0.121 nm.

4.5 The metallic bond

In contrast to the ionic and covalent bonds the properties of the metallic bond can neither be derived nor inferred from the nature of bonding in isolated molecules. The bond may, however, in some respects, be regarded as being intermediate in character between that of an ionic and of a covalent bond. Whereas in the ionic bond there is transfer of electrons to form anions and cations and a sharing of electron pairs in covalent bonding, in the metallic bond the valence electrons may be regarded as being shared by all the atoms in the crystal. We may, therefore, visualise molecular orbitals extending throughout a

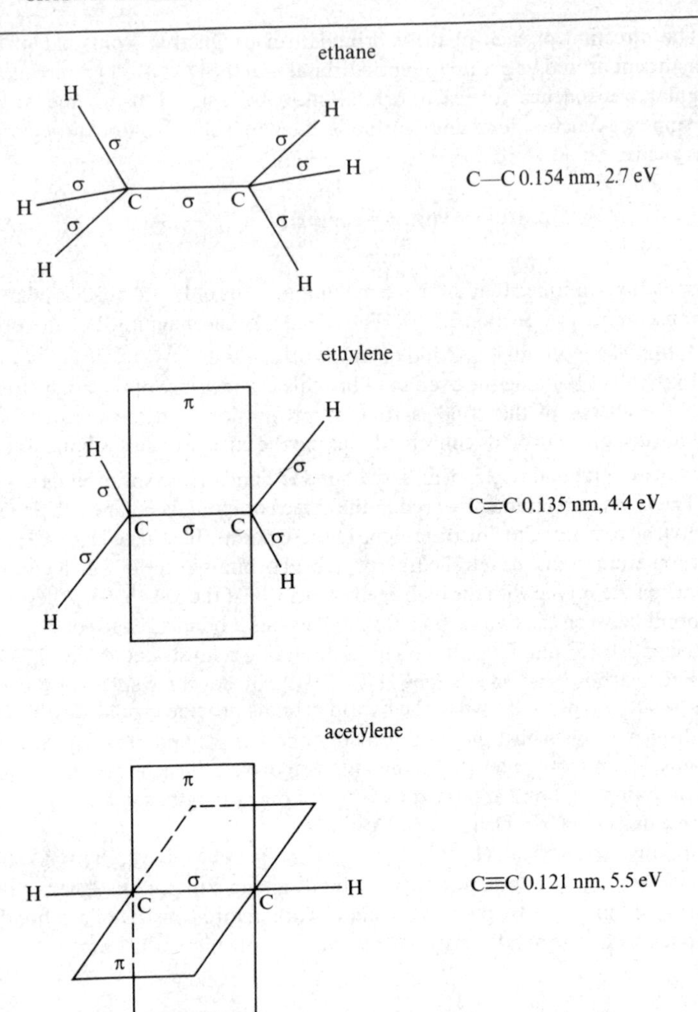

Fig. 4.16 Covalent bonds in ethane, ethylene and acetylene.

metallic crystal and these would be obtained as combinations of many atomic orbitals. For the crystal as a whole a band of orbitals is obtained and the valence electrons in a metal will be occupying those of lowest energy. Such aspects of the behaviour of electrons will be developed further in Chapter 7. A less rigorous approach can, however, give a qualitative explanation of the nature of bonding in metals.

Properties such as mechanical ductility, optical opaqueness and reflectivity, and good conductivity of heat and of electricity are all typical metallic characteristics. They are exhibited, in varying degrees, by a substantial proportion of the elements. Many alloy phases are also metallic. Many typical metals have ground state electronic configurations such that there are one or two electrons in the outer s orbital. These are the valence electrons and they are responsible for giving to elements their metallic character. However, the trivalent element aluminium and the quadrivalent elements tin and lead are also metallic, although many of the elements near the middle of the Periodic Table exhibit covalent bonding and are, consequently, non-metallic. The transition metals are also metals although less typical in their behaviour because of the presence of the incompletely filled d orbitals.

Since the valence electrons are shared by all the atoms in a crystal they may be regarded as an electron gas permeating the whole crystal. The metal atoms minus the valence electrons form a regular array of positive ions on the crystal lattice and the bonding is essentially an inter-attraction between the array of positive ions and the electron gas.

Metals crystallise typically in one of three structures, namely, the body-centred cubic, the face-centred cubic and the hexagonal closed packed structures. These are illustrated in Figs. 3.3, 3.6 and 3.10. In the b.c.c. and f.c.c. structures a single atom is associated with each lattice point. The h.c.p. structure has a simple hexagonal space lattice with a basis of atoms at 000 and $\frac{2}{3}\frac{1}{3}\frac{1}{2}$ and it should be noted that the atom positions in the structure do not themselves form a space lattice.

These three structures are typified by a high degree of packing of atoms and an atomic arrangement whereby each atom has many near neighbours, i.e. the structures have a high coordination number. This is 8 for the b.c.c. structure and 12 for the f.c.c. and h.c.p. structures. Slight departures from ideality in the h.c.p. structures in metals such as zinc and cadmium lead to 6 of the 12 atoms having slightly smaller separation distances than the remaining 6. Both the f.c.c. and the ideal h.c.p. structures correspond to the closest packing of spheres and this leads to a fraction of 0.74 of the total available volume being occupied. The arrangement of atoms in the basal (0001) planes of the h.c.p. structure is identical with that in a (111) plane of the f.c.c. structure and the two structures differ only in the repeat pattern of identical layers. This is of the form ABABA... for the h.c.p. and ABCABC... for the f.c.c. structure. Both structures are thus similar and crystal transformations between the two are observed to occur, e.g. in the metal cobalt. In the b.c.c. structure the atoms are in contact only along the body diagonal of the cube and the packing fraction is, therefore, slightly lower at 0.68.

These three metallic structures should be compared with the much more open diamond structure (packing of 0.34), which has a coordination number 4, and with the rhombohedral structure of the elements arsenic, antimony and bismuth which has a coordination number 3. The strongly directed bonds of the covalent structure of low coordination number are replaced by the more general binding of the electron gas in metallic structures, these exhibiting closer packing and higher coordination numbers.

The cohesive energies and interatomic spacings of some of the more important metallic elements are compared in Table 4.5 with those for a few covalently bonded solids. Of interest are the low cohesive energies of the alkali metals which are associated with a large atomic spacing and a small density of atoms. The ionic radii of these elements, derived from data on ionic compounds, are much less than the corresponding radii in the solid metals. This provides strong evidence that in the solid metal the electron cores do not overlap and that therefore the valence electrons lie primarily outside these core regions. However, for other metals, such as gold, the atomic and ionic radii are nearly equal and the structure is then well represented by a close packing of spherical ions.

The binding of the atoms in a solid metal is the result of a balance between a number of repulsive and attractive energy components. The magnitude and relative importance of the different components varies from one metal to another and as can be seen from Table 4.5 the cohesive energies also differ appreciably. First, there is a repulsive energy component due to mutual repulsion of the ion cores. In the alkali metals, the alkaline earth metals and in aluminium the electron cores are virtually unchanged in the solid and repulsion between them is purely electrostatic. In other metals there is some overlap of neighbouring electron cores giving Pauli exclusion repulsion, e.g. the 3d sub-shells of copper overlap.

Table 4.5 Cohesive energies and interatomic spacings of selected elements†

Element	**Cohesive Energy /eV atom^{-1}**	**Structure**	**Lattice Constants**		**Coordination number**	**Closest distance of approach /nm**
			***a*/nm**	***c*/nm**		
Li	1.4	b.c.c.	0.351		8	0.303
Na	1.0	b.c.c.	0.428		8	0.371
K	0.84	b.c.c.	0.533		8	0.462
Cu	3.3	f.c.c.	0.361		12	0.255
Ag	2.8	f.c.c.	0.409		12	0.289
Au	3.4	f.c.c.	0.408		12	0.288
Zn	1.3	h.c.p.	0.266	0.495	6,6	0.266
Cd	1.1	h.c.p.	0.298	0.562	6,6	0.298
Mg	1.5	h.c.p.	0.321	0.521	6,6	0.321
Fe	3.9	b.c.c.	0.287		8	0.248
Co	4.2	h.c.p.	0.251	0.409	6,6	0.251
Ni	4.1	f.c.c.	0.352		12	0.249
Al	3.1	f.c.c.	0.405		12	0.286
C	6.4	diamond	0.357		4	0.154
Si	3.6	diamond	0.543		4	0.235
Ge	3.3	diamond	0.566		4	0.245

† *The values refer to room temperature*

The other principal repulsive energy component is that arising from the translational kinetic energies of the valence electrons. As we shall see in the next chapter the valence electrons in a crystal can be represented on the free electron theory by a system of plane travelling waves having a range of values of propagation constant k, and the electrons on this model will possess kinetic energies ranging from zero up to a maximum value termed the Fermi energy E_F. The mean energy of a valence electron is $3/5E_F$ so that in the metal sodium for example the average energy is found to be 1.9 eV. Typical values of average energies for other metals lie in the range 1–7 eV. Essentially the value of E_F is governed by the number of free electrons being accommodated in unit volume of

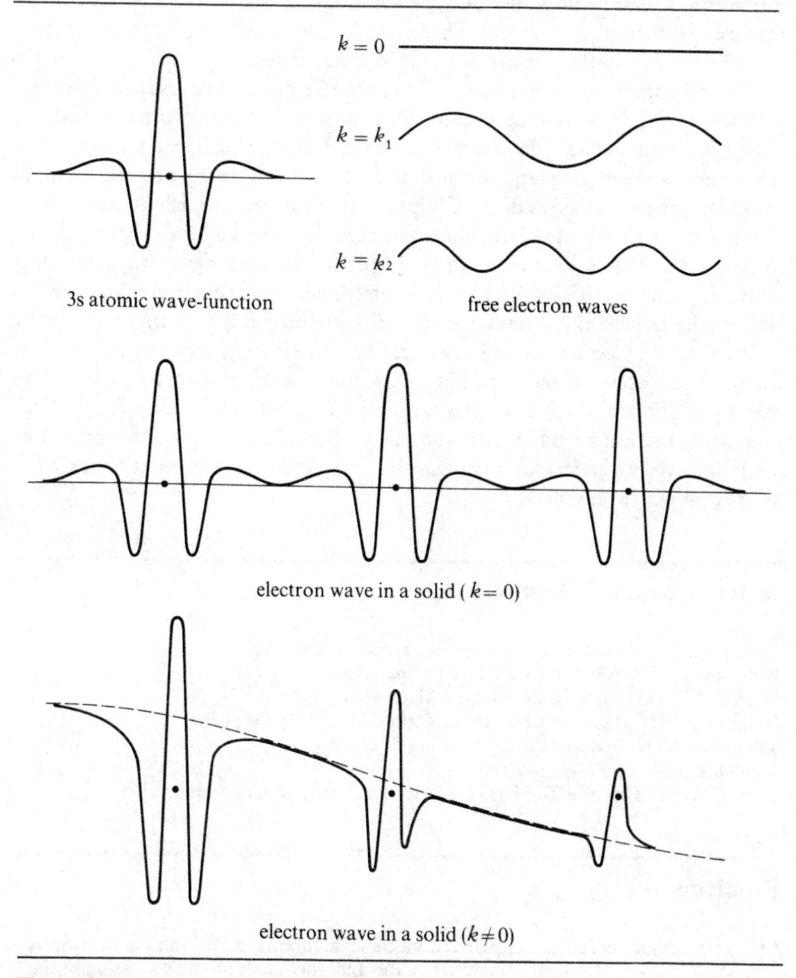

Fig. 4.17 Wave-functions for electrons in a solid.

the crystal and this is determined by the valency and the interatomic spacing. The repulsive energy arising out of this kinetic energy component will thus increase with decreasing atomic separation.

In order to see how the attractive energy components arise, the bonding in the metal sodium can be used as an example. As can be seen from Fig. 2.14 the charge cloud of the 3s electron in the free atom greatly exceeds that of the electron core which has a radius of only about 0.1 nm. The size of the free atom of sodium greatly exceeds that of a sodium ion. In the solid metal the spacing of the ions is such that a valence electron can never be further than 0.18 nm from a nucleus and its average distance from the nucleus is *decreased* as compared with the free atom situation. Consequently, there is an overall fall in the potential energy of the valence electron in the crystal. This provides an attractive energy component which increases as the separation of the ions decreases.

The other main attractive component in binding arises because of a difference in the form of the wave-functions of electrons in a crystal as compared with the isolated atomic orbitals. The former can be regarded as being of an intermediate character between the plane waves of the free electron theory and the localised atomic orbitals discussed in Chapter 2. Near each nucleus the waves approximate to the atomic orbitals whereas outside the cores they resemble plane waves. This is illustrated in Fig. 4.17. The kinetic energy at any point on a wave is given by $-(\hbar^2/2m_e)\nabla^2\psi$, i.e. it is determined by the curvature of the wave. If now the atomic and crystal orbitals are compared it is seen that outside the cores there is a reduction in curvature, i.e. the kinetic energy of the wave is reduced and this enhances binding. The total electron energy of the lowest energy orbital (for which $k = 0$), works out to be about 3 eV lower in energy than that of the free atom orbital in sodium. For other values of k the difference is less and the increased energy for $k \neq 0$ may be interpreted as the translational kinetic energy previously discussed.

References and further reading

Azároff, L. V. (1960) *Introduction to Solids*, McGraw-Hill.
Blakemore, J. S. (1969) *Solid State Physics*, Saunders.
Dekker, A. J. (1967) *Solid State Physics*, Macmillan.
Holden, A. (1971) *Bonds Between Atoms*, Oxford University Press.
Kittel, C. (1971) *Introduction to Solid State Physics*, Wiley.
Sproull, R. L. (1964) *Modern Physics*, Wiley.
Wert, C. A. and **Thomson, R. M.** (1970) *Physics of Solids*, McGraw-Hill.

Problems

4.1 In a sodium chloride ion pair the mean separation of a Na^+ and a Cl^- ion is 2.5×10^{-10} m. Compare the electrostatic and gravitational forces existing in the molecule and comment on the significance of the result.

4.2 Show that for a linear array of equally spaced ions of alternating sign the Madelung constant has a value of $2 \ln 2$.

4.3 If atoms are represented by rigid spheres show that they occupy 74 per cent of the available volume in a f.c.c. structure, 68 per cent in a b.c.c. structure and 34 per cent in a diamond structure.

4.4 Many ionic solids crystallise in the cesium chloride structure (Fig. 3.17). Assuming that cations and anions have radii r_1 and r_2 respectively show that the cube edge for the structure is given by

$$a = \frac{2}{\sqrt{3}}(r_1 + r_2).$$

Investigate this relationship for some of the alkali halides using values of ionic radii and for lattice parameters from tables.

4.5 The closest distance of approach of ions in the cesium chloride structure occurs along a cube diagonal. Show that contact between the ion of radius r_1 at the cube centre with the ions of radius r_2 at the cube corners will cease if $r_2/r_1 > 1.37$. This gives the limiting ionic radius ratio for the stability of this structure.

4.6 The repulsive energy component in equation (4.5) can be replaced by a term of the form $A \exp(-r/\rho)$ where the characteristic length ρ is a fraction of the equilibrium spacing r_0 and is a measure of the range of repulsive interaction. Following the procedure used for deriving equation (4.6) show that $\rho = r_0/n$. From the data given in the text determine the range of repulsive interaction in a sodium chloride crystal.

4.7 Using the expression for the lattice energy of an ionic crystal given by equation (4.5) determine the factor by which the equilibrium separation of the ions would be changed if the electronic charge were doubled. Determine also the factor by which the lattice energy is changed.

4.8 Show that provided the wave-functions ψ_A and ψ_B in equation (4.9) are normalised the normalisation constant N_S in the equation is given by

$$N_S = \frac{1}{(2+2S)^{\frac{1}{2}}}$$

where S is the value of the overlap integral $\int_V \psi_A \psi_B \, dV$. Deduce also the value of the normalisation constant N_{AS} in the equation and confirm the form of Fig. 4.11.

4.9 Equations (4.11) give the composition of the four sp^3 orbitals. Select any two of these orbitals and show by using the orthogonality condition as in equation (1.111) that the pair chosen are independent of each other. This is an extension of Problem 2.18 and use may be made of the conclusions drawn in that problem.

4.10 Show that any sp^3 orbital in equation (4.11) is normalised.

4.11 Show that the maximum value of the angular dependence function of the sp^3 orbital

$$\psi_1 = \tfrac{1}{2}(s + p_x + p_y + p_z)$$

is defined by $\phi = 45°$ and $\theta = \tan^{-1}\sqrt{2}$.

Confirm that the diamond lattice is consistent with tetrahedral bonds oriented at $2\theta = 2\tan^{-1}\sqrt{2}$.

4.12 Show that the maximum values of the angular dependences of s, p and sp^3 orbitals are in the ratio $1:\sqrt{3}:2$.

4.13 What is the wave-function for an sp^3 hybrid orbital which has its maximum in the z direction?

(*Hint:* Note that an sp^3 orbital must have a 3:1 weighting of p to s characteristic and that the squares of the appropriate coefficients give the relative proportions.)

4.14 Using the results in the previous question determine the form of the other sp^3 orbitals given that one of these lies in the yz plane.

(*Hint:* Consider the sign of the p_z components of the three orbitals, the symmetry of the equations and the requirement for the orbitals to be tetrahedrally oriented.)

4.15 An sp^2 orbital directed along the z-axis is represented by

$$\psi_1 = \frac{1}{\sqrt{3}}s + \sqrt{\frac{2}{3}}p_z.$$

Bearing in mind that an sp^2 orbital has a 2:1 weighting of p to s characteristic determine the form of the other two orbitals and confirm that they are normalised.

4.16 Show that the three sp^2 orbitals in the previous question are oriented at 120° to each other and therefore must lie in one plane. Select any pair of the three orbitals and confirm that they are orthogonal.

4.17 Show that the maximum value of the angular dependence function of an sp^2 orbital is 1.992 times greater than that of an s orbital.

5. The free electron theory of metals

5.1 The Drude–Lorentz theory

The concept of electrons in a metal being free to move about as if they constituted an electron gas was first put forward by Drude in 1900 soon after the discovery of the electron. The theory was later extended by Lorentz who applied classical kinetic theory of gases to the free electron gas in a metal. According to the theory, the free electrons would have the same type of distribution of speeds or energies as apertains to classical distinguishable particles, namely the Boltzmann distribution.

Although in the Drude theory, the behaviour of free electrons in a metal is in many ways analogous to that of gaseous molecules, there is one significant difference between the two cases. In the free electron theory, collisions between electrons are neglected and one is concerned only with collisions of the electrons with the metal ions. This concept, in a rather modified form, is retained in later developments of the theory. It should, however, be noted that the coulombic attractions between the electrons and the fixed ions are neglected. Drude also assumed that the mean free path of an electron between collisions is governed by the lattice spacing in the crystal and is independent of the speed of the electron.

5.2 Electrical conductivity and mobility of electrons

The electric intensity or field strength $\boldsymbol{E}$ is the force on a unit positive electric charge placed in that field. An alternative definition of field strength is that given by the relationship

$$E_x = -\frac{dU}{dx}, \qquad \ldots(5.1)$$

the potential U varying with position along the x-axis. The unit of field strength is thus the volt/metre.

The electrical conductivity σ of a solid is the rate at which charge is transported across unit area of a solid due to unit applied electric field. If the current/unit area or current density is J_x then the conductivity σ is given by

$$\sigma = \frac{J_x}{E_x}. \quad \ldots(5.2)$$

Before we can derive a value for the electrical conductivity of a metal we must first obtain a qualitative model of the effect of an electric field on the assembly of free electrons in a metal. Under thermal equilibrium conditions as many electrons are moving in any one direction as in the opposite, so that considering all the valence electrons in a metal sample their average velocity is zero. There is no net flow of charge in any direction; there is no electric current. When an

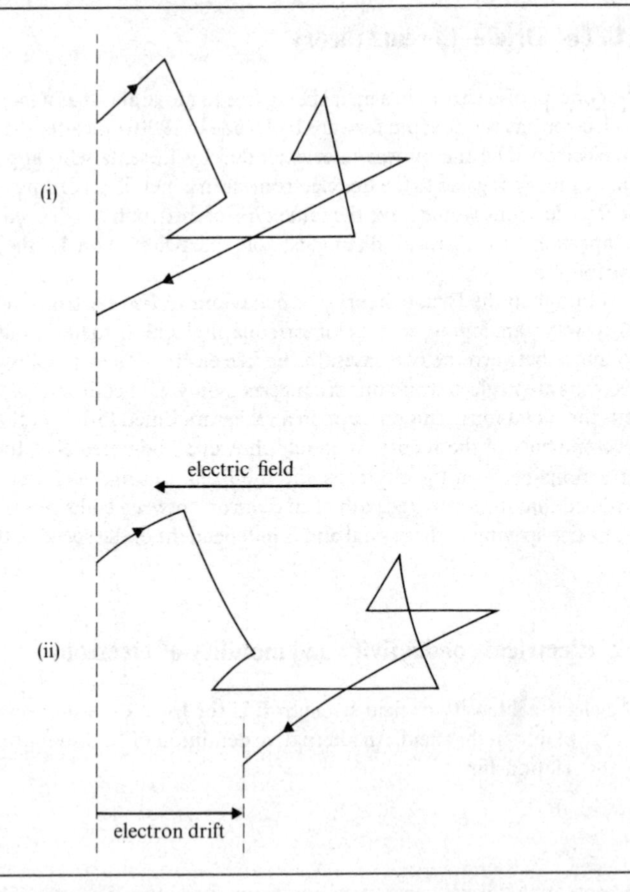

Fig. 5.1 (i) Random electron movements. For a metal at room temperature the mean electron velocity is ~ 10^6 m s^{-1}, and the mean free path is ~ 10 nm. (ii) Electron drift in applied electric field. Note assumption that similar collisions occur as in (i). The drift velocity is shown very much magnified.

electric field is applied across the sample the electrons will be accelerated during their free periods between collisions. This acceleration is in a direction directly opposite to that of the field since the charge on the electron is negative. Simultaneously collisions of the electrons with the ions will tend to restore the condition whereby all the electron velocities are random. In equilibrium the situation is equivalent to the valence electrons in the sample all possessing a common drift velocity due to the effect of the applied electric field (see Fig. 5.1). The electron drift constitutes the electric current. The drift velocity is superimposed on the random electron velocities and its value will be many orders of magnitude less than that of the random electron velocities. The situation is analogous to that obtaining in a gas flowing along a tube due to a pressure difference applied across its ends.

If we let v_d be the mean drift velocity of the electrons due to an applied field $\boldsymbol{E}$, then the rate at which charge is transported across unit area, i.e. the current density J, is given by

$$J = nev_d, \qquad \ldots(5.3)$$

where n is the number of electrons in unit volume of the metal and e is the magnitude of the electronic charge.

From (5.2) and (5.3) it follows that

$$\sigma = ne\frac{v_d}{E} \qquad \ldots(5.4)$$

or

$$\sigma = ne\mu, \qquad \ldots(5.5)$$

where μ is the electron drift velocity in unit field, termed the *mobility* of the electron. The concept of mobility is of importance and is retained in later theories of electrical conduction. Electron mobility values range rather widely, e.g. it is 3×10^{-3} m^2 V^{-1} s^{-1} for copper compared with 8 m^2 V^{-1} s^{-1} for the semiconductor indium antimonide.

5.3 Mean free time τ

We may now treat the electron collision process on a classical model and thereby obtain an expression for the drift velocity of the electrons. Although detailed knowledge of the type of collisions made by electrons is not required, it is necessary in the theory to make a number of assumptions regarding the collision processes. These assumptions are essentially the same as in the theory of gaseous collisions. Firstly, that the probability of an electron colliding in a particular time interval is independent of the time which has elapsed since a previous collision, and secondly that the probability of a collision in any time interval is directly proportional to the length of that time interval.

Let us consider a group of n_0 electrons at time $t = 0$, and let the probability of any single electron making a collision in unit time be $1/\tau$. The quantity τ is a constant for the system and will have the dimensions of time. If after a time t the number of electrons remaining uncollided is n then the change dn in n due to collisions in a further time dt is

$$dn = -\frac{n}{\tau}dt$$

which yields on integration

$$n = n_0 \exp(-t/\tau). \qquad ...(5.6)$$

The fraction uncollided at time t is n/n_0 which also gives the probability of an electron not making a collision in time t.

The mean free time of an electron is the total life of all the n_0 electrons divided by n_0 so that we have

$$\text{Mean free time of an electron} = \int_0^\infty \frac{t\,dn}{n_0}$$

$$= -\frac{1}{\tau}\int_0^\infty t \exp\left(-\frac{t}{\tau}\right) dt$$

$$= \tau.$$

Thus the quantity τ may be identified with the mean free time between collisions for an electron. The mean free path l for an electron may then be defined by the relationship

$$l = \tau\bar{v}, \qquad ...(5.7)$$

where $\bar{v}$ is the mean speed of the electrons.

We may now return to the concept of a drift velocity. During their free times the electrons will have an acceleration of eE/m_e in the field $\mathbf{E}$, where m_e is the mass of the electron. Since an electron observed at a random instant of time will, on the average, have been free for a time τ, the average drift velocity v_d achieved by the electrons in the field will be given by

$$v_d = \frac{eE\tau}{m_e}. \qquad ...(5.8)$$

As pointed out in the text by Shockley, although an electron observed at random time will on the average have been free for a time τ and will also remain free for an additional time τ this should not be taken to imply that the mean free time between collisions is 2τ rather than τ. When the mean free time τ is defined each electron free time is given equal weight. A sampling of free times taken at random instants, however, automatically favours long free times and yields a mean free time of 2τ. A sampling of all possible free times, i.e. all possible path lengths yields the correct value of τ.

Since the mobility is defined as the drift velocity in unit field we have from equation (5.8) that

$$\mu = \frac{e\tau}{m_e}. \qquad \text{...(5.9)}$$

By substituting numerical values in this equation an estimate of the mean free time between collisions may be obtained. For example, for copper $\mu = 3 \times 10^{-3}$ $m^2\ V^{-1}\ s^{-1}$ and a value for τ of 1.7×10^{-14} s is obtained. This should, however, only be regarded as an order-of-magnitude calculation. For example, as will be discussed in Chapter 7, the *effective mass* of an electron in a metal is not equal to the free electron mass. Furthermore, a more detailed treatment of the theory shows that the mean free time should not be regarded as a constant, and that it is more satisfactory to represent τ as a function of electron energy. τ is then termed the relaxation time.

5.4 Electrical and thermal properties of metals in terms of classical theory

The results of the last two sections have not been essentially altered as a result of applying quantum theory to the free electron gas in a metal. Before dealing with this aspect of the theory it is of interest to see what classical theory predicted for the properties of metals and in what respects it failed to agree with experiment.

5.4.1 *The electrical conductivity of metals*

An expression for the electrical conductivity of a metal on the classical free electron theory is obtained directly from (5.5) and (5.9) which yield

$$\sigma = \frac{ne^2\tau}{m_e} \qquad \text{...(5.10)}$$

or, using (5.7), the alternative form

$$\sigma = \frac{ne^2 l}{m_e \bar{v}}. \qquad \text{...(5.11)}$$

It is of interest to examine how the electrical conductivity of a metal as given by (5.11) will vary with temperature. First, the variation of the mean electron speed $\bar{v}$ with temperature can be obtained by making use of results from the kinetic theory of gases. We have, for an ideal gas, the pressure p related to the root mean square speed v_r of the molecules by the equation

$$p = \tfrac{1}{3}\rho v_r^2, \qquad \text{...(5.12)}$$

where ρ is the density of the gas. Putting k as Boltzmann's constant

$$pV = kT \qquad \text{...(5.13)}$$

referred to a single molecule. It follows that the mean energy of a gas molecule is given by

$$\tfrac{1}{2}mv_r^2 = \tfrac{3}{2}\,kT. \qquad \ldots(5.14)$$

Applying this relationship to the electron gas we have that the mean electron energy is $\frac{3}{2}kT$ and therefore that the mean electron speed is proportional to $(kT/m_e)^{\frac{1}{2}}$. The numerical value of the proportionality constant depends on the averaging procedure used to obtain the electron speed $\bar{v}$, and also on whether the electron gas is assumed to follow the Maxwell–Boltzmann velocity distribution apertaining to gaseous molecules. We therefore put

$$\sigma \propto \frac{ne^2 l}{(m_e kT)^{\frac{1}{2}}}. \qquad \ldots(5.15)$$

It is found experimentally that many metals exhibit, for a range of temperatures about room temperature, an electrical conductivity dependent upon $1/T$. At lower temperatures the temperature dependence is more complex and at a sufficiently low temperature the conductivity of a metal sample is determined by the concentration of impurity. The conductivity is then temperature independent. In the Drude theory the electrons are assumed to collide with the ions on the lattice so that their mean free path l could be expected to be proportional to the mean lattice spacing and therefore to increase with temperature. Expression (5.15) for the conductivity of a metal, derived on classical theory, does not therefore give the correct temperature dependence for the electrical conductivity of a metal.

5.4.2 *The heat capacity according to classical theory*

It is instructive when dealing with the theory of specific heats of solids to compare the behaviour of equivalent amounts of all solids. It is usual, therefore, to discuss the theory of specific heats in terms of a molar heat capacity C_V related to the specific heat c of the solid by the relationship

$$C_V = M_r c.$$

where M_r is the relative molecular mass of the solid. One is thus dealing with N_A atoms of each solid, where N_A is the Avogadro number.

In the previous section we saw that the mean energy of a molecule of an ideal gas is $\frac{3}{2}kT$. Since each molecule possesses three degrees of freedom, the average energy per degree of freedom is $\frac{1}{2}kT$. This is a particular case of a general theorem, termed the *Principle of equipartition of energy*. For a system of particles interacting with harmonic forces both the kinetic and the potential energies of the particles have to be taken into account. Each energy term will have an average value of $\frac{1}{2}kT$ so that the total mean energy for each particle is kT.

Let us now see how this principle is applied to the problem of determining the heat capacity of a solid. When solids are heated there is an increase in the vibrational energies of the atoms and the internal energy of the solid is thus increased. Assuming the atoms in a solid behave as harmonic oscillators the internal energy U_m of one mole of solid which contains N_A atoms is

$$U_m = 3N_A kT = 3RT, \quad \ldots(5.16)$$

where R is the gas constant for one mole.

The molar heat capacity C_V is the rate of change with temperature at constant volume of the internal energy of one mole of solid, i.e.

$$C_V = \left(\frac{\partial U_m}{\partial T}\right)_V = 3R. \quad \ldots(5.17)$$

In addition to this contribution to the molar heat capacity classical theory predicts for a metal a further component of heat capacity due to the free electron gas. The mean electron energy is equal to $\frac{3}{2}kT$, so that for the assembly of N_A electrons in one mole of a monovalent metal the total electron energy is $\frac{3}{2}N_A kT$. The corresponding heat capacity is $\frac{3}{2}R$ so that the total heat capacity of a metal should be $\frac{9}{2}R$ compared with $3R$ for an insulator.

These results are not in accord with experiment. Heat capacities of solids are always found to be functions of temperature, varying with T^3 at low temperatures and approaching a limiting value of $3R$ at a sufficiently high temperature which differs from one solid to another. The predicted additional electronic contribution of $\frac{3}{2}R$ to the heat capacity of a metal is not observed experimentally.

5.4.3 *The thermal conductivity of metals*

A well-known result from the kinetic theory of gases is that the thermal conductivity K of an ideal gas is given by

$$K = \frac{C\bar{v}l}{3}, \quad \ldots(5.18)$$

where $\bar{v}$ and l are, respectively, the mean speed and mean free path of the gas molecules. C is here the heat capacity referred to unit volume of gas.

The classical expression for the thermal conductivity of a metal is obtained by applying equation (5.18) to the electron gas in a metal. Since C refers to unit volume we have

$$C = \tfrac{3}{2}nk, \quad \ldots(5.19)$$

where n is the number of electrons per unit volume of the metal. Putting the mean electron speed $\bar{v}$ proportional to $(kT/m_e)^{\frac{1}{2}}$ and using (5.7) the thermal conductivity K of a metal is then obtained in the form

$$K \propto \frac{k^2}{m_e} n\tau T. \quad \ldots(5.20)$$

In addition using (5.10) we have that the ratio of thermal to electrical conductivity of a metal is given by

$$\frac{K}{\sigma} \propto \frac{k^2}{e^2}T. \qquad \ldots(5.21)$$

It is noted that neither n nor τ appears in this equation so that the ratio K/σ should be the same for all metals at the same temperature. The relationship is one previously put forward as an empirical expression and known as the Wiedemann and Franz Law. Although its derivation was an important success of the classical free electron model, the theory failed to yield the correct temperature dependence or the correct magnitudes of the electrical and thermal conductivities separately.

5.5 The quantum-mechanical free electron theory

We have up to now treated the free electrons in a metal in a manner analogous to that used in the kinetic theory of gases. On this basis we have treated electrons as classical distinguishable particles having velocities or energies governed by the Maxwell–Boltzmann distribution law. No restriction was placed on the number of electrons which could possess any particular value of energy. In accordance with the Principle of equipartition of energy the mean energy of an electron was put as $\frac{3}{2}kT$, so that at the absolute zero of temperature all the electrons would possess zero energy. The successes and limitations of the theory have already been discussed.

In the remainder of this chapter the application of quantum theory and wave mechanics to the assembly of free electrons in a metal will be discussed. The theory, first put forward by Sommerfeld in 1928, postulated that the valence electrons in a metal obey the Fermi–Dirac quantum statistics. In accordance with the theory the electrons obey the Pauli exclusion principle and, in contrast to classical theory, they must be regarded as indistinguishable particles.

The application of the Pauli expulsion principle to the electrons in isolated atoms was discussed in Chapter 2. It was then seen that an electron state in an atom is defined by four separate quantum numbers, and that no two electrons in an atom can be in the same state as defined by these quantum numbers. The Pauli exclusion principle when applied to the valence electrons in a crystal may be stated in the general form—no two electrons may have the same state of motion and position simultaneously. The state of motion of an electron is defined by its momentum and spin. We shall see in the succeeding sections how quantum numbers are introduced in the theory.

The method of approach is to determine first, using the principles of wave mechanics, the number of possible or allowed quantum states for electrons in crystals. The derivation of the number of states as a function of energy is of fundamental importance in the development of the theory of metals and semiconductors and will be treated fully in the sections immediately following. Following this, the mode of occupation of the quantum states by the valence electrons in a metal will be considered. Here the electrons must be considered as indistinguishable particles obeying the Pauli principle and governed by Fermi–Dirac quantum statistics. The electron density is high and the electron gas in a metal is termed degenerate. When, however, the electron gas density is low, as for example in a semiconductor, the mode of occupancy of the quantum states is essentially governed by the classical Maxwell–Boltzmann statistics. The Pauli restriction is not then of significance in determining the occupation of the states since with low electron density the probability of two electrons occupying the same state is very small.

5.6 Quantum states on the free electron model

There are two different methods which may be used to determine the number of allowed quantum states for the free electrons in a metal. One possible approach is to determine which *stationary electron waves* are possible in the crystal taking into account appropriate boundary conditions. This is analogous to the solution of the problem of determining the number of possible stationary waves on a string. Alternatively, we may deal with a system of *travelling waves* in the crystal, in which case we must determine which waves can be propagated without attenuation through the crystal. Both methods are instructive and will be given in the following sections.

5.6.1 *The stationary wave approach: one-dimensional model*

In the Sommerfeld model the electrons are assumed to move in a field free space within the metal, i.e. coulombic interactions between the electrons and the ions on the lattice are neglected. Electrons experience strong forces which tend to prevent them from leaving the metal, and this behaviour is represented on the model by a potential barrier at the surface.

The simplest model of this type is a one-dimensional model in which the electron is contained in a box $x = 0$ to $x = L$ by infinite potential barriers (see Fig. 5.2). In this model the potential energy of an electron is assumed to be the same everywhere within the box. It is mathematically convenient to set this potential energy as zero. This is permissible, since as we have already seen only *differences* in energy can be measured and have physical significance. We therefore put the potential energy V as

$$V = 0 \text{ for } 0 < x < L; \quad V = \infty \text{ for } 0 \geqslant x \geqslant L \qquad \ldots(5.22)$$

The potential energy binds the electron within the box and there is zero probability of the electron being found outside the box.

We now wish to determine which stationary electron waves are possible. As discussed in Chapter 1 the allowed electron wave-functions will be solutions of

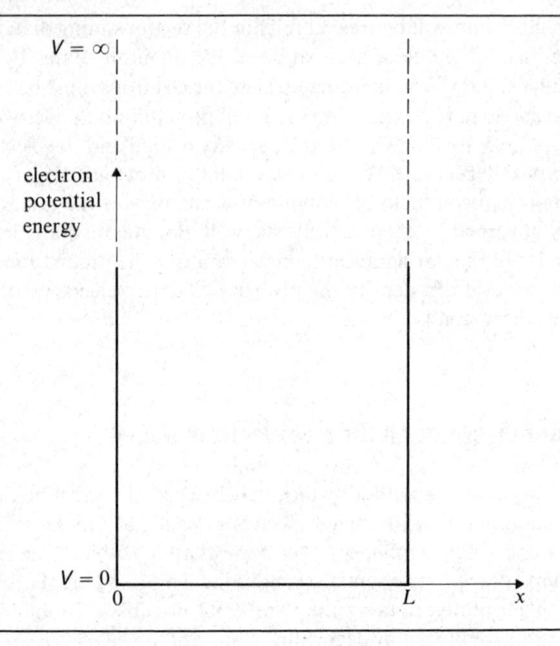

Fig. 5.2 Potential energy of electron in a square well having infinite sides—one-dimensional model.

the one-dimensional time-independent Schrödinger wave equation which represents stationary states, i.e.

$$\frac{d^2\psi}{dx^2}+\frac{2m_e}{\hbar^2}(E-V)\psi = 0. \qquad \ldots(1.104)$$

A general solution of this equation is

$$\psi = A\sin kx+B\cos kx, \qquad \ldots(5.23)$$

where k is a *quantum number* and A and B are constants whose values are to be determined.

We recall from Section 1.8.1 that for problems involving stationary states the value of the quantity ψ^2 at a point is a measure of the probability of finding the electron at that point. Since the potential barriers at $x = 0$ and $x = L$ are of infinite height, then ψ^2 must be zero outside the box. Thus, if in equation (5.23) we set $\psi = 0$ at $x = 0$ we find $B = 0$, and for $\psi = 0$ at $x = L$, the condition

$$kL = n\pi \qquad (n = 1, 2, 3, \ldots). \qquad \ldots(5.24)$$

It should be noted that the solution $n = 0$ would give $\psi = 0$ at all points. This is not a physically acceptable solution since by definition the electron must lie within the box.

The allowed wave-functions ψ_n are thus obtained in the form

$$\psi_n = A_n \sin\left(\frac{n\pi x}{L}\right) \qquad (n = 1, 2, 3, \ldots) \quad \ldots(5.25)$$

by substituting condition (5.24) in equation (5.23). Equation (5.25) illustrates how the application of the boundary conditions has led to a restriction in the possible electron wave-functions. n is a *quantum number*. The constants A_n can be determined from the normalisation condition (see Section 1.8.1) that the probability of finding the electron within the box is 1, i.e.

$$\int_0^L |\psi_n|^2 \,dx = \int_0^L A_n^2 \sin^2\left(\frac{n\pi x}{L}\right) dx = 1. \qquad \ldots(5.26)$$

It is readily verifiable that on integration this gives the values

$$A_1 = A_2 = \ldots = A_n = \left(\frac{2}{L}\right)^{\frac{1}{2}}. \qquad \ldots(5.27)$$

The final solution for the allowed wave-functions is therefore

$$\psi_n = \left(\frac{2}{L}\right)^{\frac{1}{2}} \sin kx = \left(\frac{2}{L}\right)^{\frac{1}{2}} \sin \frac{n\pi x}{L}. \qquad \ldots(5.28)$$

Each wave-function defined by a particular integral value of n will have its own specific value of total energy E, which may be determined by substituting for ψ_n in the Schrödinger wave equation (1.104). This gives

$$E = \frac{\hbar^2 k^2}{2m_e} = \frac{\pi^2}{2} \frac{n^2 \hbar^2}{m_e L^2}. \qquad \ldots(5.29)$$

This gives, in accordance with equation (1.78), the kinetic energy of a free electron, and it indicates that the quantum number k can be equated with the propagation constant k. We note also for future reference that, by combining equations (5.28) and (5.29), the wave-function is given in terms of energy E as

$$\psi_n = \left(\frac{2}{L}\right)^{\frac{1}{2}} \sin\left(\frac{(2m_e E)^{\frac{1}{2}}}{\hbar} x\right). \qquad \ldots(5.30)$$

If the length L of the box is made large compared with the atomic spacing in a crystal the allowed values of electron energy given by equation (5.29) all become very closely spaced and the variation of E with k is quasi-continuous. Each integral value of n defines a possible value of the propagation constant k and a corresponding electron wavelength λ given by

$$\lambda = \frac{2\pi}{k} = \frac{2L}{n}. \qquad \ldots(5.31)$$

The electron momentum p is given by

$$p = \hbar k = \frac{n\pi\hbar}{L}. \qquad \ldots(5.32)$$

To summarise, one notes that the application of the Pauli exclusion principle to the assembly of free electrons in a metal has led to the quantisation of electron energy and momentum. The picture therefore is that of a large number of possible standing waves, analogous to standing waves on a string, each particular wave-function (or eigenfunction) having its own specific energy (or eigenvalue). Sketches of the allowed wave-functions together with a representation of their energies are given in Figs. 5.3 and 5.4.

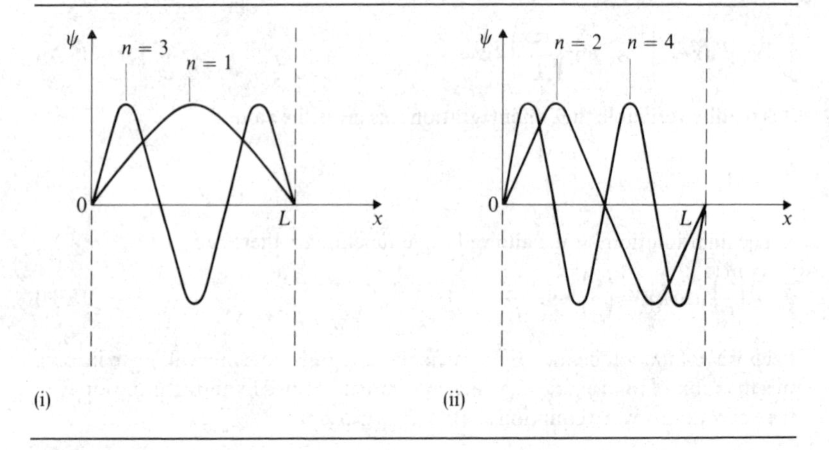

Fig. 5.3 Allowed wave-functions for the one dimensional model: (i) the symmetric wave-functions for $n = 1$ and $n = 3$; (ii) the antisymmetric wave-functions for $n = 2$ and $n = 4$.

We note that wave-functions corresponding to odd values of n have an odd number of half wavelengths fitting into the box and are therefore symmetrical about the centre of the box. They are therefore symmetric wave-functions and those corresponding to n even are antisymmetric.

Let us briefly consider the implications of the energy level spacings in Fig. 5.4. If we take a box of atomic dimensions it may be regarded as a convenient model for an isolated atom and the energy levels those of electrons in an atom. The energy levels derived from the model are clearly not in agreement with those obtained by the use of a coulombic potential, as for example, in the Bohr model, and which are verified experimentally. The model therefore fails to provide the correct sequence of energy levels for an atom although, as may be readily verified, the levels do lie within the correct range of energy. Thus for a box of length 0.3 nm the transition for $n = 2$ to $n = 1$ gives an energy = 12.5 eV, which would correspond to an optical spectral line at 100 nm.

If, on the other hand, the model is used to represent a solid metal, e.g. putting

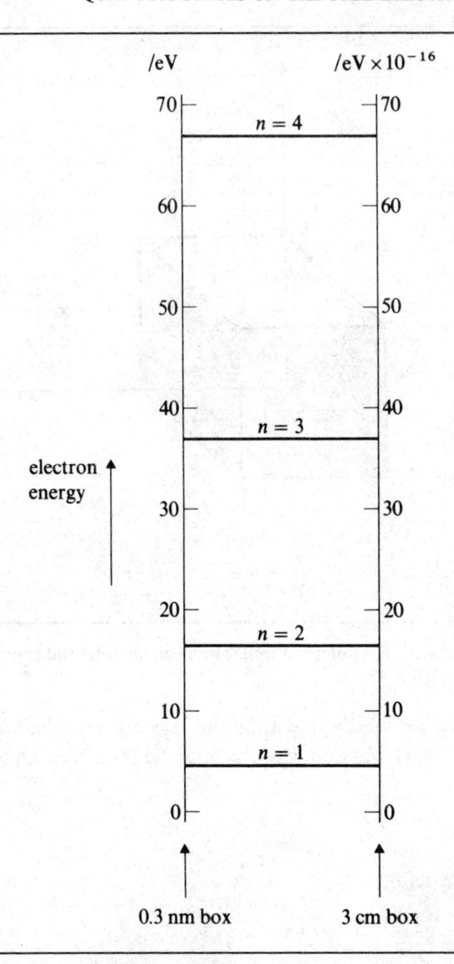

Fig. 5.4 Energy levels for electrons in a one-dimensional square potential well having infinite sides.

$L = 3$ cm, the levels are then much more closely spaced and the quasi-continuous spectrum obtained approximates to the continuously variable energy predicted on a classical model of a solid. The quantisation of energy does, however, lead to important differences between the quantum and classical models and these will be discussed later.

5.6.2 *The three-dimensional model*

A closer approximation to the representation of the behaviour of free electrons in a metal than the model given is that of an electron in a three-dimensional box. As before, we assume infinite potential barriers at the faces of the box and put the

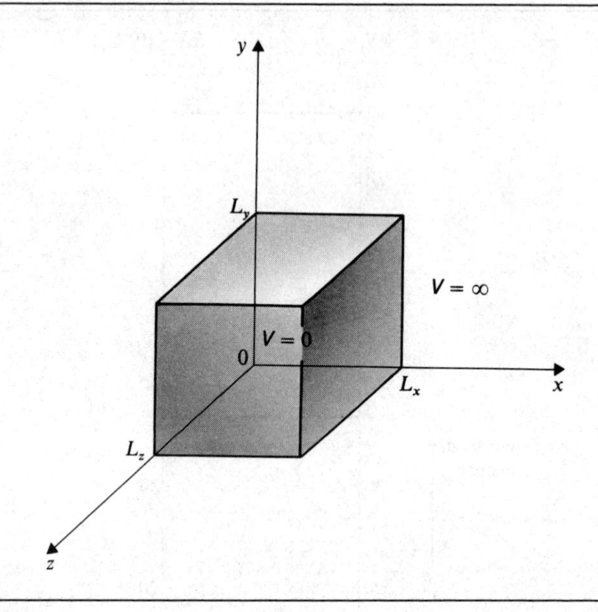

Fig. 5.5 The three-dimensional potential well. The electron potential energy is zero within the box and infinite outside.

potential energy of an electron within the box as zero (see Fig. 5.5). The appropriate Schrödinger wave equation is now the three dimensional form

$$\nabla^2\psi + \frac{2m_e}{\hbar^2}E\psi = 0 \qquad \ldots(1.105)$$

and the box may be defined by

$$0 \leqslant x \leqslant L_x; 0 \leqslant y \leqslant L_y; 0 \leqslant z \leqslant L_z. \qquad \ldots(5.33)$$

Putting in the appropriate boundary conditions and using the normalisation condition that the electron lies within the box it may be verified that the appropriate wave-functions are given by

$$\psi_{xyz} = \left(\frac{8}{V}\right)^{\frac{1}{2}} \sin\left(\frac{n_x\pi}{L_x}x\right) \sin\left(\frac{n_y\pi}{L_y}y\right) \sin\left(\frac{n_z\pi}{L_z}z\right), \qquad \ldots(5.34)$$

where n_x, n_y and n_z are quantum numbers and where $V = L_xL_yL_z$ is the volume of the box.

As for the one-dimensional case, a zero set of values of (n_x, n_y, n_z) is not acceptable since this leads to $\psi_{xyz} = 0$ everywhere. Furthermore, since we are essentially interested in determining the total number of different wave-functions, one may restrict (n_x, n_y, n_z) to positive values. Negative values of the quantum numbers do not yield any additional different wave-functions.

By differentiating (5.34) and substituting in the Schrödinger wave equation, the energy E corresponding to each wave-function is obtained as

$$E = \frac{\pi^2 \hbar^2}{2\, m_e}\left[\left(\frac{n_x}{L_x}\right)^2 + \left(\frac{n_y}{L_y}\right)^2 + \left(\frac{n_z}{L_z}\right)^2\right]. \qquad \ldots(5.35)$$

We note that each set of values of the integers defining a wave-function give different values of the energy. If, however, the symmetry of the system is increased by making the box a cube of side L, the energy E will be given by

$$E = \frac{\pi^2}{2} \frac{\hbar^2}{m_e L^2}(n_x^2 + n_y^2 + n_z^2) \qquad \ldots(5.36)$$

and each value of E can now be obtained from three different sets of values of (n_x, n_y, n_z), e.g. (3, 2, 2), (2, 3, 2) and (2, 2, 3). The three wave-functions have a common energy and the energy level is termed to be threefold degenerate.

5.6.3 *The density of states*

The results derived in (5.6.2) may now be used to determine the number of electrons which may be accommodated in the box for any particular range of energy values. We must, however, bear in mind that for each wave-function there will be two electron states, one for each electron spin. The general picture is as follows. At small values of energy, i.e. small (n_x, n_y, n_z), for a given incremental energy range there are many fewer possible wave-functions than at larger energies. Since only positive integral values of the triplet (n_x, n_y, n_z) need be considered, the total number of wave-functions for the cubic box for all energies up to a specific value E will be given by the volume of the octant of a sphere of radius r, where

$$r^2 = n_x^2 + n_y^2 + n_z^2. \qquad \ldots(5.37)$$

We note that this will be strictly accurate only for fairly large values of the quantum triplet. Taking into account that each wave-function gives rise to two electron states the total number of states N_s for energies up to E is given by

$$N_s = \tfrac{1}{8}\tfrac{4}{3}\pi(n_x^2 + n_y^2 + n_z^2)^{\frac{3}{2}} \times 2. \qquad \ldots(5.38)$$

Using equation (5.36) and putting $V = L^3$ this becomes

$$N_s = \frac{1}{3\pi^2}\left(\frac{2m_e}{\hbar^2}\right)^{\frac{3}{2}} V E^{\frac{3}{2}}. \qquad \ldots(5.39)$$

This is the number of states for energies up to a maximum value E. The number of electron states dN_s for the energy range E to $E + dE$ may be denoted by $S(E)\,dE$ and is obtained by differentiating equation (5.39).

This gives

$$dN_s = S(E)\,dE = \frac{1}{2\pi^2}\left(\frac{2m_e}{\hbar^2}\right)^{\frac{3}{2}} V E^{\frac{1}{2}}\,dE. \qquad \ldots(5.40)$$

The function $S(E)$ given by

$$S(E) = \frac{dN_s}{dE} = \frac{1}{2\pi^2}\left(\frac{2m_e}{\hbar^2}\right)^{\frac{3}{2}} VE^{\frac{1}{2}} \qquad \ldots(5.41)$$

is termed the density of states and is shown as a function of energy in Fig. 5.6.

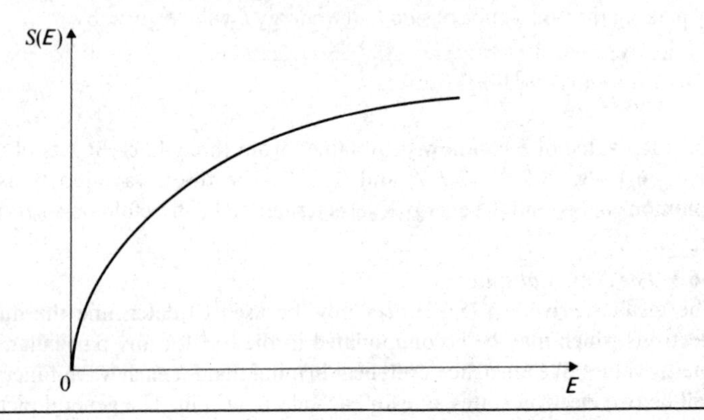

Fig. 5.6 The parabolic density of states function $S(E)$, $(=dN_s dE)$, of the free electron theory.

Before discussing the implications of this important relationship it is instructive to consider its derivation by an alternative method.

5.7 The travelling wave method

In many problems involving electron waves it is convenient to deal with travelling waves instead of stationary waves. In addition to the present use of the method we shall come across further examples of its use in Chapters 6 and 7. The boundary conditions of the problem are now introduced in a different manner, termed the method of periodic boundary conditions. In the model an infinite crystal is assumed, which is subdivided into a large number of cubic cells of side L and volume V. If the wave-function repeats itself exactly whenever x, y or z is increased by L, e.g.

$$\psi(x, y, z) = \psi(x+L, y, z)$$

or $\qquad \ldots(5.42)$

$$\psi(x, y, z) = \psi(x+L, y+L, z+L)$$

then the electron wave is propagated unchanged through the crystal.

In Chapter 1 we saw that a wave travelling in three dimensions can be represented by

$$\Psi = A\exp[ik(lx+my+nz)-i\omega t]. \qquad ...(1.27)$$

A wave-function of this form satisfying condition (5.42) is

$$\Psi = A\exp\left[i\frac{2\pi}{L}(n_x x+n_y y+n_z z)-i\omega t\right], \qquad ...(5.43)$$

where the quantum numbers n_x, n_y, and n_z can take on integral positive, integral negative and zero values. Putting

$$k_x = \frac{2\pi}{L}n_x;\ k_y = \frac{2\pi}{L}n_y;\ k_z = \frac{2\pi}{L}n_z, \qquad ...(5.44)$$

the wave-function can be written in the form

$$\Psi = A\exp[i(k_x x+k_y y + k_z z)-i\omega t]. \qquad ...(5.45)$$

Since we are interested in solutions which are independent of time we may omit the time dependence from this equation and look for solutions of the stationary wave equation (1.105). It is also convenient to write the wave-function in vector form

$$\psi = A\exp(i\boldsymbol{k}\cdot\boldsymbol{r}), \qquad ...(5.46)$$

where

$$|\boldsymbol{k}| = k_x^2+k_y^2+k_z^2)^{\frac{1}{2}} \qquad ...(5.47)$$

is the magnitude of the wave-fector $\boldsymbol{k}$. This wave-vector therefore defines the wave-function in both magnitude and direction. Proceeding as for the stationary wave model we find the energy of the wave given by the alternative forms

$$E = \frac{2\pi^2\hbar^2}{m_e L^2}(n_x^2+n_y^2+n_z^2) = 2\pi^2\frac{n^2\hbar^2}{m_e L^2}, \qquad ...(5.48)$$

where n^2 values are positive integral. Alternatively,

$$E = \frac{\hbar^2}{2m_e}(k_x^2+k_y^2+k_z^2) = \frac{\hbar^2 k^2}{2m_e}. \qquad ...(5.49)$$

This relationship is illustrated in Fig. 5.7.

It should be noted that any particular value of electron energy given by equation (5.48) may be obtained from a number of different combinations of the integral numbers (n_x, n_y, n_z). We may take as a specific example the energy level corresponding to $n^2 = 14$. This value of n^2 is given by the (n_x, n_y, n_z) values of $(\pm 1, \pm 2, \pm 3)$ which will give 8 degenerate energy levels. Since in addition there are 6 possible permutations of n_x, n_y, and n_z the total number is increased to 48. Finally, for each triplet defining a wave-function we may have two quantum states corresponding to the two possible electron spins, so that a total of 96

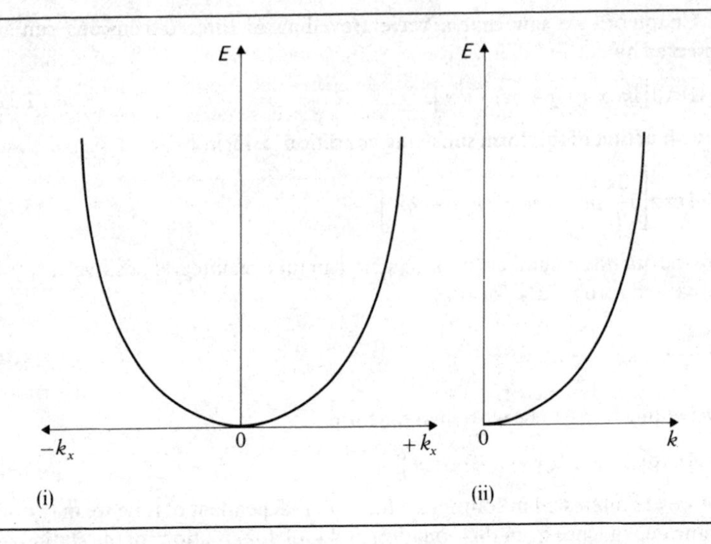

Fig. 5.7 (i) Variation of electron energy E with propagation constant k for waves travelling along the positive and negative direction of the k_x-axis. (ii) Variation of electron energy with magnitude k of wave-vector $\boldsymbol{k}$.

electrons in the crystal may have the value of energy corresponding to $n^2 = 14$. The physical interpretation is that a number of electron waves travelling in different crystal directions can have the same value of energy.

The density of states may now be determined in a similar manner to that used in the stationary wave model. Since electron waves may travel in any crystal direction both positive and negative values of n_x, n_y, and n_z are allowed. The number of possible wave-functions for energies up to E is therefore given by the assembly of values (n_x, n_y, n_z) lying within the whole sphere of radius r defined by equation (5.37). The number of quantum states N_s will be double this number and will therefore be given by

$$N_s = \tfrac{4}{3}\pi(n_x^2 + n_y^2 + n_z^2)^{\frac{3}{2}} \times 2. \qquad \ldots(5.50)$$

Using equation (5.48) and putting $L^3 = V$ we obtain

$$N_s = \frac{1}{3\pi^2}\left(\frac{2m_e}{\hbar^2}\right)^{\frac{3}{2}} V E^{\frac{3}{2}}, \qquad \ldots(5.39)$$

which is in agreement with the result obtained using the stationary wave model. It follows that the travelling wave model will give the same density of states $S(E)$ as obtained previously, i.e.

$$S(E) = \frac{\mathrm{d}N_s}{\mathrm{d}E} = \frac{1}{2\pi^2}\left(\frac{2m_e}{\hbar^2}\right)^{\frac{3}{2}} V E^{\frac{1}{2}}. \qquad \ldots(5.41)$$

It is of interest to examine the procedure adopted in deriving these equations. Each set of values (n_x, n_y, n_z) gives a corresponding set (k_x, k_y, k_z) defining the wave-vector $\boldsymbol{k}$ of the travelling wave. In a $\boldsymbol{k}$-space representation each point (k_x, k_y, k_z) occupies a volume $(2\pi/L)^3$ and can accommodate two electrons (see Fig. 5.8). Hence, for a volume V of crystal, a unit volume of $\boldsymbol{k}$-space will

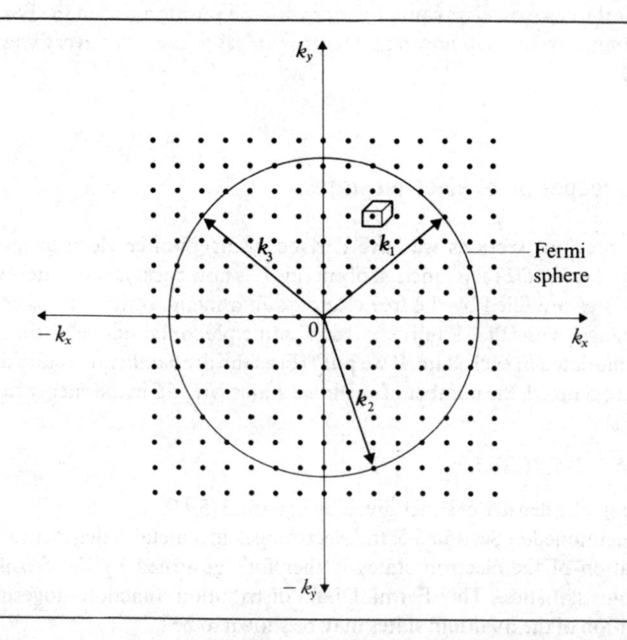

Fig. 5.8 A section of $\boldsymbol{k}$-space showing the cells of volume $(2\pi/L)^3$. Cells of equal energy are equidistant from the origin. Cells within the Fermi sphere are all occupied at $T = 0$ K. Wave-vectors such as $\boldsymbol{k}_1$, $\boldsymbol{k}_2$ and $\boldsymbol{k}_3$ lying on the Fermi sphere have equal magnitude k_F.

accommodate $V/4\pi^3$ electrons. The available states up to a given energy E will be contained within a spherical volume in $\boldsymbol{k}$-space and will be given by

$$N_s = \frac{V}{4\pi^3} \cdot \tfrac{4}{3}\pi k^3 = \frac{k^3}{3\pi^2} V. \quad \text{...(5.51)}$$

This condition is readily shown to be equivalent to that given in equation (5.39) and it yields a density of states in $\boldsymbol{k}$-space of

$$\frac{dN_s}{dk} = \frac{k^2 V}{\pi^2}. \quad \text{...(5.52)}$$

An alternative approach is to represent the momenta components of a wave-function on a momentum diagram. Application of the Heisenberg uncertainty principle then shows that the uncertainty in momentum corresponds to a cell of volume h^3/L^3 in momentum space. The results already derived then follow.

Finally, one may note one significant difference between the stationary wave and the travelling wave models of the free electron. The boundary conditions assumed in the stationary wave model leads to the result that the probability of finding the electron at any point varies with the position within the box. On the travelling wave model, however, the probability is constant everywhere in the crystal.

5.8 Occupation of electron states

In the previous sections we have derived, using the free electron model, the density of states $S(E)$ as a function of energy. It is now necessary to determine how these states are filled by the free electrons in a metal, bearing in mind that in accordance with the Pauli exclusion principle only one electron can be accommodated in each state. If we put $f(E)$ as the probability of a state of energy E being occupied, the number of occupied states $N(E)\,dE$ in the energy range E to $E+dE$ is

$$N(E)\,dE = S(E)f(E)\,dE, \qquad \ldots(5.53)$$

$S(E)$ being the density of states given by equation (5.41).

As mentioned in Section 5.5, the electron gas in a metal is degenerate and the occupation of the electron states is therefore governed by the Fermi–Dirac quantum statistics. The Fermi–Dirac distribution function governing the occupation of the quantum states may be shown to be

$$f(E) = \frac{1}{\exp[(E-E_F)/kT]+1}. \qquad \ldots(5.54)$$

The quantity E_F, which has the dimensions of energy, is termed the Fermi level or Fermi energy and its value is determined by the number of particles in the system considered. The concept of Fermi level is of great interest in the theory of metals and semiconductors and its properties will be developed in this section and are summarised in the next section.

Substituting for $S(E)$ and $f(E)$ in (5.53) we have

$$N(E)\,dE = \frac{1}{2\pi^2}\left(\frac{2m_e}{\hbar^2}\right)^{\frac{3}{2}} V \frac{E^{\frac{1}{2}}}{\exp[(E-E_F)/kT]+1}\,dE \qquad \ldots(5.55)$$

and since the total number of occupied states is equal to the total number of free electrons in the whole crystal then

$$n = \frac{1}{2\pi^2}\left(\frac{2m_e}{\hbar^2}\right)^{\frac{3}{2}} \int_0^\infty \frac{E^{\frac{1}{2}}}{\exp[(E-E_F)/kT]+1}\,dE, \qquad \ldots(5.56)$$

where n is the number of free electrons per unit volume of crystal. The analytical evaluation of this integral is difficult except in a few special cases which will be discussed presently.

5.8.1 *Occupation of the states at absolute zero*

By substituting $T = 0$ K in equation (5.54) it is readily verified that the distribution function then has the following values

$$f(E) = 0 \text{ for } E > E_F; \; f(E) = 1 \text{ for } E < E_F \qquad \ldots(5.57)$$

i.e. for a metal at 0 K all the states up to the Fermi level E_F are filled and all those above E_F are empty (see Fig. 5.9). Since the position of the Fermi level in a metal is dependent on temperature (although only to a slight extent), to be accurate we should denote the value of the Fermi level at absolute zero by $E_F(0)$. The integral in equation (5.56) may now be evaluated by substituting $f(E) = 1$ and changing the upper limit of the integral to $E_F(0)$. This yields

$$E_F(0) = (3\pi^2 n)^{\frac{2}{3}} \frac{h^2}{2m_e}. \qquad \ldots(5.58)$$

The average electron energy at 0 K will be given by

$$E_{AV}(0) = \frac{\int_0^{E_F} E\, N(E)\, dE}{\int_0^{E_F} N(E)\, dE} = \frac{\int_0^{E_F} E^{\frac{3}{2}}\, dE}{\int_0^{E_F} E^{\frac{1}{2}}\, dE} = \tfrac{3}{5} E_F(0). \qquad \ldots(5.59)$$

The value of $E_F(0)$ for a typical metal such as sodium may be readily calculated from equation (5.58) provided it is assumed that each atom of the metal contributes one free electron. Calculated values for the position of the Fermi energy for a number of metals are given in Table 5.1 from which it may be noted that the values are typically of a few electron-volts. Thus, in contrast to the classical Maxwell–Boltzmann distribution on which all the electrons would have zero energy at absolute zero, the application of the Pauli exclusion principle has been such as to compel the electrons in a metal to have energies of a few electron volts.

The occupation of the electron states may be looked at in a slightly different manner. In k-space all states up to a maximum value of k equal to k_F are filled whilst all states for $k > k_F$ are empty. The surface in k-space separating the filled and unfilled states is termed the Fermi surface (see Fig. 5.8). In accordance with

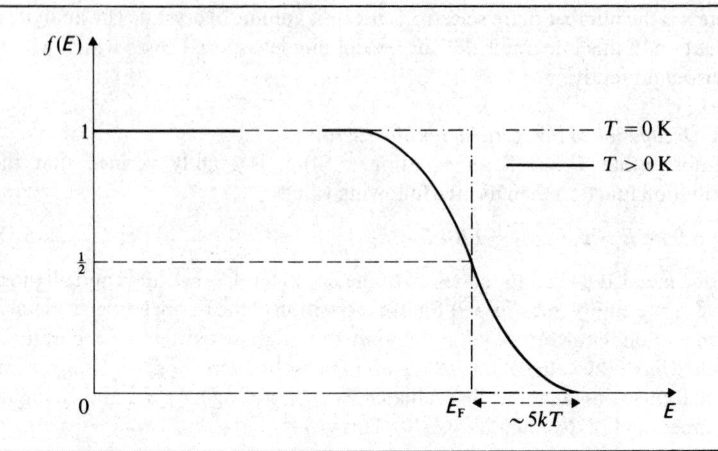

Fig. 5.9 The Fermi–Dirac distribution function for $T = 0$ K and $T > 0$ K.

equation (5.51) the number of electrons within this surface is $k_F^3 V/3\pi^2$ so that putting n as the number of electrons per unit volume the radius of the Fermi sphere is given by

$$k_F = (3\pi^2 n)^{\frac{1}{3}}. \qquad \ldots(5.60)$$

The corresponding value of the Fermi energy is

$$E_F = \frac{\hbar^2 k_F^2}{2m_e}. \qquad \ldots(5.61)$$

We should bear in mind that E in the Schrödinger wave equation denotes the total electron energy. Since the potential energy of the electron was set as zero in the preceding theory the quantity E representing the electron energy, e.g. in equation (5.49), may be equated to the translational kinetic energy of the electron. We therefore obtain, for example, that the velocity u_F of an electron at the Fermi level is given by

$$\tfrac{1}{2} m_e u_F^2 = E_F. \qquad \ldots(5.62)$$

The value of u_F for copper is 1.5×10^6 m s^{-1}.

5.8.2 *Occupation of the states at ordinary temperatures*

We may next consider how the free electrons in a metal are filling the available quantum states at a temperature well above absolute zero, e.g. at room temperature. Consideration of the Fermi–Dirac distribution function (5.54) makes it clear that, except for states having energies within a few kT's of E_F, the probability of occupation of the states is essentially unaltered as compared with the situation at the absolute zero of temperature. Since the value of kT at room temperature is only about 0.025 eV, compared with, say, $E_F = 5$ eV for a typical

Table 5.1 Properties of metals[†]

Metal	Relative atomic mass	Density $10^{-3}\ \rho$/kg m^{-3}	Fermi energy E_F/eV	Work function Φ/eV	Electrical conductivity $10^{-7}\ \sigma/\Omega^{-1}$ m^{-1}	Thermal conductivity K/W m^{-1} K^{-1}
Li	6.94	0.53	4.7	2.5	1.1	71
Na	23.0	0.97	3.1	2.3	2.1	138
K	39.1	0.86	2.0	2.2	1.4	97
Cs	132.9	1.87	1.5	1.8	0.50	—
Cu	63.5	8.9	7.0	4.5	5.9	385
Ag	107.9	10.5	5.5	4.5	6.2	418
Au	197.0	19.3	5.5	4.9	4.6	310
Mg	24.3	1.74	7.1	3.7	2.3	154
Ca	40.1	1.54	4.7	3.2	2.8	—
Ba	137.3	3.5	3.7	2.5	0.26	—
Zn	65.4	7.1	9.4	4.3	1.7	113
Cd	112.4	8.7	7.5	4.1	1.4	96
Al	27.0	2.7	11.6	4.2	3.7	242
Fe	55.8	7.9	—	4.4	1.1	71
W	183.9	19.3	—	4.5	1.9	185

[†] *Density and electrical and thermal conductivity values are for room temperature. The Fermi energy values are calculated from equation (5.58) taking into account the valency of each metal.*

metal, it is apparent that only a comparatively small part of the total electron population, namely that near the top of the distribution, is affected. The form of the distribution function at room temperature is shown in Fig. 5.9, and that of the occupation of quantum states in Fig. 5.10. As a result of the increase in temperature some electrons have been raised from energy levels just below E_F to

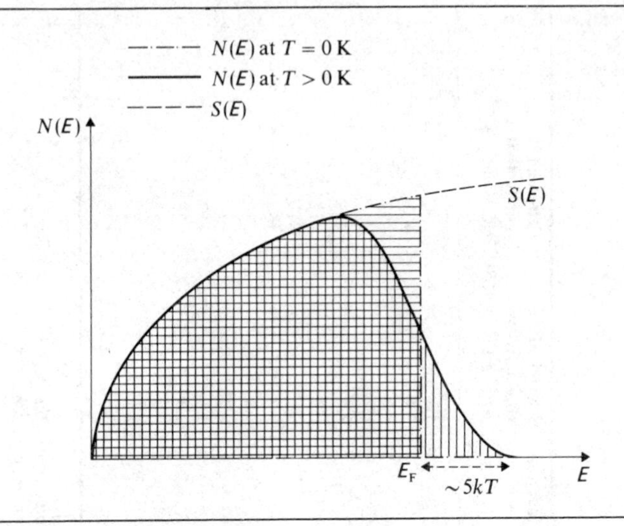

Fig. 5.10 Occupation of electron states at $T = 0$ K and $T > 0$ K.

levels slightly above E_F. This has, of course, yielded unoccupied states for the corresponding region below the Fermi level. Note that the changes in the form of the distribution are shown much magnified near the Fermi level on both diagrams. The mean electron energy will not be appreciably changed and will have a value of a few electron-volts. This may be compared with the result on classical theory that the mean electron energy will be given by $\frac{3}{2}kT$ (see equation (5.14)) which is only 0.037 eV at 300 K. The Fermi–Dirac and Maxwell–Boltzmann distributions are compared in Fig. 5.11. In conclusion it may also be pointed out that the description given here for the mode of occupancy of the

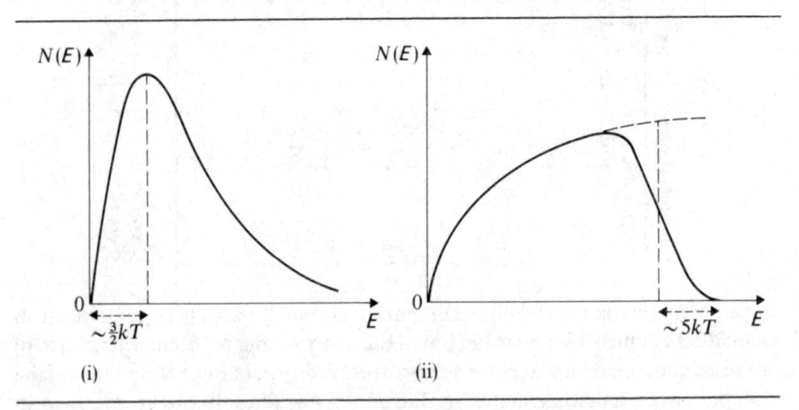

Fig. 5.11 Comparison of (i) classical and (ii) Fermi–Dirac distributions.

states is applicable for $E_F \gg kT$. This condition will always be valid for solid metals and for which the free electron model is applied.

5.9 The Fermi level

It is useful here to collect together and summarise the properties which are associated with the concept of the Fermi level. Although the concept has been developed in the context of the free electron theory of metals it is also extensively used in semiconductor theory (see Chapter 8).

We have seen that at absolute zero the Fermi level is that energy level which in a metal separates the filled and unfilled states. It is thus governed by the number of electrons to be accommodated. At any temperature above absolute zero we find from the Fermi–Dirac distribution function (5.54) that for $E = E_F$, the value of $f(E) = \frac{1}{2}$, i.e. the Fermi level is that energy level which has a 50 per cent probability of being occupied.

The Fermi level may also be considered from a different viewpoint, namely that of contact potential difference between solids. If two dissimilar solids are placed in contact with each other then, provided that they are in thermal equilibrium, the Fermi level for both solids must lie at the same value of energy. That this must be so follows from the fact that on joining the two solids electrons will flow from the states of higher energy in one to the unfilled lower-energy states in the other. This electron flow would continue until electron levels at the same value of energy in both solids have the same probability of occupation. The Fermi levels must therefore be equalised and a contact difference in potential is established between the two solids. This principle is of importance in the understanding of the behaviour of p-n junctions.

The discussion of the Fermi level up to this point has been quite general and it applies to both metals and semiconductors. We may now return to the free electron theory, which is a good approximation to the behaviour of some metals, but which is not applicable to semiconductors except in the special case when their conductive properties become metallic.

The primary effect of a change of temperature on the free electrons is, as has been discussed, that of a redistribution of the electrons between states which are within $\sim kT$ of the Fermi level. However, the position of the Fermi level, defined as that level which has a 50 per cent probability of being occupied, will itself be dependent on temperature. The change in its position with temperature is, however, small and the reason for the temperature dependence is as follows. The form of the distribution function $f(E)$ is symmetrical about E_F at all temperatures. The density of states function $S(E)$, however, increases with energy, and as the temperature rises an increasing proportion of the electron population will be occupying states of higher energy. Therefore, since the total number of electrons remains constant, i.e. the area under the curve in Fig. 5.10 remains constant, the level for which $f(E) = \frac{1}{2}$ must move to a lower energy with increasing temperature. It would only remain constant if the density of states $S(E)$ was constant for all values of E.

The exact position of the Fermi level at any temperaure will be given by the solution of the integral in equation (5.56). Provided it is assumed that $E_F \gg kT$, which assumption is valid for a solid metal, the integral may be solved and yields

$$E_F = E_F(0)\left[1 - \frac{\pi^2}{12}\left(\frac{kT}{E_F(0)}\right)^2\right]. \quad \ldots(5.63)$$

Since kT is always much less than $E_F(0)$ for a metal the change in Fermi level with temperature is small.

The quantity E_F may be considered to define a *Fermi temperature* T_F corresponding to the Fermi energy kT_F. Typical values for metals, which may be calculated from E_F values in Table 5.1, are $\approx 10^4$ K. A normal gas obeys Maxwell–Boltzmann statistics and has a mean energy per molecule $\sim kT$. Consequently T_F is the temperature at which the mean energy of the electron gas would approach that of an ordinary gas. At this temperature the Fermi–Dirac distribution approaches that given by Maxwell–Boltzmann statistics and T_F is referred to as the degeneracy temperature. A metal is always, of course, at a temperature much lower than T_F and the electron gas is degenerate.

5.10 Electrical conductivity according to the free electron theory

The relationship

$$\sigma = \frac{ne^2\tau}{m_e} \quad \ldots(5.10)$$

was derived on classical theory by considering the collision processes taken part in by the free electrons in a metal. We must now consider how the application of the Pauli exclusion principle will modify these collision processes. The general principle involved is that a particular electron collision is only allowed provided there is an unoccupied state appropriate to the position and momentum of the electron after collision.

As discussed in Section 5.7, electron states can be represented by cells on a three-dimensional momentum diagram, the distance of a cell from the origin giving a measure of the energy of that state. On the free electron model the electrons at the absolute zero of temperature will be occupying states lying within a sphere about the origin of momentum space. The boundary of the occupied states is the *Fermi surface* (see Fig. 5.12). At ordinary temperatures this boundary surface lies at energy E_F and is slightly less sharp than at 0 K. Partial occupation of electron states extends over a region $\sim 5kT$ about E_F.

The collisions in which electrons take part are very nearly elastic so that any

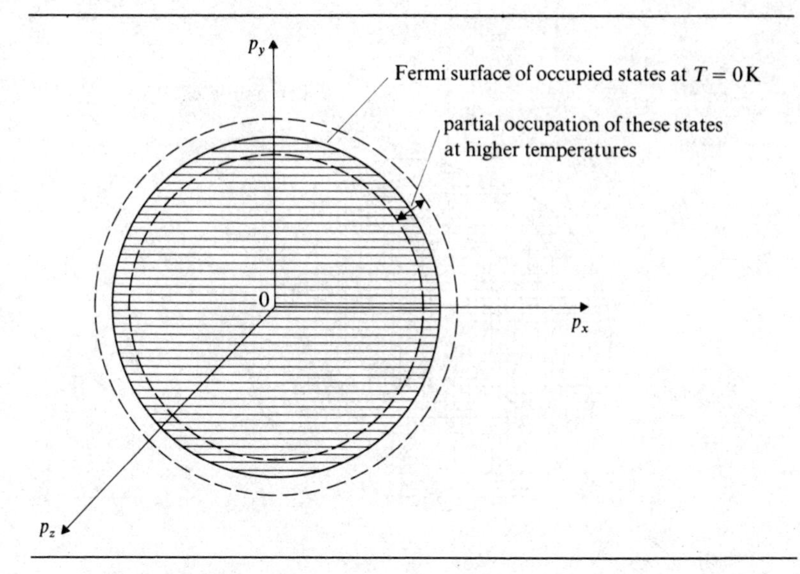

Fig. 5.12 The momentum diagram showing the Fermi surface of occupied states corresponding to the Fermi level E_F. Note the partial occupation of states having energies within $\sim 5kT$ of E_F.

changes in electron energy which result are very small. On the three-dimensional momentum diagram an electron involved in a collision is thus scattered to an electron state very nearly equidistant from the origin. Since in a metal all the electron states up to within $\sim 5kT$ of E_F are filled it follows that electron collisions involving such states are forbidden by the Pauli exclusion principle. An electron colliding must always have an empty state to receive it. Electrons occupying states close to the Fermi level can, however, take part in many collisions, and these will have the effect of changing the wave-vectors of the electrons.

When an electric field is applied the electrons are accelerated during their free times, and this is represented in Fig. 5.13 by a very slight displacement of the Fermi sphere of occupied states in a direction opposite to that of the field. We note that each electron has its wave-vector changed by a small amount by the field, and that the process is only possible since the changes are systematic. The electrons are thus all moving to fill the states vacated by other electrons. It is clear that we may define a mean drift velocity of all the electrons in the crystal as was done in Section 5.2 and from this point of view all the free electrons may be considered to be taking part in the electrical conduction process.

However, it is only electrons occupying states near E_F which suffer collisions, namely those which are not prevented by the Pauli exclusion principle. In particular electrons within states in the region A are able on collision to fill states in region B, which we note will have nearly equal energy. Such collisions will tend to restore the equilibrium electron distribution, which condition will be

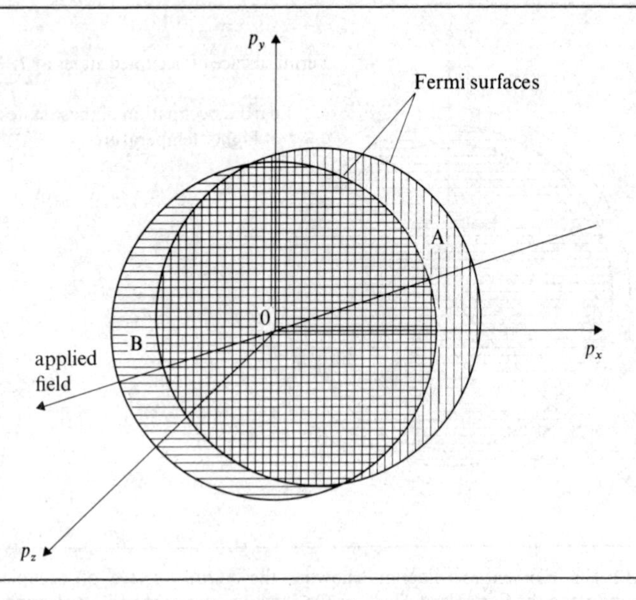

Fig. 5.13 Effect of applied electric field on distribution of electron momenta. Horizontal hatching shows the undisturbed distribution in zero field. Vertical hatching shows the distribution in applied field equivalent to all electron velocities increased by an increment. (cf. Fig. 5.1).

approached with a characteristic time termed the relaxation time. This relaxation time will be the mean free time τ_F of the electrons having energy close to E_F. We therefore obtain the interesting result that the mean free time τ_F governs the drift velocity achieved by the whole assembly of free electrons in the crystal. The value of the electrical conductivity σ according to the quantum theory is therefore

$$\sigma = \frac{ne^2\tau_F}{m_e}. \qquad \ldots(5.64)$$

The value of τ_F for any particular metal may be calculated directly from equation (5.64) provided the conductivity is known. Typical values are $\sim 10^{-14}$ s. It is governed by the collision processes in which the electrons take part, and as will be discussed in Section 6.5.2 the main scattering process is that due to lattice vibrations. The scattering increases with increasing temperature and the theory shows that τ_F is inversely proportional to T at ordinary and high temperatures. It follows that the electron mobility μ and electrical conductivity σ decrease with increasing temperature according to $1/T$ (see Section 5.4.1).

The electrons in a metal may also be scattered by impurity atoms and by crystal imperfections such as dislocations and grain boundaries. These scattering processes are nearly independent of temperature so that they are of particular significance at low temperatures when the lattice scattering is small.

To summarise, we have found that the conductivity of a metal depends on the electron concentration and on the electron mobility. The carrier concentration is large, being of the same order as the number of atoms per cubic metre in the metal. It remains essentially constant with temperature whereas the mobility decreases with increasing temperature because of the changes in lattice scattering. As we shall see in Chapter 8, the carrier concentration in a semiconductor is very much less than in a metal, and the temperature variation of the conductivity of an intrinsic semiconductor is essentially governed by changes in carrier concentration rather than by mobility changes.

5.11 Electron emission processes

We have already discussed in Section 1.6.1. the emission of electrons from metals on irradiation with light of appropriate wavelength and have seen how this emission is explained in terms of quantum theory. In addition to this photoelectric emission there are two other main processes by which electrons may be emitted from the surface of a metal. These are thermionic emission and field emission.

5.11.1 *Thermionic emission*

When a metal is heated electrons are emitted from its surface. This process is termed thermionic emission and is found to be strongly temperature-dependent. It forms the basis of the use of heated tungsten and thoriated tungsten filaments in electronic valves, and these filaments have typical operating temperatures for electron emission of about 2500 K and 1900 K, respectively. Emission is also observed for certain metallic oxides such as barium oxide and these are used in the form of indirectly heated oxide coated cathodes having a much lower operating temperature of about 1000 K.

For an understanding of the thermionic emission process to be obtained we need to reconsider the model of Section 5.6.1. This assumed infinitely high potential barriers at the surface of a metal and would thus not allow of any process of electron emission. Let us therefore assume instead a finite barrier height W but retain as in the previous model a constant potential $V = 0$ within the metal (Fig. 5.14). We are interested in solutions for which the electron is bound to the well, i.e. for $E < W$. The general solution obtained in equation (5.23) of the previous model is still valid for the region of the well, although the constants A and B will now differ. Solutions also exist for the regions on either side of the well and these may be shown to be in the form of waves dying exponentially away from the boundary. As illustrated in Fig. 5.15, the different wave solutions must merge at the boundary and this governs the exact form of the solutions including the values of the constants involved.

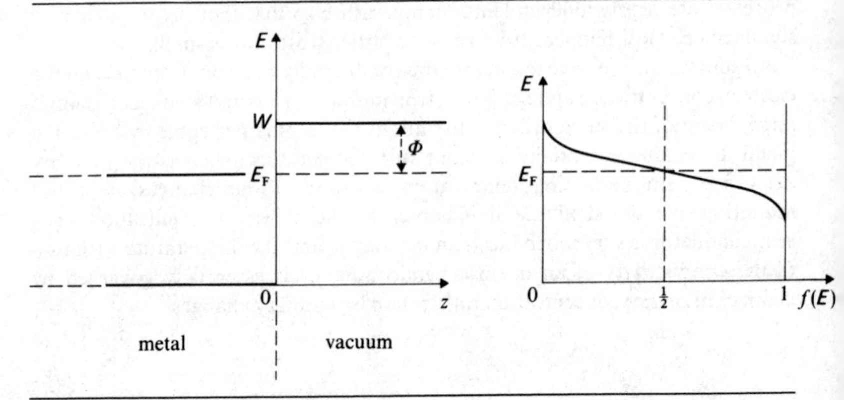

Fig. 5.14 Simplified model illustrating thermionic emission. Electrons in the solid are distributed in energy about E_F and require energy $>(E_F+\Phi)$ to leave the solid.

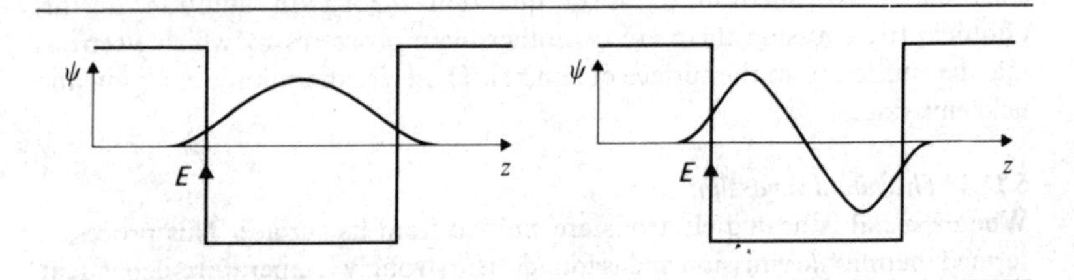

Fig. 5.15 Wave-functions for $n = 1$ and 2 are superimposed on a one-dimensional well. Penetration of wave-functions outside the well should be noted.

The model therefore leads, as before, to a range of possible electron energy states and the distribution of occupied states relative to the Fermi level E_F is governed by the temperature of the metal. An electron must possess energy E greater than the barrier height W in order that it may leave the solid. Thus denoting $(W-E_F)$ by Φ the condition for emission is that

$$E \gg E_F + \Phi. \quad \ldots(5.65)$$

Φ is termed the work function and differs for different metals. Values of a few electron-volts are found for the work functions of many metals so that at a sufficiently high temperature some electrons in the distribution will have sufficiently high energy to overcome the barrier and thermionic emission will then occur.

The principal interest in the phenomenon is that of investigating the variation of thermionic emission current density with temperature. This is represented by the Richardson–Dushman equation in the form

$$J = A_0 T^2 \exp\left(-\frac{\Phi}{kT}\right), \qquad \ldots(5.66)$$

where A_0 is a universal constant having the value 1.2×10^6 A m^{-2} K^{-2}. This equation will now be derived on the basis of the simple model outlined.

As well as possessing sufficient energy, an electron leaving the metal must also have the appropriate velocity component normal to the surface so that it may overcome the barrier forces. If the z direction is normal to the surface of the metal then the minimum component of momentum p_{zc} in this direction which allows the electron to escape will be given by

$$p_{zc} = (2m_e E)^{\frac{1}{2}} = \left[2m_e(E_F + \Phi)\right]^{\frac{1}{2}} \qquad \ldots(5.67)$$

in accordance with equation (5.65). This equation defines a plane in momentum space above which all electrons have sufficient energy and momentum to escape.

Since Φ is of the order of a few electron-volts (e.g. 4.5 eV for tungsten) one is concerned with electron energies in the distribution at a level considerably higher than the Fermi level and to attain emission the temperature must be high. Only the tail of the distribution function (5.54) is then of importance and the function may be approximated to

$$f(E) = \frac{1}{\exp[(E - E_F)/kT]} = \exp\left(\frac{E_F}{kT}\right)\exp\left(-\frac{E}{kT}\right), \qquad \ldots(5.68)$$

which is the classical Boltzmann distribution.

This equation gives the probability of a state of energy E being occupied when $E \gg E_F$ and is a good approximation provided that $E - E_F > 2kT$. Since one is primarily interested in the values of momentum required to overcome the barrier this distribution function can be written in the form

$$f(E) = \exp\left(\frac{E_F}{kT}\right)\exp-\left(\frac{p_x^2 + p_y^2 + p_z^2}{2m_e kT}\right) \qquad \ldots(5.69)$$

to give the probability of being occupied of a state having momentum components p_x, p_y and p_z.

It is also necessary to know the density of states in terms of the momenta components. From Section 5.7 we have the result that for a unit volume of crystal, a volume h^3 of momentum space will accommodate two electrons, one of each spin. The number of states dN_s with momenta in the range p_x to $p_x + dp_x$, p_y to $p_y + dp_y$ and p_z to $p_z + dp_z$ will thus be given by

$$dN_s = \frac{2}{h^3} dp_x dp_y dp_z. \qquad \ldots(5.70)$$

Relationship (5.69) gives the fraction of these states which are occupied.

To find the current density J_z in the direction normal to the surface we use relationship (5.4) so that for electrons of momentum p_z this will be given by

$$J_z = \frac{nep_z}{m_e}. \qquad \ldots(5.71)$$

Using equations (5.69) and (5.70) this can be put in the form

$$J_z = \frac{e}{4\pi^3 m_e \hbar^3} p_z \exp\left(\frac{E_F}{kT}\right) \exp\left(-\frac{p_x^2+p_y^2+p_z^2}{2m_e kT}\right) dp_z dp_y dp_z. \qquad \ldots(5.72)$$

The total current density J_z of thermionic emission will be obtained by integrating this expression noting that all values of p_x and p_y are allowed but only values of $p_z \geqslant p_{zc}$. We therefore have

$$J_z = \frac{e}{4\pi^3 m_e \hbar^3} \int_{-\infty}^{\infty} \exp-\left(\frac{p_x^2}{2m_e kT}\right) dp_x \int_{-\infty}^{\infty} \exp-\left(\frac{p_y^2}{2m_e kT}\right) dp_y \times$$

$$\int_{p_{zc}}^{\infty} \exp\frac{(2m_e E_F - p_z^2)}{2m_e kT} p_z \, dp_z. \qquad \ldots(5.73)$$

Using standard integrals this expression yields

$$J_z = A_0 T^2 \exp-\left(\frac{\Phi}{kT}\right),$$

where $\qquad \ldots(5.74)$

$$A_0 = \frac{em_e k^2}{2\pi^2 \hbar^3},$$

which is the Richardson–Dushman relation.

Emission is strongly influenced by the exponential dependence on temperature and on the value of the work function Φ. Typical values are given in Table 5.1. The work function varies according to the crystal planes exposed on the surface of the sample and also itself shows some temperature dependence due to thermal expansion. The emission is very sensitive to any surface contamination or oxidation and is reduced if a space charge is allowed to build up, as will occur in the absence of a sufficiently high anode potential.

5.11.2 *Field emission*

When an electron is leaving the surface of the metal there is a redistribution of the charge on the surface and this is such as to maintain the surface at a uniform potential. This condition can only be achieved by the lines of force from the charge intersecting the surface normally and this type of system is electrostatically equivalent to an image positive charge inside the metal as illustrated in Fig. 5.16. During emission the electron is subject to a coulombic force governed by the separation ($2z$) of the electron and its image, i.e. a force proportional to $(1/2z)^2$. The potential energy of the electron, instead of showing an abrupt step at

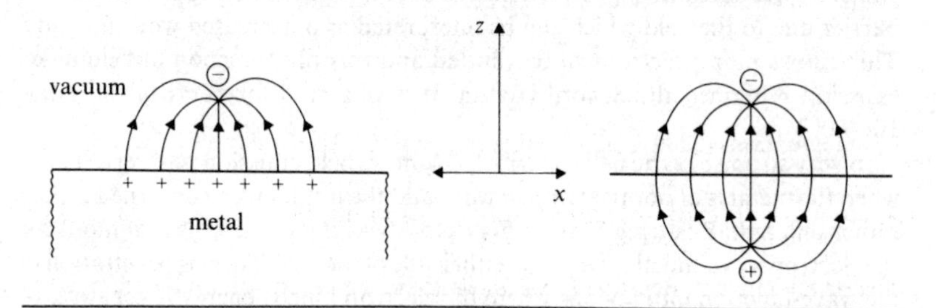

Fig. 5.16 The electric field lines are perpendicular to the metal surface since it is at constant potential. The electric field is equivalent to that of the electron and an image charge $+e$ within the metal.

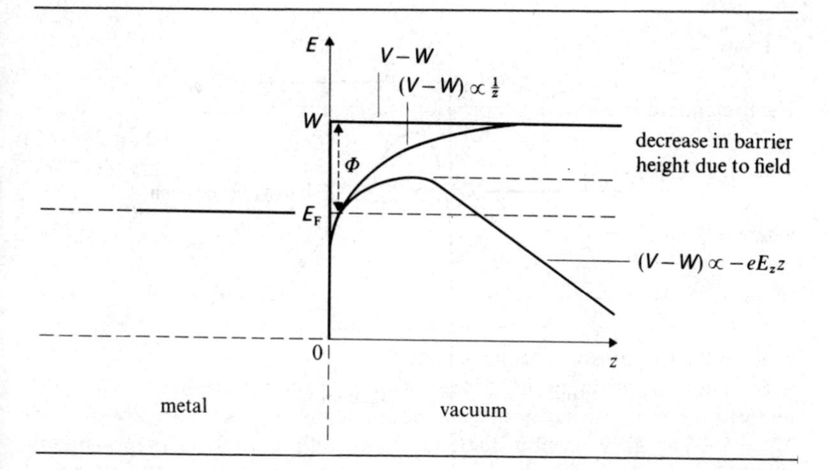

Fig. 5.17 Thermionic emission. Form of barrier is modified by taking into account the presence of the image force– Schottky effect. The electric field decreases the effective height of the barrier.

the surface as in the model, will thus vary according to $1/z$, as illustrated in Fig. 5.17.

Suppose now that the potential of a collecting anode is increased so that there exists a substantial accelerating field E_z above the surface of the metal. The electron will experience a force eE_z in this field and it will have a potential energy component due to the field which *decreases* away from the surface according to $eE_z z$. If we also take into account the modification of the barrier arising out of the image force the potential barrier V will be of the form

$$V = E_F + \Phi - \frac{e^2}{16\pi\varepsilon_0 z} - eE_z z, \qquad \ldots(5.75)$$

which is illustrated in Fig. 5.17. There is thus a slight lowering of the effective barrier due to the field which can be interpreted as a decreased work function. This allows more electrons to be emitted and this phenomenon of field-aided emission is termed the Schottky effect. It is observed for fields in the range 10^3–10^6 $V\,m^{-1}$.

In very strong electric fields, ~ 10^6–10^9 V m^{-1}, field emission is observed even when the metal is at normal temperatures and the mechanism concerned is then rather different. Referring back to Fig. 5.15 it was then noted that solutions for the electron wave-functions exist on either side of the well. There is penetration of the wave-function into regions where the electron kinetic energy is negative. In terms of classical mechanics this would of course be forbidden. As illustrated in Fig. 5.18, under the action of a very strong field, the barrier width at the Fermi

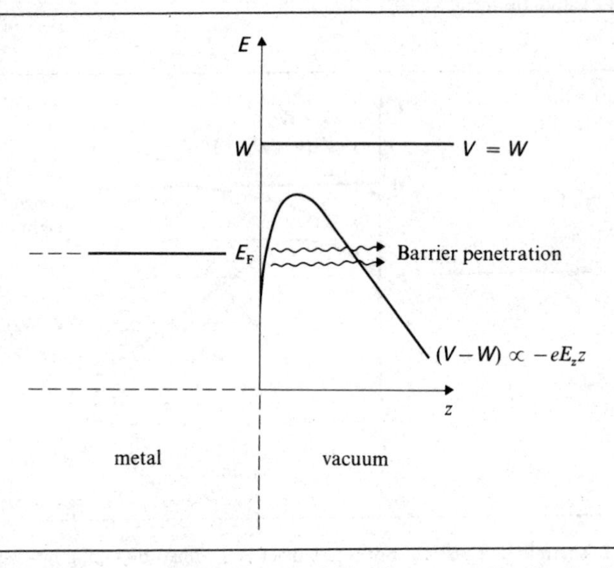

Fig. 5.18 Electron emission through the potential barrier occurs when the electric field is very large (10^6–10^9 V m^{-1}) and the barrier width is then <1 nm. The process is termed tunnelling.

level becomes very narrow (<1 nm). Under these conditions the electron can penetrate through the barrier and field emission occurs. The phenomenon is termed tunnelling and it finds application in the field-emission microscope. The emission current density is given by the Fowler–Nordheim relation in the form

$$J = B\exp\left(-\frac{\beta}{E}\right), \qquad \ldots(5.76)$$

where B and β are constants involving the work function. A strong dependence of emission on field strength is apparent and the similarity to the Richardson–Dushman relation for thermal emission should be noted.

References and further reading

Blakemore, J. S. (1969) *Solid State Physics*, Saunders.
Dekker, A. J. (1967) *Solid State Physics*, Macmillan.
Hart-Davis, A. (1975) *Solids, an Introduction*, McGraw-Hill.
Kittel, C. (1971) *Introduction to Solid State Physics*, Wiley.
Rosenberg, H. M. (1975) *The Solid State*, Clarendon.
Shockley, W. (1950) *Electrons and Holes in Semiconductors*, Van Nostrand.
Solymar, L. and **Walsh, D.** (1970) *Lectures on the Electrical Properties of Materials*, Clarendon.
Wilkes, P. (1973) *Solid State Theory in Metallurgy*, Cambridge University Press.

Problems

5.1 Show that the relationship between field strength and potential in equation (5.1) is in accord with the concept of potential difference being the work done in the transfer of charge between two points.

5.2 A potential difference of 3.3 V is applied across a 6 m length of copper wire of diameter 3×10^{-4} m. Given that the resistivity of copper is $1.72 \times 10^{-8} \Omega$ m, determine the current and the current density in the wire.

Is this a feasible current?

5.3 The element sodium has density 0.97×10^3 kg m^{-3}, relative atomic mass 23 and electrical conductivity of $2.1 \times 10^7\ \Omega^{-1}$ m^{-1}. Determine the mobility of electrons in sodium.

5.4 The density of the metal barium is 3.5×10^3 kg m^{-3} and its relative atomic mass is 137. Assuming the metal to be divalent determine the radius of the Fermi sphere and the corresponding value of the Fermi energy.

5.5 Lithium has electrical conductivity of $1.05 \times 10^7\ \Omega^{-1}$ m^{-1} and there are 4.8×10^{28} atoms per m^3. Determine the drift velocity of the electrons in the metal for an applied field of 100 V m^{-1} and compare the value obtained with the speed of electrons at the Fermi level.

5.6 Using the data from the previous example, determine the speed of an electron which has the average electron energy in the crystal. The ratio of this speed to u_F is 0.77. Explain why this ratio should be obtained.

5.7 Determine the value of the Fermi energy for copper and show that the change in the value of Fermi energy for the range 0–300 K amounts to only 0.001 per cent. Copper has density 8.93×10^3 kg m^{-3} and relative atomic mass 63.5.

5.8 Use the Heisenberg uncertainty principle to show that the uncertainty in momentum of an electron confined in a cube of side L corresponds to a cell of volume h^3/L^3. Assuming that each cell in momentum space can accommodate two electrons, show that the number of states dN_s in unit volume of a crystal for the range p to $p + dp$ is given by

$$dN_s = \frac{8\pi}{h^3} p^2\, dp.$$

Confirm that this result is in agreement with equation (5.41).

5.9 Use the density of states relationship in the previous question to show that for a metal at 0 K the average electron speed is three-quarters of the speed of an electron at the Fermi level. Compare the electron speed with that of an electron possessing the average energy.

5.10 The following values are obtained for the thermionic emission current density J (in $A\,m^{-2}$) of tunsten for the temperatures indicated (in K):

T	2000	2250	2500	2750
J	10.1	234	2970	24150

By drawing an appropriate graph determine the work function for tungsten and the value of the A-factor. Comment on the accuracy in the determination of the A-factor and compare the value obtained with that given in the text for A_0.

Is the difference outside the experimental error?

5.11 Determine the thermionic emission current from a tantalum wire of length 0.05 m and diameter 10^{-3} m which is maintained at a temperature of 2000 K. The work function for tantalum is 4.19 eV and the A-factor is 0.55×10^6 $\mathrm{Am^{-2}\,K^{-2}}$.

5.12 Show that in field-aided emission the maximum barrier height occurs at a distance

$$z_0 = \left(\frac{e}{16\pi\varepsilon_0 E_z}\right)^{\frac{1}{2}}$$

from the surface of a sample.

6. The thermal properties of solids

This chapter will be concerned with the theories which are used to explain the main thermal properties of solids, namely thermal conductivity, thermal expansion and heat capacity. It is convenient to discuss first the theories which have been proposed to explain the observed heat capacities of solids. It will then be shown how theory accounts for the thermal expansion of solids. The development of the theory of thermal conductivity of solids will be briefly discussed.

6.1 The lattice heat capacity of a solid

We have seen in Section 5.4.2 that classical theory predicts a molar heat capacity C_V of $3R$ due to the vibration of the atoms about their mean lattice position. This value of 24.9 J K^{-1} mol^{-1} is observed experimentally for many solids at room temperature. Notable exceptions, however, are the hard materials such as diamond and silicon which have values of C_V of 6.1 and 20.7 J K^{-1} mol^{-1}, respectively, at room temperature. At a sufficiently high temperature, which is as high as 2000 K for diamond, the heat capacities of all solids attain the classical value of $3R$. At low temperatures all solids show a heat capacity which decreases towards zero as the temperature decreases. This decrease is not predicted on classical theory. Typical experimental heat capacity data is shown in Fig. 6.1.

6.1.1 *The Einstein theory*

The first application of the quantum theory to the problem of the heat capacity of solids was due to Einstein. In the theory it is postulated that each atom in a solid with its three possible components of vibration may be replaced by three linear harmonic oscillators, i.e. the vibration of each atom can be represented by a combination of three simple harmonic motions. A solid containing N atoms will thus require a total of $3N$ harmonic oscillators to represent the total possible vibrations of all the atoms. The oscillators are assumed in the theory to be independent of each other, but to have the same frequency of vibration ν. The energy E of an oscillator governs its amplitude of oscillation, and following

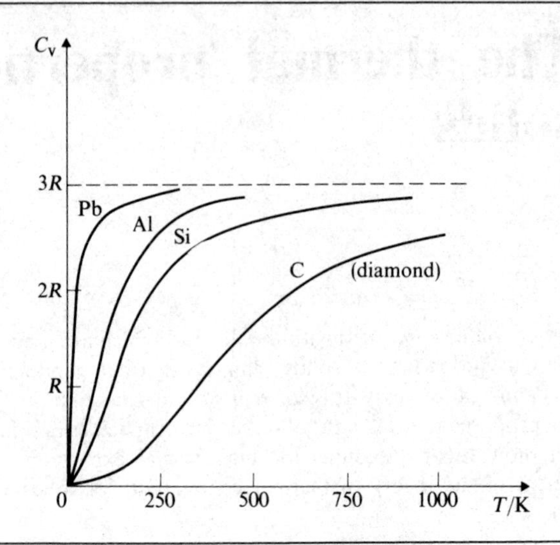

Fig. 6.1 Experimental molar heat capacities of solids.

Planck's theory, it is postulated that it can only take on the values

$$E = nh\nu \qquad (n = 0, 1, 2 \ldots). \qquad \ldots(6.1)$$

In order to determine the heat capacity of a solid it is first necessary to determine the total energy of the solid as a function of temperature. The value of the most probable energy of an oscillator at a particular temperature is thus required, i.e. the mean energy $\bar{E}$ of all the oscillators in the system. On classical theory the oscillators would have a mean energy kT.

The Boltzmann distribution which governs the equilibrium arrangement of particles in a system according to their energies has been mentioned in Section 5.1. Assuming that the Boltzmann distribution is applicable to the oscillators, then the number of oscillators δs whose energies lie between E and $E+\mathrm{d}E$ is given by

$$\delta s = A \exp(-E/kT)\,\mathrm{d}E. \qquad \ldots(6.2)$$

A is a constant whose value depends on the total number of oscillators in the system. Hence if s is the number of oscillators having energy $E = 0$, then from (6.2) the numbers having energies $h\nu$, $2h\nu \ldots$, are $s\exp(-h\nu/kT)$, $s\exp(-2h\nu/kT)$ .... The mean energy $\bar{E}$ of an oscillator will be given by the sum of the energies of all the oscillators divided by the total number of oscillators in the system, i.e.:

$$\bar{E} = \frac{h\nu s\exp(-h\nu/kT) + 2h\nu s\exp(-2h\nu/kT) + \ldots}{s + s\exp(-h\nu/kT) + s\exp(-2h\nu/kT) + \ldots}. \qquad \ldots(6.3)$$

This expression yields

$$\bar{E} = \frac{h\nu}{\exp(h\nu/kT) - 1}. \qquad \ldots(6.4)$$

The total energy U_m of the N_A atoms in a mole of solid is therefore given by

$$U_m = \frac{3N_A h\nu}{\exp(h\nu/kT) - 1}. \qquad \ldots(6.5)$$

Let us consider the limiting values of this expression. At high temperatures $h\nu/kT$ is small and terms higher than $(h\nu/kT)^2$ in the expansion of the exponential term can be neglected. The value of U_m at high temperatures is then readily shown to be

$$U_m = 3N_A(kT - \tfrac{1}{2}h\nu). \qquad \ldots(6.6)$$

We note here that the value of U_m does not converge to the classical value of $3N_AkT$ $(=3RT)$ at high temperature. Convergence to this value is, however, obtained if it is assumed that an oscillator has energy $h\nu/2$ at absolute zero of temperature, i.e. if equation (6.1) is modified to

$$E = (n + \tfrac{1}{2})h\nu \qquad (n = 0, 1, 2 \ldots). \qquad \ldots(6.7)$$

This additional energy is termed zero point energy. It should be pointed out that whereas the above derivation gives the zero point energy as *approximately* $h\nu/2$ a rigorous quantum treatment shows this value to be exact.

At low temperatures $\exp(h\nu/kT) \gg 1$ so that

$$U_m = \frac{3N_A h\nu}{\exp(h\nu/kT)}. \qquad \ldots(6.8)$$

To obtain the molar heat capacity we put, as in Section 5.4.2,

$$C_V = \left(\frac{\partial U_m}{\partial T}\right)_V. \qquad \ldots(6.9)$$

For high temperatures the total energy U_m is $3N_AkT$ $(=3RT)$ so that C_V is $3R$, in accord with the classical value. The value of the molar heat capacity at low temperatures is obtained from equation (6.8) which yields

$$C_V = 3R\left(\frac{h\nu}{kT}\right)^2 \exp\left(-\frac{h\nu}{kT}\right). \qquad \ldots(6.10)$$

The molar heat capacity thus tends to zero as the temperature approaches absolute zero, which is as required by experiment. It should be noted that the exponential factor dominates expression (6.10) at low temperatures, but that accurate comparison with the experimental data shows the exponential variation predicted gives too rapid a fall of heat capacity with temperature. Experimentally it is found that, if the solid is an insulator, the component of the heat capacity associated with the lattice approaches zero as T^3. For metals, a small additional component of heat capacity, which is due to the free electrons and which is proportional to T, is obtained. This component only becomes important at very low temperatures (see Section 6.3).

The complete expression for C_V may readily be derived from equation (6.5) and is given by

$$C_V = 3R\left(\frac{\Theta_E}{T}\right)^2 \left[\frac{\exp\left(\frac{\Theta_E}{T}\right)}{\left(\exp\left(\frac{\Theta_E}{T}\right)-1\right)^2}\right], \qquad \ldots(6.11)$$

where

$$\Theta_E = \frac{h\upsilon_E}{k}. \qquad \ldots(6.12)$$

Θ_E is known as the Einstein characteristic temperature. A plot of this equation is shown in Fig. 6.2 and a fair agreement for the temperature variation of C_V for a given solid can be obtained by choice of suitable ν or Θ_E. Numerical values of Θ_E for most solids lie within the range 100–300 K, and the corresponding characteristic frequencies υ_E from 2×10^{12} to 6×10^{12} Hz. A notable exception is diamond, which has atoms very firmly bound in the crystal lattice, and for this material the characteristic frequency is much higher ($\sim 3 \times 10^{13}$ Hz), giving a correspondingly high characteristic temperature ($\Theta_E \sim 1320$ K).

We may note in concluding the discussion of the Einstein theory that the basic assumption that the atoms in the solid are all performing independent oscillations of the same frequency is unacceptable. The motions of atoms on a crystal lattice are strongly coupled together.

6.1.2 *The Debye theory*

The Einstein theory postulated a single characteristic frequency for all the vibrations of the atoms in any particular solid. In the Debye theory the atomic displacements are considered to be capable of being represented by a system of transverse and longitudinal elastic waves within a continuous uniform solid. The system of waves would cover a wide spectrum of frequencies. Debye assumed that all the waves whatever their frequencies had the normal velocities appropriate to transverse and longitudinal waves in the solid, i.e. possible dispersion of the waves was neglected.

For the N atoms in a solid the total number of modes of vibration must be the same as in the Einstein theory, i.e. $3N$. Corresponding to each longitudinal mode of vibration of a given frequency there will be two transverse modes of vibration polarised at right angles to each other. Accordingly, the system of waves will consist of N longitudinal and $2N$ transverse modes. Each wave will possess its own specific frequency and the restriction on the total number of modes implies that there must exist a certain limiting cut off frequency ν_{m} for the system of waves.

In order to determine the total energy of the system of waves it is necessary to determine first the number of possible waves in the solid as a function of frequency, and to assign to each wave an appropriate mean energy. The total energy is then obtained by integrating the energies for all the waves up to that corresponding to the limiting frequency ν_{m}.

The number of possible waves in the solid for a given range of frequencies may be determined by considering stationary wave systems and inserting the appropriate boundary conditions. Alternatively, travelling wave solutions may be employed using the method of periodic boundary conditions (see Section 6.2.1).

Let us consider the possible stationary waves in a solid cube of side L, the edges of the cube coinciding with the coordinate axes. We recall from Chapter 1 that the standing wave solutions for a string of length L attached at both ends are of the form

$$\Psi = 2a \sin kx \cos \omega t, \qquad \ldots(1.38)$$

where

$$k = \frac{n\pi}{L}. \qquad \ldots(1.42)$$

Standing wave solutions of the three-dimensional wave equation

$$\nabla^2 \Psi = \frac{1}{v^2}\frac{\partial^2 \Psi}{\partial t^2} \qquad \ldots(1.26)$$

are required for the cube and these will be of the form

$$\Psi = A \sin\left(\frac{n_x \pi x}{L}\right) \sin\left(\frac{n_y \pi y}{L}\right) \sin\left(\frac{n_z \pi z}{L}\right) \cos \omega t, \qquad \ldots(6.13)$$

where A is a constant and n_x, n_y and n_z are positive integers. The similarity of this solution to that derived in Section 5.6.2 for electron waves in a box should be noted. Substituting in equation (1.26) we find

$$\frac{\pi^2}{L^2}\left(n_x^2 + n_y^2 + n_z^2\right) = \frac{\omega^2}{v^2}. \qquad \ldots(6.14)$$

The theory assumes that there is no dispersion of the waves, i.e. the wave velocity v is constant and putting ν as the frequency of the wave we have

$$n_x^2 + n_y^2 + n_z^2 = \frac{4L^2\nu^2}{v^2}. \qquad \ldots(6.15)$$

The number of possible waves $\mathrm{d}N_{\mathrm{s}}$ of any one polarisation having frequencies less than ν is given by the number of sets of values (n_x, n_y, n_z) lying within the octant of the sphere of radius $(n_x^2+n_y^2+n_z^2)^{\frac{1}{2}}$, so that

$$\begin{aligned} dN_s &= \tfrac{1}{8}\tfrac{4}{3}\pi\left(\frac{4L^2\nu^2}{v^2}\right)^{\frac{3}{2}} \\ &= \tfrac{4}{3}\pi\left(\frac{L\nu}{v}\right)^3. \end{aligned} \qquad \ldots(6.16)$$

For a range of frequencies ν to $\nu+d\nu$ the number of possible waves is then given by

$$dN_s = \frac{4\pi V\nu^2}{v^3}d\nu, \qquad \ldots(6.17)$$

where V is the volume of the box.

This result may be shown to be generally valid whatever the shape of the box. In addition, since for any longitudinal wave there will correspond two transverse waves polarised at right-angles to each other, equation (6.17) may be put in the general form

$$dN_s = 4\pi V\left(\frac{2}{c_t^3}+\frac{1}{c_l^3}\right)\nu^2\, d\nu, \qquad \ldots(6.18)$$

where c_t and c_l are the transverse and longitudinal wave velocities in the solid. The limiting frequency ν_m of the set of waves may be determined by equating the total number of possible vibrations to $3N$, i.e.

$$\int_0^{\nu_m} 4\pi V\left(\frac{2}{c_t^3}+\frac{1}{c_l^3}\right)\nu^2\, d\nu = 3N \qquad \ldots(6.19)$$

which yields

$$\nu_m = \left(\frac{9N}{BV}\right)^{\frac{1}{3}}, \qquad \ldots(6.20)$$

where

$$B = 4\pi\left(\frac{2}{c_t^3}+\frac{1}{c_l^3}\right). \qquad \ldots(6.21)$$

It is postulated in the theory that the energies of the waves are governed by quantum theory, so that the mean energy $\bar{E}$ of a wave of frequency ν is given by equation (6.4). The total energy U_m of the N_A atoms in a mole of solid will therefore be given by

$$\begin{aligned} U_m &= 3N_A\bar{E} \\ &= \int_0^{\nu_m} \frac{h\nu}{\exp\left(\dfrac{h\nu}{kT}\right)-1} BV\nu^2\, d\nu. \end{aligned} \qquad \ldots(6.22)$$

Putting $x = h\nu/kT$, this expression yields

$$U_m = 9N_A kT\left(\frac{1}{x_m^3}\right)\int_0^{x_m} \frac{x^3}{\exp x - 1}\,dx \qquad \ldots(6.23)$$

where

$$x_m = \frac{h\nu_m}{kT} = \frac{\Theta_D}{T}. \qquad \ldots(6.24)$$

Θ_D has the dimensions of temperature and is termed the Debye temperature of the solid.

We may now determine what equation (6.23) predicts for the lattice heat capacity. First of all, at high temperatures $x \to 0$ so that $(\exp x - 1) \to x$. The definite integral in (6.23) then reduces to $\frac{1}{3}x_m^3$ giving

$$U_m = 3N_A kT \qquad \ldots(6.25)$$

and, as before, a molar heat capacity of $3R$ at high temperatures.

At low temperatures $x_m = h\nu_m/kT \to \infty$ so that the upper limit of the definite integral in equation (6.23) can be replaced by ∞. The integral is then standard and has a value of $\pi^4/15$. Using equation (6.24) the total energy is

$$U_m = 9R\frac{T^4}{\Theta_D^3}\frac{\pi^4}{15} \qquad \ldots(6.26)$$

so that the lattice molar heat capacity at low temperatures is given by

$$C_V = \frac{12}{5}\pi^4 R\left(\frac{T}{\Theta_D}\right)^3. \qquad \ldots(6.27)$$

The T^3 variation of the lattice component of C_V is observed for a wide range of solids at low temperatures ($T < \Theta_D/20$). As already pointed out there is in metals an additional component due to the free electrons which becomes dominant at very low temperatures (<4 K). Although Θ_D can be calculated from equation (6.24), using values of the velocity of sound waves in the solid, a fit to the observed C_V values is best obtained by using Θ_D as an adjustable parameter. Values of Θ_D normally lie in the range 100–500 K for most solids, beryllium (1160 K) and diamond (1860 K) being exceptions. For the lowest temperatures a limiting value of Debye temperature Θ_D is normally quoted. Deviations in the behaviour of a solid from that predicted on the Debye theory can conveniently be represented in the form of a variation of Θ_D with temperature. The values of C_V as a function of T/Θ_D on the Debye theory are given in Table 6.1 and are compared in Fig. 6.2 with those obtained on the Einstein theory.

The Debye theory as outlined here neglects zero point energy. If this is included, the allowed energies of a vibrational mode of frequency ν are given by equation (6.7) and its mean energy by

$$\bar{E} = \tfrac{1}{2}h\nu + \frac{h\nu}{\exp(h\nu/kT) - 1}. \qquad \ldots(6.28)$$

Table 6.1 The variation of molar heat capacity with temperature†

Θ/T	T/Θ	**Debye theory** $C_V/\text{J K}^{-1}\text{ mol}^{-1}$	**Einstein theory** $C_V/\text{J K}^{-1}\text{ mol}^{-1}$
0	∞	24.94	24.94
1	1.0	23.7	22.9
2	0.5	20.6	18.1
3	0.33	16.5	12.4
4	0.25	12.6	7.58
5	0.20	9.19	4.26
6	0.17	6.63	2.24
7	0.14	4.76	1.12
8	0.125	3.45	5.36×10^{-1}
9	0.11	2.53	2.49×10^{-1}
10	0.10	1.89	1.13×10^{-1}
15	0.067	5.75×10^{-1}	1.72×10^{-3}
20	0.050	2.43×10^{-1}	2.10×10^{-5}
25	0.040	1.36×10^{-1}	
50	0.020	1.55×10^{-2}	
100	0.010	1.94×10^{-3}	
200	0.005	2.43×10^{-4}	

† *Values refer to $C_V : T/\Theta_E$ for the Einstein theory and $C_V : T/\Theta_D$ for the Debye theory. The low temperature approximations given in the text are valid for $\Theta_E/T \geqslant 7$ and $\Theta_D/T \geqslant 50$.*

The second term here corresponds to the mean energy $\bar{E}$ in equation (6.4) and is the mean excess energy of the mode above the zero point energy $h\nu/2$. If the heat capacity is derived using equation (6.28) instead of (6.4) it is readily verified that the value of C_V is unchanged, since the temperature-independent term involved in the expression for the total energy of the vibrational modes disappears on differentiation.

It has also been assumed in the theory that the maximum cut off frequency ν_m is the same for both the transverse and longitudinal waves. This limit is, as we have seen, essentially due to the atomic structure of the solid. A pictorial representation of the vibrational modes makes it clear, however, that it is the *shortest wavelengths*, governed by the separation of the atoms, which should be common to both the transverse and longitudinal modes. Since the transverse and longitudinal velocities c_t and c_l are unequal it is therefore necessary to use two different values of ν_m in the theory corresponding to the different modes. This leads to an expression for C_V containing two constants of the Debye type. This type of expression can, in some cases, give improved agreement with experimental data.

The Debye theory assumes that there is no dispersion of the waves, i.e. that there is no variation of the phase velocity of the waves with wavelength. If dispersion of the waves is taken into account then theory shows that the frequency distribution of the vibration modes is much more complex than in the Debye theory and sharp maxima occur in the frequency distribution. The theory

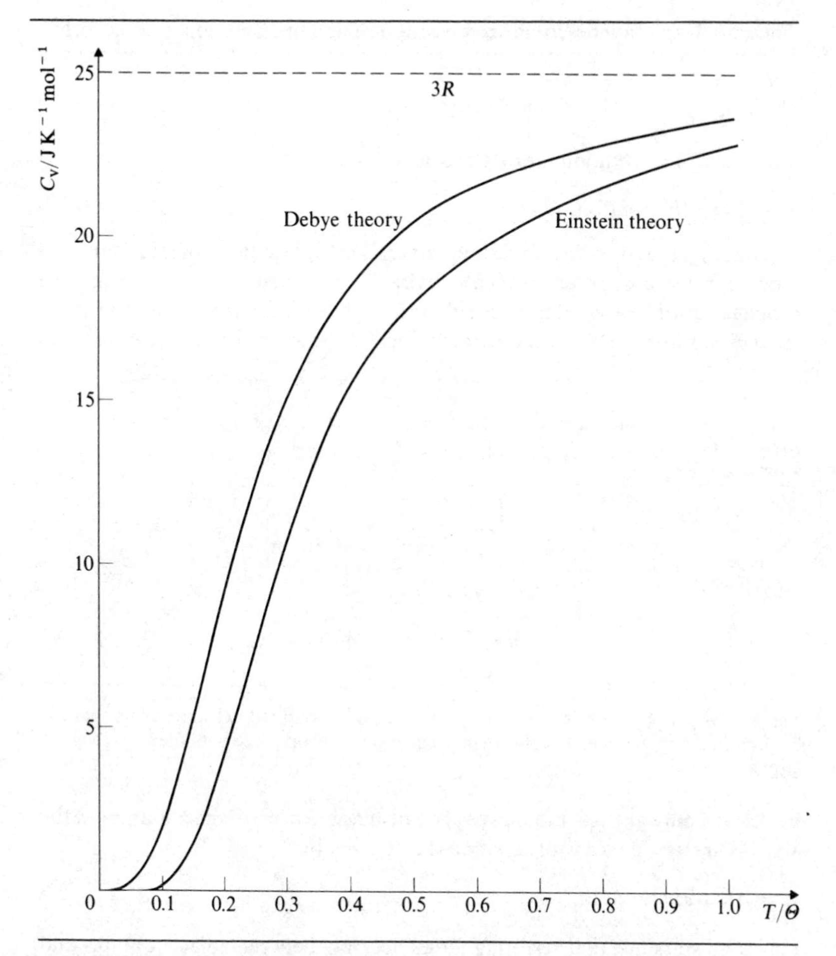

Fig. 6.2 The variation of molar heat capacity C_V given by the Einstein theory (C_V vs. T/Θ_E) and the Debye theory (C_V vs. T/Θ_D).

of lattice vibrations and the reasons for the dispersion of lattice waves will be considered further in the following section.

6.2 Lattice vibrations

A complete treatment of the modes of vibration of a crystal can be approached by considering first the propagation of waves along a one-dimensional array of atoms, i.e. along a linear lattice. We may recall from Chapter 1 that the general differential equation describing a wave motion in one dimension is of the form

$$\frac{\partial^2 \Psi}{\partial x^2} = \frac{1}{v^2} \frac{\partial^2 \Psi}{\partial t^2} \qquad \ldots(1.23)$$

which has for its solution the displacement

$$\Psi = A \exp \mathrm{i}(kx - \omega t). \qquad \ldots(1.22)$$

Assuming that there are no restrictions on the displacements of the atoms then both transverse and longitudinal waves are involved. Let us consider the propagation of a longitudinal wave along an infinite linear array of atoms each of mass m and spaced a apart. Referring to Fig. 6.3 the wave displacement is seen to

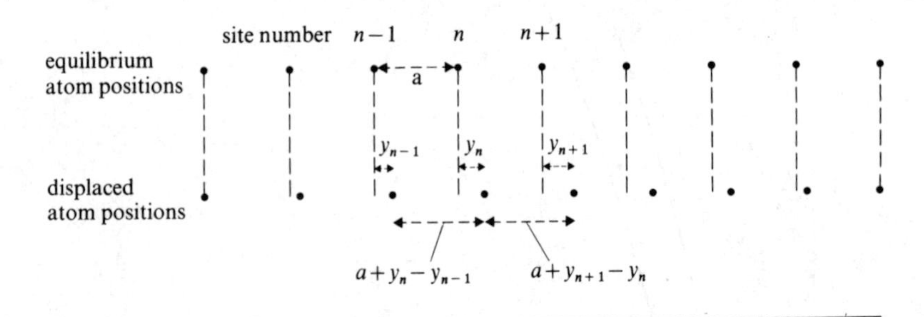

Fig. 6.3 Displacements of a linear array of atoms by a longitudinal wave. Note that the longitudinal displacements are shown much exaggerated compared with the spacing of the atoms.

be defined only at each atomic site. The displacement y_n of the nth atom in the array is, in accordance with equation (1.22), given by

$$y_n = A \exp \mathrm{i}(kna - \omega t). \qquad \ldots(6.29)$$

Let us now assume that restoring forces act only between adjacent atoms and that the atoms behave as a line of spring-coupled masses, i.e. we assume Hooke's law is obeyed. In accordance with equation (6.29) the restoring force F_n on the nth atom is given by

$$F_n = m\frac{\mathrm{d}^2 y_n}{\mathrm{d}t^2} = -m\omega^2 y_n \qquad \ldots(6.30)$$

and from Fig. 6.3 is seen to be

$$-m\omega^2 y_n = f(y_{n+1} - y_n) - f(y_n - y_{n-1}), \qquad \ldots(6.31)$$

where f is the force constant for the system. This may be put in the form

$$-m\omega^2 y_n = f y_n[\exp(\mathrm{i}ka) - 1] - f y_n[1 - \exp(-\mathrm{i}ka] \qquad \ldots(6.32)$$

by use of equation (6.29) and it simplifies to

$$\omega^2 = 4\left(\frac{f}{m}\right)\sin^2\left(\frac{ka}{2}\right). \quad \ldots(6.33)$$

Positive and negative values of k correspond to waves travelling in the positive and negative directions along the x-axis. Since only positive values of the angular frequency ω have physical significance the condition for propagation can be written in the form

$$\omega = \omega_c\left|\sin\left(\frac{ka}{2}\right)\right| \quad \ldots(6.34)$$

where

$$\omega_c = 2\left(\frac{f}{m}\right)^{\frac{1}{2}} \quad \ldots(6.35)$$

is the maximum value of the angular frequency. It defines a limiting cut off frequency ν_c for waves which may be propagated without attentuation along the linear lattice. The form of the variation of ω with k for the longitudinal waves is shown in Fig. 6.4, from which it is noted that it is a periodic function of period $2\pi/a$. The form of the function for transverse waves is similar.

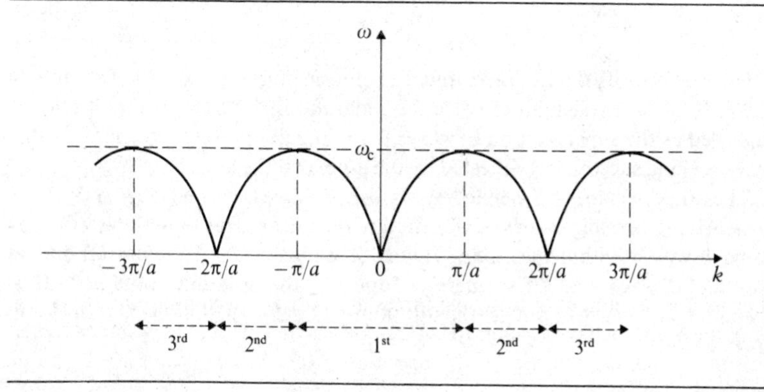

Fig. 6.4 Variation of ω with k for longitudinal waves along a linear array of atoms. The ranges of values of k defining the 1st, 2nd and 3rd Brillouin zones are indicated.

The longitudinal modes of vibration of the infinite linear array of atoms are thus described by equation (6.30) and the associated condition of equation (6.34) relating to ω and k. Let us consider some of the implications of this relationship. We recall from Section 1.5.2 that the phase velocity of a wave motion is given by ω/k and the group velocity by $d\omega/dk$. Dispersion of the waves occurs if the phase velocity is a function of wavelength, i.e. if the relationship between ω and k is not linear.

It is clear, therefore, from equation (6.34) that in the present problem there is dispersion of the waves travelling along the linear lattice.

The values of the phase and group velocities of the waves for certain limiting cases may be readily obtained. Using equations (6.34) and (6.35) the phase velocity $v = \omega/k$ may be put in the form

$$v = \left(\frac{f}{m}\right)^{\frac{1}{2}} \frac{\sin\left(\frac{ka}{2}\right)}{\left(\frac{ka}{2}\right)}. \qquad \ldots(6.36)$$

For long wavelengths $k = 2\pi/\lambda \to 0$ so that the phase velocity is given by

$$v = \left(\frac{f}{m}\right)^{\frac{1}{2}} \qquad \ldots(6.37)$$

which is independent of wavelength. Consequently the long wavelength waves are propagated without dispersion with equal phase and group velocities.

Let us next consider the propagation of the waves along the lattice at the limiting frequency ν_c, i.e. when $k = \pi/a$ or $-\pi/a$. The corresponding phase velocity v_c may be found from equation (6.36) and its value is

$$v_c = \frac{2}{\pi}\left(\frac{f}{m}\right)^{\frac{1}{2}}. \qquad \ldots(6.38)$$

The group velocity $d\omega/dk$ is zero and the solution must then consist of a standing wave. It has a wavelength of twice the separation of the atoms and it may be regarded as the superposition of waves travelling in opposite directions, i.e. the wave is being successively reflected by the atoms on the lattice.

The range of values of k defined by $-\pi/a < k \leqslant \pi/a$ is termed the *first Brillouin zone*. All the possible modes of vibration of the linear array can be described by using k values within this zone. Values of k outside the zone correspond to wavelengths $< 2a$ and these merely duplicate the solutions obtained. It is therefore customary to restrict the solutions to k values within the first Brillouin zone.

6.2.1 *Boundary conditions for the linear array*

There is one further important property of the solutions which has not yet been discussed, namely the result that only certain specific values of k lead to allowed solutions. This result may be derived by considering a finite array of atoms and inserting the boundary condition that the end atoms of the array are fixed. Alternatively, periodic boundary conditions may be employed.

Let us consider first a solution of the stationary wave type for a length L of a linear array of N atoms having interatomic separation a. The solutions must represent the sum of two identical waves travelling in opposite directions and as shown in Section 1.5.1 can therefore be of the form

$$y = (A \sin kx + B \cos kx) \exp(-i\omega t). \qquad \ldots(6.39)$$

Since the two end atoms of the array are fixed, the boundary conditions are

$$y = 0 \text{ when } x = 0;\ y = 0 \text{ when } x = (N-1)a = L \qquad \ldots(6.40)$$

The first condition gives $B = 0$ so that equation (6.39) becomes

$$y = A \sin kx \exp(-i\omega t) \qquad \ldots(6.41)$$

and, on substituting the second condition, the allowed values of k of

$$\frac{\pi}{L}, \frac{2\pi}{L}, \frac{3\pi}{L}, \ldots \frac{(N-2)\pi}{L}$$

are obtained. It may be noted from equation (6.41) that $k = 0$ and $k = (N-1)\pi/L = \pi/a$ do not allow motion of any of the atoms of the array, and, as discussed previously, larger values of k do not describe any additional different modes of vibration of the atoms. There are, therefore, a total of $(N-2)$ solutions and this corresponds to the number of atoms allowed to vibrate.

In applying the method of periodic boundary conditions travelling wave solutions are used together with the condition that any solution must be periodic over a sufficiently large number N of atomic spacings. If this repeat distance Na is equal to L then the periodicity condition can be written in the form

$$y(na) = y(na + L) = y[a(n+N)]. \qquad \ldots(6.42)$$

Inserting this condition in equation (6.29) it is readily verified that the allowed values of k are given by

$$\exp(ikL) = 1. \qquad \ldots(6.43)$$

This equation yields

$$k = 0, \pm\frac{2\pi}{L}, \pm\frac{4\pi}{L}, \ldots \pm\frac{N\pi}{L} \quad \text{for } N \text{ even}$$

and

$$k = 0, \pm\frac{2\pi}{L}, \pm\frac{4\pi}{L}, \ldots \pm(N-1)\frac{\pi}{L} \quad \text{for } N \text{ odd.}$$

It may be noted that, in accordance with the result of the previous section, the solutions $\pm N\pi/L = \pm\pi/a$, obtained for N even, correspond to a stationary wave, i.e. the waves travelling in opposite directions combine to form a single solution. The total number of solutions for either N odd or N even is, therefore, equal to N. The allowed values of k are spaced twice as far apart as for the stationary wave method, but, since both positive and negative k values are allowed, the total number of solutions for a given range of values of ω is the same as that for N particles free to oscillate using the stationary wave method. The solutions correspond to waves travelling in either direction along the linear array.

Summarising, the application of the boundary conditions to the problem leads to the result that only certain values of k are allowed, i.e. the ω vs. k curves shown in Fig. 6.4 consist of a large number of closely spaced points. The stationary wave and travelling wave solutions are illustrated in Fig. 6.5 for an array of seven

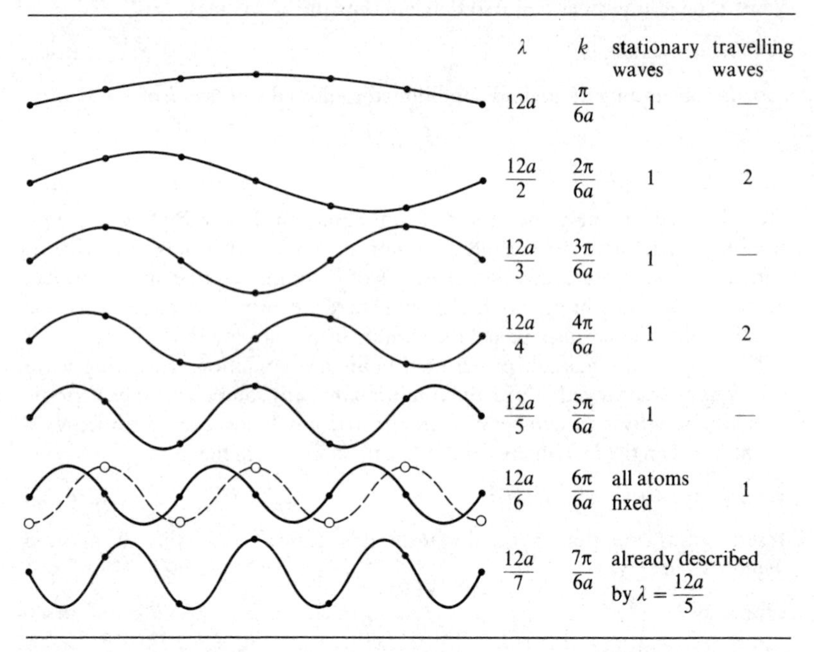

Fig. 6.5 Allowed stationary waves and travelling waves for a linear array of atoms.

atoms. The number of standing wave solutions is 5, corresponding to the number of atoms oscillating. For the travelling wave, the repetition grouping of atoms is 6. Two pairs of travelling wave solutions are obtained together with a solution for $k = 0$, which corresponds to all the atoms moving together in phase, and the solution for $k = \pi/a$ in which the movement of alternate atoms are exactly out of phase.

6.2.2 *Vibrational modes of a crystal*

The theory developed can be extended and applied to the solution of problems concerning real crystals. For a linear array of atoms both longitudinal and transverse waves are possible. The transverse vibrational modes can be represented by the superposition of two independent modes confined to different planes and this gives rise in a real crystal to two transverse branches which may or may not coincide.

In addition, longitudinal modes of vibration of the atoms will occur and the limiting velocity c_l for long wavelength longitudinal waves in a crystal will always differ from the transverse velocity c_t. As a consequence there will be separate longitudinal and transverse branches of the ω vs. k curves for a crystal, as is illustrated in Fig. 6.6. These branches are termed *acoustical modes* since, in the

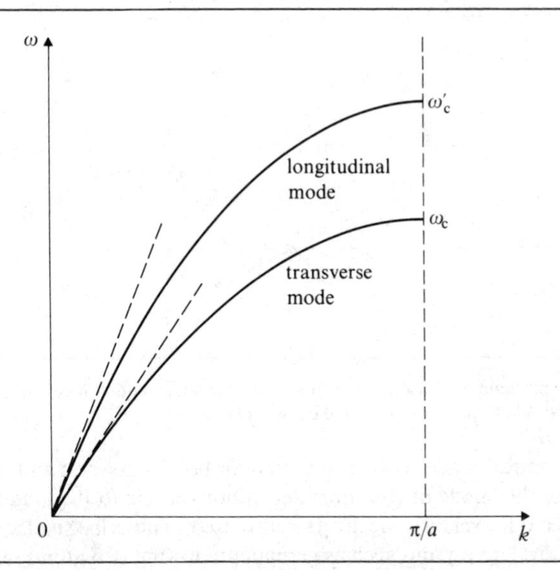

Fig. 6.6 Dispersion curves for the transverse and longitudinal modes for a linear array of atoms. The force constant f differs for transverse and longitudinal modes giving two different limiting velocities at $k = 0$ and two different limiting angular frequencies at $k = \pi/a$.

long wavelength limit, the waves travel in the crystal with the velocity of sound appropriate to the particular polarisation.

If, however, a linear array is considered in which atoms of two different masses are arranged alternately, it is found that, in addition to the transverse and longitudinal acoustical modes, vibrational modes are obtained in which the displacements of the dissimilar atoms are always opposite to each other. The displacement of atoms in transverse modes of the two types is illustrated in Fig. 6.7.

Transverse waves of the second type can be excited by long wavelength infrared radiation in ionic crystals since the two dissimilar atoms in the unit cell in the crystal carry opposite electric charges and will, therefore, be displaced in opposite directions in the transverse electric field of the electromagnetic radiation. The excitation is most pronounced for $k \to 0$ which we note from Fig. 6.8 corresponds to an angular frequency $\omega_0 \neq 0$. Infrared radiation of frequency ω_0 is strongly absorbed in the crystal and the mode of vibration of the crystal is

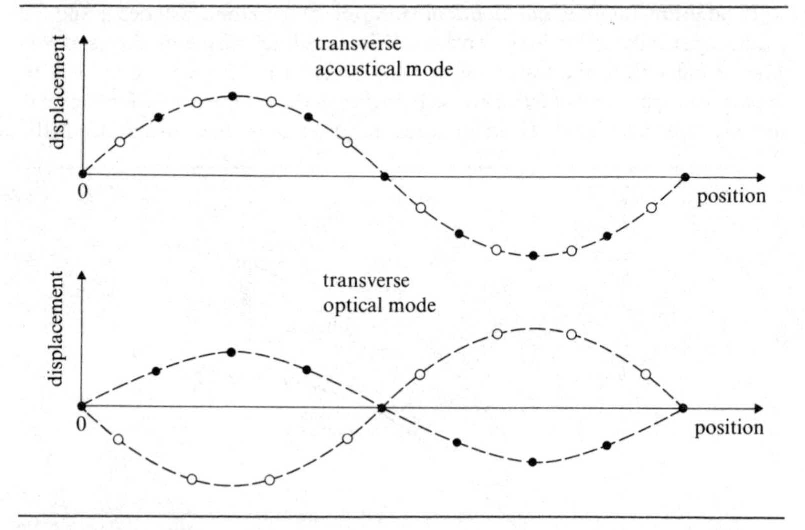

Fig. 6.7 Displacement of atoms in transverse acoustical and optical modes. The two different types of atoms are represented by • and ∘.

termed an optical mode. This term is used for both transverse and longitudinal branches of this mode of vibration, and is not specific to the long wavelength transverse mode excited optically. It is also used to describe similar vibrational modes in covalent crystals such as germanium having two atoms in a unit cell, although there is then no associated charge separation.

Referring again to Fig. 6.8 the acoustic and optical branches are seen to be separated by a forbidden band of frequencies. If the two atom masses are much different this band is wide and the optical branch is flat, i.e. it covers a narrow frequency range. In addition the presence of two types of atoms has doubled the effective periodicity of the lattice and has halved the width of the Brillouin zone.

A full discussion of the linear crystal with two types of particle is given in Chapter 3 of *Wave Mechanics of Crystalline Solids* by Smith (1963).

6.2.3 *Heat capacity from the theory of lattice vibrations*

As discussed in the previous sections the vibrational modes of a linear array consist of both transverses and longitudinal modes which obey a dispersion relationship of the form given by equation (6.34). This equation may be put in terms of the frequency ν, i.e.

$$\nu = \nu_c \left| \sin\left(\frac{ka}{2}\right) \right|. \qquad \ldots(6.44)$$

It was also shown that only certain specific values of k were allowed, and using the stationary wave model (and thus restricting k to positive values) these values

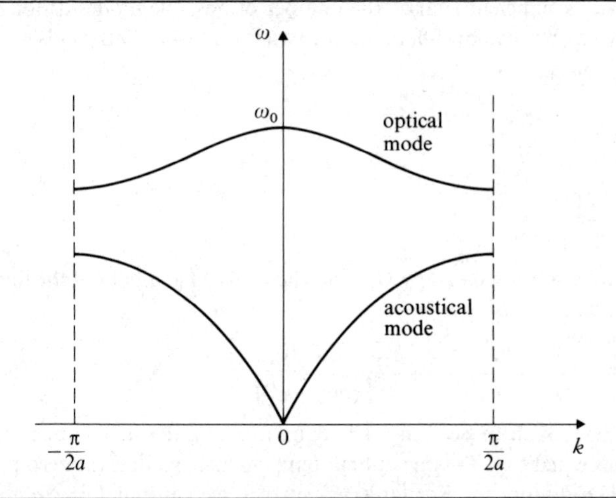

Fig. 6.8 Dispersion curves for the acoustical and optical modes for a linear array with alternate atoms of different masses. Note that the boundary of the 1st Brillouin zone now lies at $\pi/2a$ and that the limiting frequency for the optical branch is not zero for $k = 0$.

were shown to be given by $s\pi/L$, where s is integral. The possible vibrational modes of a given type are therefore given by the frequencies

$$\nu = \nu_c \sin \frac{s\pi a}{2L} \qquad [s = 1, 2, 3, \ldots (N-2)] \quad \ldots(6.45)$$

with a limiting upper frequency ν_c.

In order to derive the total energy of all the modes it is necessary to determine the mean energy of a mode of frequency ν and the number of modes in a given frequency range. Let us first confine attention to a single transverse mode of a particular frequency ν given in (6.45) for a specific value of s. The allowed energy values of this mode are restricted to the values

$$E = (n + \tfrac{1}{2})h\nu \qquad (n = 0, 1, 2 \ldots) \quad \ldots(6.7)$$

i.e. the mode may be in its lowest energy state of energy $\frac{1}{2}h\nu$, or in one of a number of excited states corresponding to the different values of n. The mean energy $\bar{E}$ of a mode is a function of temperature and its value may be determined using the Boltzmann law and following a procedure similar to that used in Section 6.1.1 for calculating the mean energy of an oscillator. The value of $\bar{E}$, which has already been quoted in 6.1.2, is

$$\bar{E} = \tfrac{1}{2}h\nu + \frac{h\nu}{\exp(h\nu/kT) - 1}. \quad \ldots(6.28)$$

Each value of s in equation (6.45) defines a different mode so that provided

$L\,(=Na)$ is sufficiently large, the number of possible modes $\mathrm{d}s$ for a frequency $\mathrm{d}\nu$ may be obtained by differentiating equation (6.45). This yields

$$\mathrm{d}s = \frac{2L\,\mathrm{d}\nu}{\pi a \nu_c \cos\dfrac{s\pi a}{2L}} \qquad \ldots(6.46)$$

$$= \frac{2L\,\mathrm{d}\nu}{\pi a(\nu_c^2 - \nu^2)^{\frac{1}{2}}}. \qquad \ldots(6.47)$$

The total vibrational energy U_s of all the transverse modes of the linear array is then given by

$$U_s = \int_0^{\nu_c} \left[\frac{h\nu}{\exp(h\nu/kT)-1} + \frac{h\nu}{2}\right] \frac{2L\,\mathrm{d}\nu}{\pi a(\nu_c^2 - \nu^2)^{\frac{1}{2}}}. \qquad \ldots(6.48)$$

As we have seen in Section 6.1.1 (equation 6.6), the first term in the bracket becomes equal to $(kT - \frac{1}{2}h\nu)$ at high temperatures so that the zero point energy term then disappears. For large values of L we can put $L = Na$ and the high temperature value of U_s is therefore

$$U_s = \int_0^{\nu_c} \frac{2NkT}{\pi} \frac{\mathrm{d}\nu}{(\nu_c^2 - \nu^2)^{\frac{1}{2}}} = NkT. \qquad \ldots(6.49)$$

If two sets of transverse modes and the additional longitudinal mode are included as discussed in Section 6.2.2 one obtains

$$U_s = 3NkT. \qquad \ldots(6.50)$$

Putting N equal to the Avogadro number N_A the limiting value of $3R$ is obtained for the heat capacity at high temperatures.

An alternative approach is to treat the lattice vibrations as an assembly of particles, the quantum of vibrational energy $h\nu$ being termed a *phonon*. Thus the change in energy for a particular vibrational mode of frequency ν when the value of n in equation (6.7) increases or decreses by 1 corresponds respectively to the absorption or emission of a phonon of energy $h\nu$. The nth excited state of a mode of frequency ν can then be represented by n phonons each of energy $h\nu$. The statistics governing these particles must therefore not place a restriction on the number occupying a given energy state, i.e. the exclusion principle does not apply to phonons. The appropriate statistics for phonons are the Bose–Einstein statistics according to which the probability of occupation $f(E)$ of a state of energy E is given by

$$f(E) = \frac{1}{\exp[(E - E_B)/kT] - 1}. \qquad \ldots(6.51)$$

The energy E_B is determined by the number of particles N in a system. When N is very large $E_B = E_0$ where E_0 is the lowest energy of the system. We may now put

$$E - E_0 = h\nu \qquad \ldots(6.52)$$

as the energy of a phonon, and equation (6.51) then gives the probability of a phonon of energy $h\nu$ being excited. The mean energy $\bar{E}$ of phonons of this frequency is obtained as the product of the phonon energy and its excitation probability so that

$$\bar{E} = \frac{h\nu}{\exp(h\nu/kT)-1}. \qquad \ldots(6.53)$$

A zero point energy of $h\nu/2$ will, in addition, be associated with the vibrational mode of this frequency so that essentially we have the same result as (6.28) for the mean energy of the mode.

Summarising, it may be seen that the lattice vibrations can be treated either as quantised waves or as a system of identical particles termed phonons. A similar duality can be shown to hold in the case of electromagnetic radiation between electromagnetic waves and photons. The frequency spectrum of the lattice vibrations was first calculated by Blackman. For a real crystal the interatomic spacings will differ in different crystallographic directions and the problem is essentially one of determining the variation of ω with k for a number of different crystal directions. Theory shows that the frequency distributions of the vibrational modes for a real crystal may consist of a number of sharp peaks instead of relationship (6.47) deduced for a linear lattice. The form of the distributions will depend on the type of crystal lattice. These have given, in a number of cases, C_V values in good agreement with experiment and have been used, for example, to explain the observed experimental temperature variation of the Debye temperature Θ_D for sodium chloride. It is also possible to determine the frequency distribution of the vibrational modes of a solid by experiment. This is achieved by measuring the variation of ω with k for a number of crystal directions using neutron scattering techniques. It should, however, be noted that the details of the frequency distributions cannot be inferred from heat capacity data. A comparison of the frequency distributions obtained in the different theories is given in Fig. 6.9.

6.3 The electronic component of the heat capacity of a metal

As discussed in Section 5.4.2 classical theory predicted a contribution of $\frac{3}{2}R$ to the molar heat capacity of a metal due to the free electrons. It is found by experiment, however, that the heat capacities of metals and insulators at high temperatures are not significantly different. We may now enquire what contribution to the heat capacity, if any, is predicted on the quantum theory by the free electrons in a metal.

Qualitatively the picture is as follows. When the temperature of a metal is changed there is a redistribution of electrons between states which are within $\approx 5kT$ of the Fermi level E_F. Since $kT \ll E_F$ at ordinary temperatures these electrons, which are able to receive thermal energy, form only a small proportion

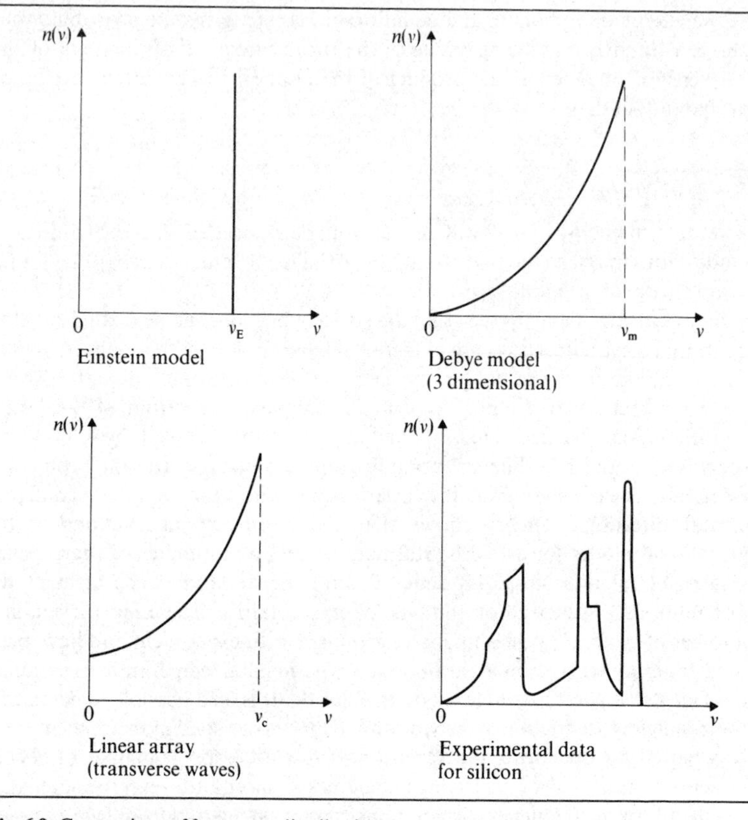

Fig. 6.9 Comparison of frequency distributions.

of the total number of free electrons. Electrons occupying states of lower energy cannot receive thermal energy since to do so they would have to move to states of higher energy and these are *already occupied.* Interchange of electrons between two occupied states does not change the system and has no physical significance. It follows that the only contribution to the heat capacity should come from the small proportion of the electrons occupying states close to the Fermi level.

In order to calculate the heat capacity of the electron gas we may use the nomenclature and results of Chapter 5 and proceed as follows.

The number of electrons in volume V having energies between E and $E+\mathrm{d}E$ is $S(E)f(E)\,\mathrm{d}E$ so that the total energy U_n of the electron gas is given by

$$U_n = \int_0^\infty E\,S(E)f(E)\,\mathrm{d}E \qquad \ldots(6.54)$$

$$= \frac{1}{2\pi^2}\left(\frac{2m_e}{\hbar^2}\right)^{\frac{3}{2}} V \int_0^\infty \frac{E^{\frac{3}{2}}}{\exp[(E-E_F)/kT]+1}\,\mathrm{d}E \qquad \ldots(6.55)$$

using equations (5.41) and (5.54). This may be shown to yield on integration

$$U_{\rm n} = \tfrac{3}{5}nE_{\rm F}(0)\left[1+\tfrac{5}{12}\pi^2\left(\frac{kT}{E_{\rm F}(0)}\right)^2\right], \qquad \ldots(6.56)$$

where n is the number of free electrons per unit volume. The factor $\frac{3}{5}nE_{\rm F}(0)$ is the internal energy per unit volume of the electron gas at absolute zero. The heat capacity per unit volume is given by the change of internal energy with temperature so that we have

$$C = \frac{{\rm d}U_{\rm n}}{{\rm d}T} = \frac{\pi^2}{2}n\frac{k^2T}{E_{\rm F}(0)}. \qquad \ldots(6.57)$$

The molar electronic heat capacity C_V will be given by the same expression but with n replaced by the Avogadro number $N_{\rm A}$. We therefore obtain

$$C_V = \frac{\pi^2}{2}R\frac{T}{T_{\rm F}}, \qquad \ldots(6.58)$$

where $E_{\rm F}(0)$ has been replaced by $kT_{\rm F}$. $T_{\rm F}$ is the Fermi temperature and metals at room temperature have values of $T_{\rm F}/T$ ranging typically between 50 and 300. The classical molar electronic heat capacity of $\frac{3}{2}R$ is thus reduced by a factor $\sim\frac{1}{3}T_{\rm F}/T$ so that at ordinary temperatures the electronic heat capacity is small compared with the lattice heat capacity of $3R$.

At low temperatures the Debye theory predicts a T^3 variation of the lattice heat capacity so that at such temperatures the total heat capacity C_V is given by

$$C_V = \alpha T + \beta T^3, \qquad \ldots(6.59)$$

where α and β are constants given by equations (6.58) and (6.27), respectively.

At a very low temperature, usually below 4 K, the first term in equation (6.59) becomes dominant and an experimental test of the theory is then possible. It is found that agreement between theory and experiment is good in the case of the alkali metals sodium, potassium and lithium, and that it may be further improved by the replacement of the free electron mass $m_{\rm e}$ in equation (6.55) by an effective electron mass $m_{\rm e}^*$. The concept of effective electron mass will be developed in Chapter 7. We note here that the values of $m_{\rm e}^*$ are different for each metal and differ from the free electron mass $m_{\rm e}$ by a numerical factor. For the transition metals iron, cobalt, nickel and manganese, however, the experimental values of the constant α exceed those given by equation (6.58) by factors greater than 10. The transition metals have incompletely filled electron shells below the outer shell and the behaviour of these electrons cannot be represented adequately by the free electron theory.

6.4 The thermal expansion of solids

Crystalline solids, with the exception of very few such as invar, expand on heating and have expansion coefficients lying in the range 10^{-4}–10^{-6} per $\rm K^{-1}$.

The measured macroscopic expansion is a manifestation of an expansion of the crystal lattice with increased temperature. The expansion coefficient for a solid will be related to the type of bonding exhibited by the crystal and hence also to its heat capacity.

To investigate the change in interatomic spacing one may recall from Chapter 4 the form of the relationship between the energy of a crystal and the interatomic spacing, namely

$$E(r) = -\frac{A}{r^n}+\frac{B}{r^m} \qquad \ldots(4.1)$$

i.e. a balance between a negative attractive energy term and a positive repulsive energy term. Let us first consider the situation at 0 K. Referring to Fig. 6.10 the

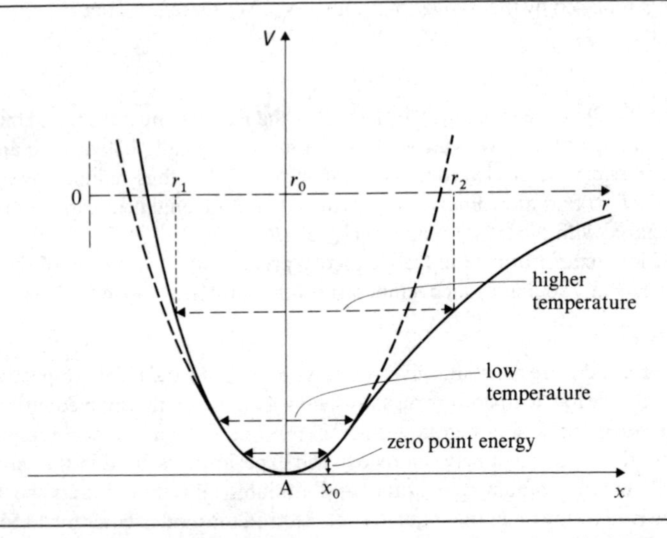

Fig. 6.10 The vibrational energy of the atoms of a solid as a function of the displacement x of the atoms from their equilibrium separation r_0.

equilibrium spacing of the atoms at 0 K corresponds to the minimum of the energy curve. The atoms of the solid will be in their lowest vibrational energy state, corresponding to the zero point energy $h\nu/2$ and, as illustrated, will possess a minimum amplitude of vibration x_0.

The energy of the solid at a higher temperature, as we have seen, can be represented on the Einstein model in terms of a single characteristic frequency ν_E and the atoms will then possess a displacement amplitude $> x_0$. At comparatively low temperatures the displacements are small and the vibrational motion may be assumed to be simple harmonic. Consequently, the potential energy V of the atom vibrations will be related to the displacement x from their mean position by

$$V = cx^2, \qquad \ldots(6.60)$$

where c is a constant. This parabolic relationship is shown dotted in Fig. 6.10, V and x being referred to the point A as origin. The increased displacement with temperature rise should be noted. However, the *mean* positions of the atoms remain unchanged, which is in accord with the experimental observation that the thermal expansion coefficients of all solids approach zero as $T \to 0$ K.

At higher temperatures the atomic displacements are larger and the vibrational potential energy V may then be described by the modified expression

$$V = cx^2 - gx^3 - fx^4, \qquad \ldots(6.61)$$

where f and g are constants. The second and third terms describe the deviation from simple harmonic behaviour and are called the anharmonic terms. Their individual effect in providing a vibrational energy consistent with the curve shown in the figure should be considered. Referring to Fig. 6.10 it is apparent that the mean atomic spacing $\frac{1}{2}(r_1 + r_2)$ at such temperatures exceeds r_0, i.e. thermal expansion of the lattice is occurring. A calculation of thermal expansion based on equation (6.61) is given by Kittel (1971) in Chapter 6 of *Introduction to Solid State Physics*, and the experimental proportionality with temperature is confirmed.

Finally it may be noted that strong binding corresponds to a rather sharp minimum in the energy curve and therefore referring to the figure one should then expect a small expansion coefficient. Thus diamond exhibiting very strong covalent bonding possesses a coefficient of expansion about 40 times less than that of the ionic crystal NaF.

6.5 The thermal conductivity of solids

The conduction of heat in solids occurs by two processes. Lattice vibrations conduct heat in all solids; free electrons may also contribute to the heat conduction, and in many metals the electronic component may be dominant yielding thermal conductivities which are one or two orders of magnitude greater than that of insulators. However, for semimetals such as, for example, bismuth and also in alloys the contribution of the lattice conductivity is important. It may be noted that the ratio of the thermal conductivities of metals to insulators is much less than the corresponding ratio of the electrical conductivities.

6.5.1 *Lattice heat conduction*

Let us first consider the lattice conduction process. The qualitative picture of heat conduction by the lattice vibrations is simple. If one part of a solid is maintained at an elevated temperature this results in an increased amplitude of the vibration modes of the atoms within that region of the solid. These more energetic vibrations are then gradually transmitted to adjacent cooler regions in the solid, so that conduction of *heat* occurs along the temperature gradient.

A quantitative theory of thermal conductivity was first put forward by Debye (1914). As in his earlier theory of the heat capacity of solids he considered thermal motion as an assembly of all the possible atomic displacements represented by a system of waves in a continuous medium. If the lattice vibrations were regarded simply as sets of simple harmonic trains of waves, then such trains of waves could be shown to pass freely across each other without modification, i.e. there would be no scattering of the waves. In order to account for thermal resistivity it is necessary to consider how lattice waves may interact with each other so that heat may be transferred. It is therefore necessary to consider the anharmonic terms in the lattice potential energy and the associated anharmonic atomic displacements. These displacements are such as to give rise to local variations in density, and owing to the dependence of the wave propagation velocity on density will therefore have the property of deflecting lattice waves. The anharmonic terms are providing a source of coupling between the lattice waves and give rise to mutual scattering of the waves. If the process is treated as one of collisions of a phonon gas one may note that such a mechanism is also necessary in order to achieve at any temperature the thermal equilibrium distribution of phonons of different frequencies.

By analogy with a classical gas we may, as in Section 5.4.3, put the thermal conductivity K of the phonon gas as

$$K = \tfrac{1}{3}Cl\bar{v}, \qquad \ldots(6.62)$$

where C is the heat capacity referred to unit volume of the solid, l the mean free path of the phonons and where for $\bar{v}$ the mean velocity of propagation of elastic waves in the solid may be used. Experimentally determined values of K, C and $\bar{v}$ allow the mean free path l of the phonons to be calculated from equation (6.62). Values of $l \sim 10^{-9}$ m are obtained for glasses, i.e. amorphous solids, which values are close to the mean separation of the molecules in these substances.

A theory of phonon–phonon collisions was developed by Peierls (1929). The wave–particle duality in the behaviour of lattice waves has already been mentioned and the particle aspects of phonons is analogous to those of photons and electrons. A phonon of frequency ν can thus be ascribed energy $h\nu\,(=\hbar\omega)$ and a momentum $\hbar\boldsymbol{k}$ where $\boldsymbol{k}$ is the wave vector of the phonon. According to the theory two phonons of frequencies ν_1 and ν_2 may collide to give a third phonon of frequency ν_3 provided that

$$h\nu_1 + h\nu_2 = h\nu_3 \qquad \ldots(6.63)$$

i.e. there is conservation of energy in the process. The corresponding equation for the conservation of momentum is represented by

$$\hbar\boldsymbol{k}_1 + \hbar\boldsymbol{k}_2 = \hbar\boldsymbol{k}_3, \qquad \ldots(6.64)$$

where $\boldsymbol{k}_1$, $\boldsymbol{k}_2$ and $\boldsymbol{k}_3$ are the wave-vectors of the three phonons. In conjunction with equation (6.63) this equation implies that the energy flow is neither changing in magnitude or direction and such collision processes cannot therefore provide any resistance to the flow of heat.

Peierls proposed, therefore, that during the phonon interaction process the lattice as a whole could absorb or donate momentum. The momentum equation is then modified to

$$\hbar \boldsymbol{k}_1 + \hbar \boldsymbol{k}_2 = \hbar \boldsymbol{k}_3 + \hbar \boldsymbol{G}, \qquad \ldots(6.65)$$

where $\boldsymbol{G}$ is a vector of the reciprocal lattice. The exchange of momentum with the lattice may be compared with that occurring during the coherent elastic scattering of X-ray photons and equation (6.65) should be compared with the Bragg reflection condition (3.29). It may also be noted that a relationship similar to (6.65) is obtained when the scattering of electrons by phonons is considered. The process described by equation (6.65) is termed an *umklapp* (i.e. a reversal) process, and, in contrast with the normal process of equation (6.64), is one which gives rise to thermal resistance.

It is of interest to examine what the theory predicts for the variation of thermal conductivity with temperature. We note from equation (6.51) that at high temperatures the probability of a phonon of frequency ν being excited is proportional to $kT/h\nu$. This is valid for phonons of all frequencies and consequently the density of phonons is proportional to temperature. It may also be shown that the mean free path of a phonon is inversely proportional to the density of all other phonons with which it is interacting. We thus have $l \propto 1/T$. Since the heat capacity C_V is a constant at high temperatures (e.g. for $T \gg \Theta_D$) we have the result that $K \propto 1/T$ at high temperatures. This result is in accord with experiment for many solids.

At rather lower temperatures the phonon spectrum is modified so that for a solid below the Debye temperature Θ_D, long wavelength phonons predominate, namely those with $\boldsymbol{k}$ values much less than $\boldsymbol{G}$. Peierls argues that for the *umklapp* process to be effective both the incident phonons must have wave-vectors $\sim \boldsymbol{G}/2$. He therefore puts the threshold excitation energy of the phonons as $k\Theta_D/2$ (corresponding to $h\nu_m/2$ on the Debye theory), and from equation (6.51) it then follows that the number of excited phonons will be proportional to $\exp(2T/\Theta_D)$. Proceeding as before we obtain

$$K \propto l \propto \exp\left(\frac{\Theta_D}{2T}\right). \qquad \ldots(6.66)$$

Fair agreement with experiment is found for an equation of this form although modified values of Θ_D are usually required. At low temperatures the mean free path l becomes large and will eventually become comparable with the dimensions of a sample and be limited by them. In accordance with equation (6.62) the thermal conductivity K will then be proportional to the smallest dimension of the sample and to the heat capacity. A dependence on T^3 is thus obtained. It may also be noted that phonons can also be scattered by impurity atoms, vacancies, dislocations and grain boundaries. Such defects in a real crystal will also provide a limit for the mean free path at low temperatures.

6.5.2 *Electronic heat conduction*

A metal will normally have a thermal conductivity possibly exceeding that of an insulator by an order of magnitude. It is concluded, therefore, that the process of heat conduction in metals is primarily due to the electrons and that phonon transport effects are much smaller. When, however, a metal is impure, its thermal conductivity is much reduced and phonon transport may then be dominant particularly at low temperatures.

The electrical conductivity of metals was discussed in Chapter V in terms of the free electron theory of metals. We may now consider what contribution may be expected to the heat conduction by the free electrons in a metal.

As we have seen in Section 5.10 only electrons having energies close to the Fermi level can take part in collisions and the relaxation time τ_F of these electrons governs the drift velocity attained by the whole assembly of free electrons in the metal. When deriving the thermal conductivity of a metal one is concerned with the mean free path of the electrons between collisions, and it is therefore appropriate to use a mean free path l defined by

$$l = \tau_F u_F, \qquad \ldots(6.67)$$

where u_F is the velocity of an electron at the Fermi level. Following classical kinetic theory of gases the thermal conductivity K of the electron gas is given by

$$K = \tfrac{1}{3} C l u_F, \qquad \ldots(6.68)$$

where C is the electronic heat capacity referred to unit volume. Using the value of C given by equation (6.57) and that of l from equation (6.67) we have

$$K = \frac{\pi^2}{6} n \frac{k^2 T}{E_F(0)} \tau_F u_F^2. \qquad \ldots(6.69)$$

Putting

$$E_F(0) = \tfrac{1}{2} m_e u_F^2 \qquad \ldots(6.70)$$

the thermal conductivity is obtained in the form

$$K = \frac{\pi^2}{3} \frac{nk^2 T}{m_e} \tau_F. \qquad \ldots(6.71)$$

In Section (5.10) it was noted that τ_F is inversely proportional to T so that equation (6.71) confirms the tendency for metals at ordinary temperatures to show little variation of thermal conductivity with temperature.

A comparison of the electrical and thermal conductivities of metals is of interest. Whereas both processes are due to electron transport they rely on different aspects, namely flow of charge and flow of energy. Referring to equation (5.64) for electrical conductivity and combining with equation (6.71) the well known Wiedmann–Franz law is obtained:

$$\frac{K}{\sigma T} = \frac{\pi^2}{3} \left(\frac{k}{e}\right)^2 = L$$
$$= 2.45 \times 10^{-8}\ \mathrm{V^2\,K^{-2}}. \qquad \ldots(6.72)$$

The constant L is termed the Lorenz number and it involves only fundamental constants. As may be confirmed from Table 5.1 many metals at room temperature yield a value of L close to this calculated value. Agreement is also obtained at higher temperatures whilst at low temperatures a decreased value is obtained.

The value of the relaxation time τ_F for the electron collisions may be derived from experimental data on thermal conductivity (from equation (6.71)) or electrical conductivity (from equation (5.64)). Values $\sim 10^{-14}$ s are typical.

Finally, we may briefly consider the main features of the scattering process for electrons in metals. The dominant mechanism at ordinary temperatures is that due to lattice waves or phonons. As discussed in Section 6.5.1 the lattice waves provide local variations in the spacing of the atoms on the lattice. Electron waves in a metal have a much higher velocity than that of the lattice waves, e.g. electrons at the Fermi level have $u_F \sim 1.5 \times 10^6$ m s^{-1} as compared with 5×10^3 m s^{-1} for lattice waves. The lattice waves are providing a stationary periodic positional displacement of the ions on the lattice for the moving electrons. Such irregularities give rise to the isotropic scattering of electrons, dominant at ordinary temperatures, and governing the electron mean free paths. We note that collisions with ion cores is not a viable mechanism for the observed scattering. In terms of a phonon representation, the phonon density at high temperatures is proportional to the temperature and a dependence of τ_F on $1/T$, already quoted, is obtained. At low temperatures the scattering caused by impurities will predominate and a thermal conductivity proportional to T is obtained.

References and further reading

Blakemore, J. S. (1969) *Solid State Physics*, Saunders.
Dekker, A. J. (1967) *Solid State Physics*, Macmillan.
Girifalco, L. A. (1973) *Statistical Physics of Materials*, Wiley.
Hart-Davis, A. (1975) *Solids, an Introduction*, McGraw-Hill.
Kittel, C. (1971) *Introduction to Solid State Physics*, Wiley.
Rosenberg, H. M. (1975) *The Solid State*, Clarendon.
Smith, R. A. (1963) *Wave Mechanics of Crystalline Solids*, Chapman and Hall.
Zemansky, M. W. (1968) *Heat and Thermodynamics*, McGraw-Hill.
Ziman, J. M. (1972) *Principles of the Theory of Solids*, Cambridge University Press.

Problems

6.1 Derive the mean energy of an oscillator, as given in equation (6.4) from expression (6.3).

(*Hint:* The sum of the geometrical progression $1 + x + x^2 + \ldots = (1 - x)^{-1}$).

6.2 Show that the Einstein theory yields the expression

$$C_V = 3R\left(\frac{\Theta_E}{T}\right)^2\left[\frac{\exp\left(\frac{\Theta_E}{T}\right)}{\left(\exp\left(\frac{\Theta_E}{T}\right)-1\right)^2}\right]$$

for the molar heat capacity of a solid.

6.3 Determine for the Einstein model the variation of C_V with Θ_E/T for the range of Θ_E/T from 0 to 50 and find the range for which the low temperature approximation stated in the text is valid. Confirm that the derived values of C_V agree with those given in Fig. 6.2.

6.4 The molar heat capacity C_V of silver at 100 K is 20.2 J K^{-1} mol^{-1}. Using the values derived in the previous question determine the Einstein characteristic temperature for silver. Determine, from data given in the text, the value of the Debye temperature Θ_D which gives the best fit to the experimental value of C_V at 100 K. Why should Θ_E be always lower than Θ_D?

6.5 Show that the value of Θ_D for silver derived in Problem 6.4 gives good agreement with the experimental value for C_V of 1.67 J K^{-1} mol^{-1} at 20 K whereas the value obtained for Θ_E yields a C_V value much lower than is observed experimentally.

6.6 Evaluate the integral in equation (6.19) and show that the Debye temperature Θ_D is given by

$$\Theta_D = \frac{h}{k}\left(\frac{9n}{B}\right)^{\frac{1}{3}},$$

where n is the number of atoms per unit volume.

6.7 The velocities of longitudinal and shear waves in aluminium are 6374 and 3111 m s^{-1}, respectively. Determine Θ_D and the limiting frequency ν_m for aluminium given that there are 6.02×10^{28} atoms of aluminium in a cubic metre.

The molar heat capacity C_V for aluminium at 100 K is 13.0 J K^{-1} mol^{-1}. Using data from the text determine the value of Θ_D which yields this value of C_V.

6.8 The following are values of the molar heat capacity C_V of tungsten at low temperatures

T/K	1	2	3	4	10
C_V/10^{-3} J K^{-1} mol^{-1}	1.36	2.90	4.82	7.23	4.30

Determine the values of the constants α and β in equation (6.59) which gives the best fit to this data and calculate the corresponding value of the limiting low temperature value for Θ_D. Assuming tungsten obeys the temperature variation given by the Debye theory, determine the value of C_V at 100 K for this Θ_D. What Θ_D would be needed to give the observed value of 16.3 J K^{-1} mol^{-1} at this temperature?

6.9 The molar heat capacity C_V of copper has the following values at the given temperatures

T/K	1	2	3	4	10
C_V/10^{-3} J K^{-1} mol^{-1}	0.76	1.8	3.4	5.8	55
T/K	15	20	50	100	300
C_V/J K^{-1} mol^{-1}	0.172	0.49	6.3	16.1	23.9

Show that for $T \leqslant 15$ K this data can be represented by equation (6.59). Determine the values of α, β and the limiting value of Θ_D and confirm that the electronic component of the heat capacity is dominant at very low temperatures. What value of Θ_D better represents the temperature variation of C_V at higher temperatures?

6.10 Given that the velocities of longitudinal and shear waves in copper are 4759 and 2325 m s^{-1} at room temperature, determine a value for Θ_D for copper. Determine also the value of the Fermi temperature T_F for copper and the constant α in equation (6.59) given that the electron density in copper is 8.45×10^{28} m^{-3}. Use these values of α and Θ_D to derive the variation of C_V for the temperature range up to 300 K and compare with the experimental values given in the previous question.

6.11 Assuming that transverse and longitudinal waves in a solid have a common velocity show that the minimum cut-off wavelength on the Debye theory is given by

$$\lambda_{min} = \left(\frac{4\pi V}{3N}\right)^{\frac{1}{3}}.$$

6.12 If in the Debye theory a minimum cut-off wavelength is defined, as for example in the previous problem, different limiting frequencies are found if the longitudinal and transverse waves have different velocities. Derive expressions for these limiting frequencies in terms of c_t and c_l. Hence show that the molar heat capacity C_V at low temperature is given by

$$C_V = \tfrac{4}{5}\pi^4 R\left[\left(\frac{T}{\Theta_l}\right)^3 + 2\left(\frac{T}{\Theta_t}\right)^3\right],$$

where Θ_l and Θ_t are modified Debye temperatures corresponding to the two modes.

6.13 Derive an expression for the ratio of the displacements of successive atoms on a linear array of spacing during passage of a longitudinal wave. Show that this ratio is not altered when k is either increased or decreased by $2\pi n/a$, where n is integral.

What is the significance of this result?

6.14 Show that for the dispersion relation in equation (6.34) the group velocity u is given by

$$u = v_0 \cos\frac{ka}{2}$$

Interpret this result for $k = 0$ and $k = \pm\pi/a$.

6.15 Use the method of periodic boundary conditions as in Section 5.7 to show that for travelling waves of the form

$$\psi = A\exp(i\boldsymbol{k}\cdot\boldsymbol{r})$$

the number of possible waves N_s for $\boldsymbol{k}$ values up to k is given by

$$N_s = \frac{k^3}{6\pi^2}V,$$

where V is the volume of the crystal. Hence confirm the result in equation (6.17) for the number of waves in the frequency range ν to $\nu + d\nu$.

7. Energy bands

7.1 Introduction

Although some metallic properties can be explained well in terms of free electron theories, detailed comparison of the theoretical predictions with experimental results show many discrepancies. The behaviour of charge carriers in a solid under the combined influence of electric and magnetic fields as in Hall effect measurements (see Section 8.4.1) may be taken as illustrative. For the alkali metals sodium and potassium there is very good agreement between the experimental values of the Hall constant and the theoretical values which are based on the assumption of one free electron per atom. However, for other metals, such as cadmium and iron, an explanation of the experimental data requires that the charge carriers be positive rather than negative. The large values of Hall constant observed for semimetals, such as arsenic and bismuth, and for semiconductors cannot be explained on the free electron theory; neither can the dependence of the sign and magnitude of the constant on the impurity content of samples as is found with semiconductors. Such experimental data require a substantially different approach for their interpretation.

The main weakness of the free electron model lies in its inability to explain why some solids are metals, some are insulators and some are semiconductors and how the differences in conductivity between these solids arise. These differences are large, e.g. as compared with copper, which has a conductivity of 5.9×10^7 $\Omega^{-1}\ \mathrm{m}^{-1}$ at room temperature, the semiconductor germanium has conductivity $2\ \Omega^{-1}\ \mathrm{m}^{-1}$ and diamond, which is an insulator, of only $\sim 10^{-6}\ \Omega^{-1}\ \mathrm{m}^{-1}$.

In the free electron theory the valence electrons in a metal are considered to move in a field-free region at a uniform potential and no account is taken of the array of ions on the lattice of the crystal. Surrounding each ion there will, however, be an electric field and the array of ions will give rise to a periodic variation in electrostatic potential. A travelling electron wave in the crystal will thus be subject to a periodic variation in its potential energy and the form of this variation will differ according to its direction in the crystal. Both the form and periodicity of this lattice potential will be governed by the type of atom and the symmetry of the crystal. There will thus be substantial differences between one solid and another and as we shall see this underlies the observed differences in properties. The influence of atomic structure on the bonding between atoms has been discussed in Chapter 4 and an approach to band theory based on consideration of free atoms will be developed first.

7.2 The free atom approach to energy bands

In studying the interactions between adjacent atoms one clear approach is to set up a simple model of an isolated atom so as to give electron wave-functions of known energy and then to see how such a model can be modified to allow for interactions between atoms. One possible model is that of the particle in the box developed in Section 5.6.1, which, as we have seen, yielded a range of possible electron wave-functions representing bound states having different energies. Another very successful model is the wave-mechanical model used in Chapter 2 for the hydrogen atom and a qualitative extension of this model was used in Section 4.4.1 to explain the nature of the bonding in the hydrogen molecule ion.

On the wave-mechanical model, the ground state of the hydrogen atom is represented by a wave-function which falls off exponentially from the nucleus. This was illustrated in Fig. 2.11. A one-dimensional model which retains this feature will now be developed and will then be extended to show the effect of interaction between adjacent atoms. The model, termed the delta well model, resembles in many respects the particle in the box model and the method of approach to its solution is similar.

The one dimensional well of width l and of depth of potential $-V$ is shown in Fig. 7.1. It has the property that the product

$$\delta = -Vl \qquad \ldots(7.1)$$

remains finite and positive as $V \to -\infty$ and $l \to 0$. A solution of the one-dimensional, time-independent, Schrödinger wave equation

$$\frac{\hbar^2}{2m_e}\frac{d^2\psi}{dx^2} + (E - V)\psi = 0 \qquad \ldots(1.104)$$

is required for an electron bound to the well. Only a solution for which $E < 0$ will conform to this condition. One valid form of solution can be expressed as

$$\psi = A\exp\alpha x + B\exp(-\alpha x), \qquad \ldots(7.2)$$

where A, B and α are constants. Since ψ must remain finite everywhere and since $V = 0$ for the regions on either side of the well the solution for these regions can be written as

$$\psi_1 = \psi_0\exp(\alpha x); \quad x < 0$$

$$\psi_2 = \psi_0\exp(-\alpha x); \quad x > 0 \qquad \ldots(7.3)$$

provided that $l = 0$.

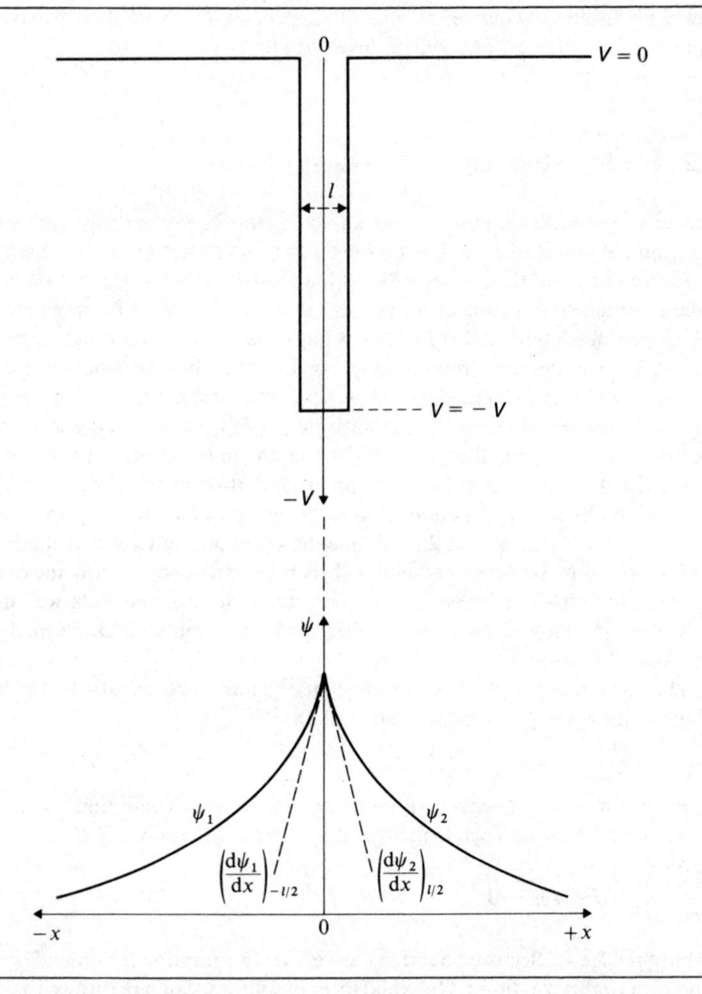

Fig. 7.1 The delta well model for a bound electron. Note that the wave-function is shown for $l \to 0$.

The constant α is necessarily positive and the equations given conform with the condition that

$$\psi_1 = \psi_2 = \psi_0 \qquad \text{...(7.4)}$$

at the centre of the well at $x = 0$.

The value of the constant α is determined by looking at the form of the wave-function at the well itself. At this point E is negligible compared with V so that the wave-function is then a solution of

$$\frac{\hbar^2}{2m_e}\frac{d^2\psi}{dx^2} = V\psi. \qquad \ldots(7.5)$$

Integrating this equation over the region of the well we have

$$\int_{-l/2}^{l/2} \frac{d^2\psi}{dx^2}\,dx = \frac{2m_e}{\hbar^2}\int_{-l/2}^{l/2} V\psi\,dx \qquad \ldots(7.6)$$

giving

$$\left(\frac{d\psi_2}{dx}\right)_{l/2} - \left(\frac{d\psi_1}{dx}\right)_{-l/2} = \frac{2m_e}{\hbar^2}Vl\psi_0$$

$$= \frac{-2m_e}{\hbar^2}\,\delta\,\psi_0 \qquad \ldots(7.7)$$

This abrupt change in the slope of the wave-function which occurs at the well is also, from equation (7.3), given by $-2\alpha\psi_0$ so that we have

$$\alpha = \frac{m_e}{\hbar^2}\delta. \qquad \ldots(7.8)$$

This is a measure of the strength of the well and it determines both the form of the wave-function of equation (7.3) and also its energy. The latter is determined by noting that for the region outside the well $V = 0$, so that from equations (1.104) and (7.3) we have

$$E = -\frac{\hbar^2}{2m_e}\alpha^2$$

$$= -\frac{m_e}{2\hbar^2}\delta^2. \qquad \ldots(7.9)$$

In contrast to the other models examined, the delta well model yields only a single solution for an electron bound to the well. For values of $E > 0$ other types of solution are, however, obtained which give a continuous distribution of values of E and which are referred to as free states. We shall not be concerned with solutions of this form.

Let us now investigate the effect on the wave-functions of bringing together two atoms represented as delta wells. As illustrated in Fig. 7.2 when the atoms are well separated there can be little interaction between them, the wave-functions are unmodified and the energies of both wave-functions are equal. Changing the sign of one wave-function has no significance since neither the potential nor the kinetic energy of the electron is changed. When the atoms are much closer together a modified wave-function will represent the behaviour of an electron in the combined field of the two nuclei. As illustrated in Fig. 7.2 there are two such possible wave-functions, these being the even, or symmetric, wave-function ψ_S

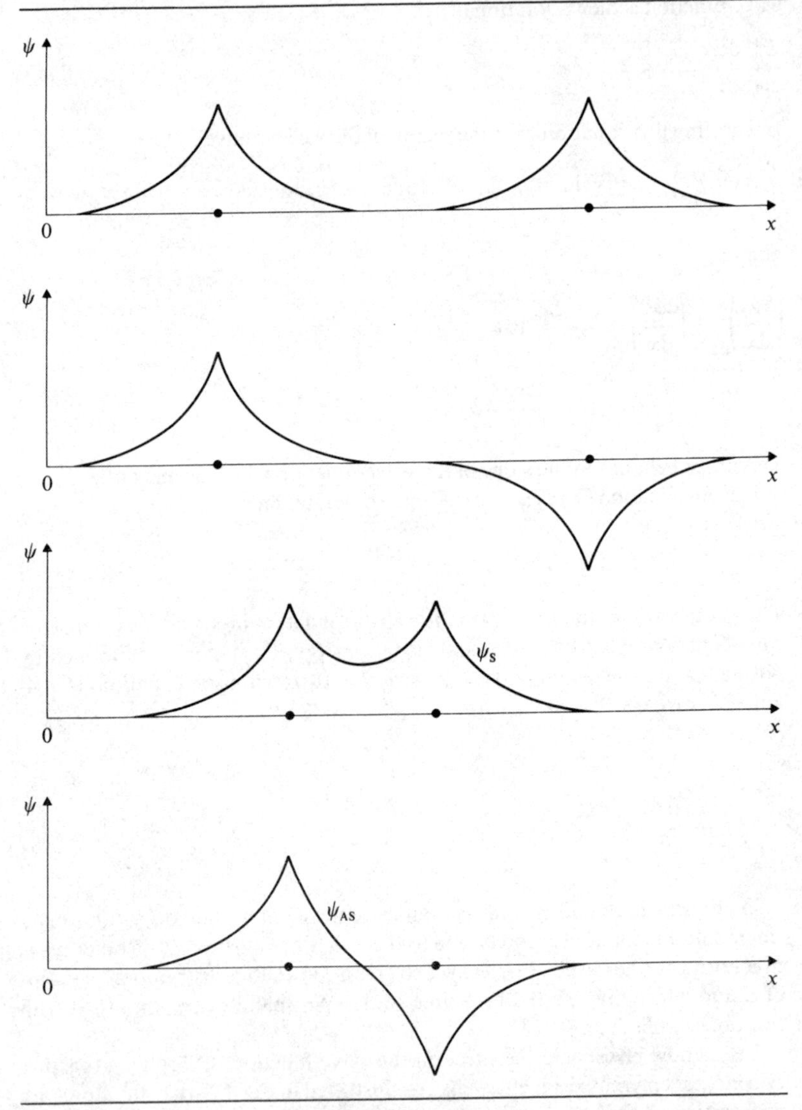

Fig. 7.2 The interaction between adjacent wave-functions showing the formation of the symmetric ψ_S and antisymmetric ψ_{AS} wave-functions when the nuclei approach each other.

and the odd, or antisymmetric, wave-function ψ_{AS}. This is the situation previously described in Section 4.4.1 when the bonding in the hydrogen molecule ion was examined and as we have seen, the energy of the ψ_S wave-function is less than that of the ψ_{AS} wave-function. The single atomic level of the isolated atoms has split into two separate energy levels whose separation is dependent on the separation of the two atoms. Each wave-function corresponds to two quantum states so that the total number of states for the two atoms has remained unaltered. In the specific example of the hydrogen molecule the two available electrons will occupy the lower energy symmetric states which are termed bonding states.

Some further insight into the splitting of the atomic level can be obtained by considering in more detail the wave-functions possible for an electron bound to the two adjacent one-dimensional delta wells. The model is further illustrated in Fig. 7.3 which shows the two wells together with the two possible wave-functions

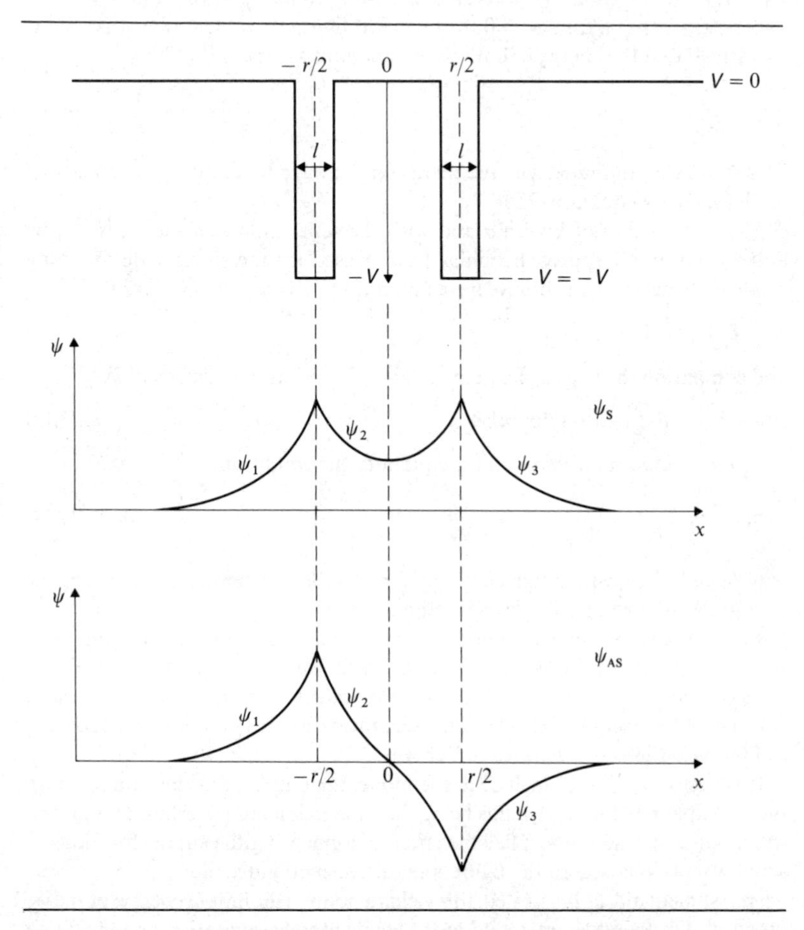

Fig. 7.3 This model, consisting of two adjacent delta wells of separation r, predicts the formation of two wave-functions ψ_S and ψ_{AS} of differing energies. Note that the two wave-functions are shown for $l \rightarrow 0$.

ψ_S and ψ_{AS}. We are primarily interested in seeing how the energies of ψ_S and ψ_{AS} are modified as the separation r of the two wells is varied.

Three regions are identified on the diagram and, bearing in mind the necessity for a wave-function to remain finite everywhere, the two possible wave-functions take the form

$$\begin{aligned}
\psi_1 &= A\exp(kx); \quad x < -r/2 \\
\psi_2 &= B[\exp(kx)+(-1)^n\exp(-kx)]; \quad -r/2 \leqslant x \leqslant r/2 \qquad \ldots(7.10) \\
\psi_3 &= (-1)^n A\exp(-kx); \quad x > r/2
\end{aligned}$$

for the three regions. $n = 1$ gives the antisymmetric form and $n = 2$ the symmetric form. Setting $V = 0$ in the Schrödinger wave equation it is readily confirmed that the energy of both wave-functions satisfy.

$$E = -\frac{\hbar^2}{2m_e}k^2. \qquad \ldots(7.11)$$

This should be compared with the energy for the wave-function for the single well model given by equation (7.9).

The relationship of k with α and with the separation r of the wells can be determined by an approach similar to that used for the single well. Omitting some of the detailed algebra we have on putting $\psi_1 = \psi_2$ at $x = -r/2$ that

$$A = B[1+(-1)^n\exp(kr)] \qquad \ldots(7.12)$$

and comparing the slopes of ψ_1 and ψ_2 at $x = -r/2$ as in equation (7.7).

$$A(k-2\alpha) = Bk[1+(-1)^n\exp(kr)]. \qquad \ldots(7.13)$$

On eliminating A and B from these equations the condition

$$\exp(kr) = \pm\frac{\alpha}{k-\alpha} \qquad \ldots(7.14)$$

is obtained, the positive sign corresponding to the symmetric, and the negative sign to the antisymmetric, wave-function.

Let us briefly examine the function (7.14). For a large separation r of the wells, $k = \alpha$, and, as pointed out previously, the wave-functions ψ_S and ψ_{AS} then have the same energy. For other values of r, two values of k are obtained so that a splitting of the energy level occurs, the separation being a function of the spacing r of the wells. This is illustrated in Fig. 7.4.

It is necessary also to include in the model the repulsive atomic interactions, the principal one for most solids being that due to ion core overlap. This gives a short range interaction and has the effect of dominating the energy functions at small atomic spacing. In the figure a repulsive energy dependence on r^{-12} has been assumed and, as illustrated, this yields a point of minimum energy for the ψ_S function. This would correspond to the stable atomic separation and the figure should be compared with Fig. 4.1. An interesting discussion of the model is given in the text by Holden (1971).

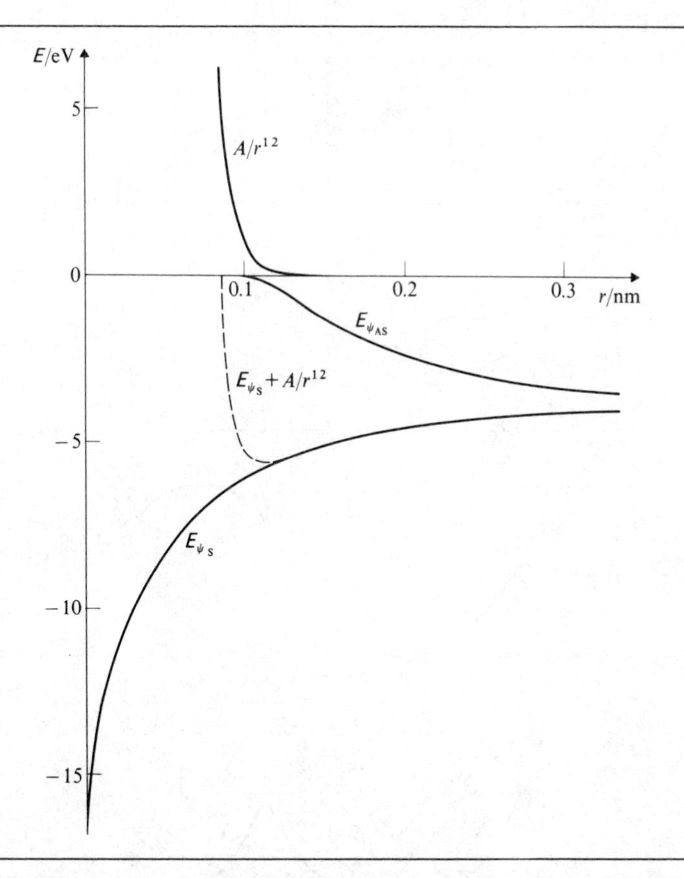

Fig. 7.4 The energies $E_{\psi S}$ and $E_{\psi AS}$ of the symmetric and antisymmetric wave-functions as a function of well separation for the model of Fig. 7.3. The constant α has been set at 10^{10} m^{-1} corresponding to wave-function energies of -3.8 eV for infinite separation of the wells. The general form of the A/r^{12} function is shown and the value of the constant A has been chosen to yield a minimum at 0.1 nm for the $E_{\psi S} + A/r^{12}$) curve.

Extension of the model to a linear array of atoms is most easily performed by a method due to Krönig and Penney which will be discussed in Section 7.4.

Qualitatively, it can be seen from Fig. 7.5 that, for an array of four atoms, there are now a total of four different wave-functions, two of which are symmetric and two antisymmetric. This illustrates the important result that the total number of quantum states available for the electrons has remained unaltered as compared with the number available in the isolated atoms. As shown later in Section 7.8 this type of result is also valid for the assembly of atoms forming a solid so that there is then a very large number of levels forming an *energy band*. The width of the band is independent of the number of atoms and is typically of a few electron-

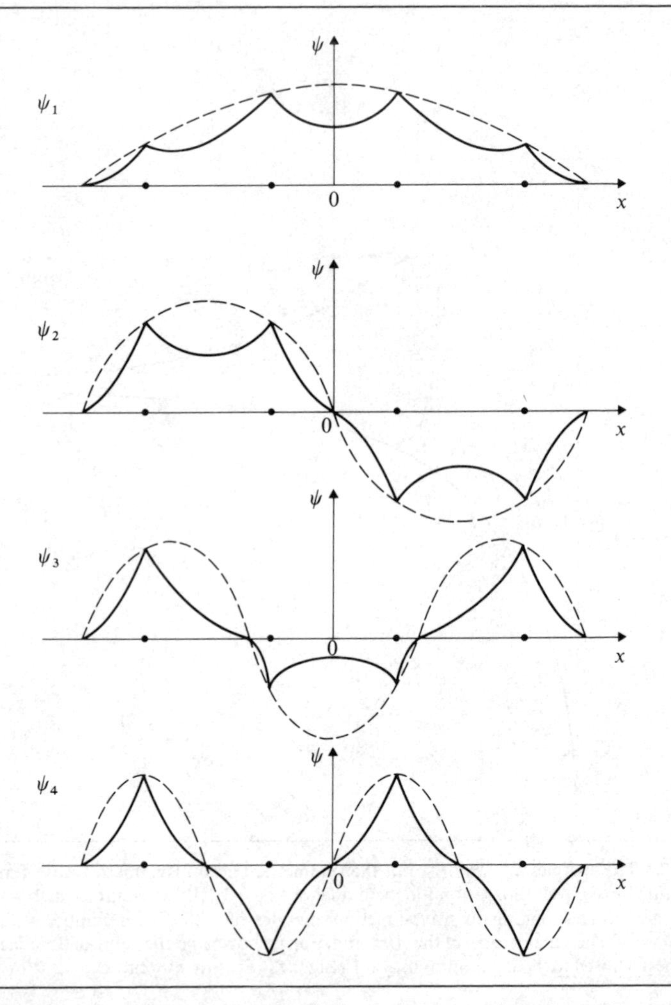

Fig. 7.5 The four wave-functions for a linear array of four atoms. Wave-functions ψ_1 and ψ_3 are symmetric; ψ_2 and ψ_4 are antisymmetric.

volts. The levels are thus very closely spaced within the band and essentially form a continuous spectrum in energy. This is illustrated in Fig. 7.6.

The simplified theory developed here may be illustrated by referring to the electronic structure of sodium. Figure 7.7 compares the band structure of sodium with the levels of the free atom. A free sodium atom in its ground state has the levels 1s, 2s, and 2p filled whilst the 3s state has one electron only. In the solid the 1s, 2s, and 2p bands are very narrow, as is evidenced by the sharpness of the X-ray spectra involving electron transitions to these bands. Electrons in these

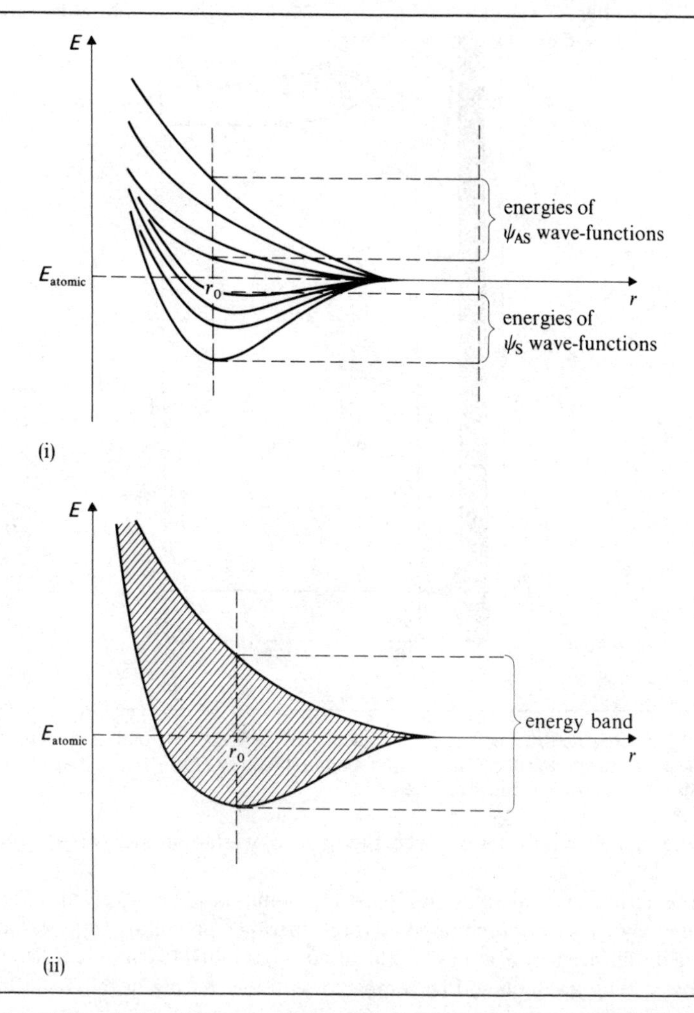

Fig. 7.6 (i) The splitting of energy levels due to the interaction between adjacent atoms for the linear array in Fig. 7.5. Note that the energies of the ψ_S wave-functions are less than, and those of the ψ_{AS} are greater than, the energy E_{atomic} of the isolated atoms. (ii) For a large number of atoms a band of energies is formed and r_0 is then the equilibrium spacing of the atoms.

inner bands are tightly bound to the nucleus and their wave-functions are essentially located on the individual atoms and are not modified appreciably by the presence of adjacent atoms. As is clear from the figure, however, the 3s and higher levels of the isolated atom have broadened out at the observed atomic spacing r_0 in the solid, to form bands of allowed energies. The extent of this broadening in

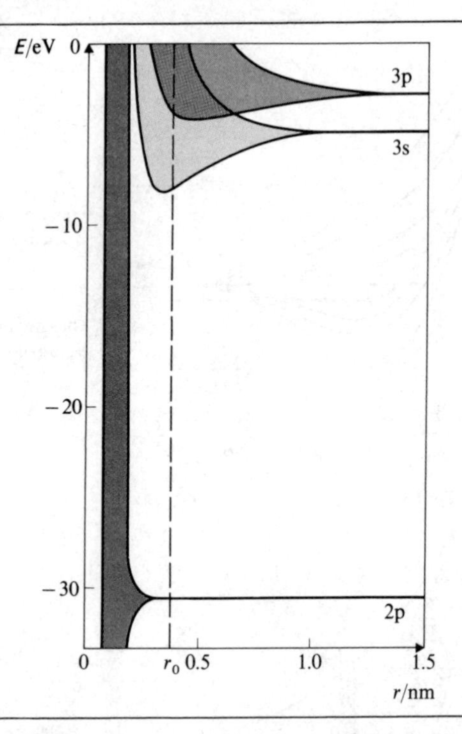

Fig. 7.7 Energy bands in sodium. The broadening of the 3s and 3p atomic levels into bands at the observed spacing of 0.37 nm in solid sodium should be noted. The overlap of the two bands is typical of a metallic structure.

sodium is such that the lower part of the 3p band overlaps in energy the top of the 3s band.

An alternative representation of the energy bands of sodium is given in Fig. 7.8 where they are shown superimposed on the periodic potential energy variation along the linear array of atoms. The localisation on individual atoms of electrons in lower lying levels should be compared with the sharing of the 3s electrons amongst all the atoms of the array.

In accordance with the concepts developed at the end of Chapter 2 an isolated sodium atom will have a single electron occupying one of the two 3s states and all other higher levels will be empty. For solid sodium containing N atoms we now find that there are $2N$ states in the 3s band and since there is only one electron per atom to be accommodated the band will only be half filled. It is this incomplete filling of the band which gives to sodium its metallic properties. The possibility of either filled or partly filled bands, the occurrence of overlapping bands and the existence of energy ranges in which there are no electron states available govern the electronic properties of solids. These concepts will be developed further in the remainder of this chapter.

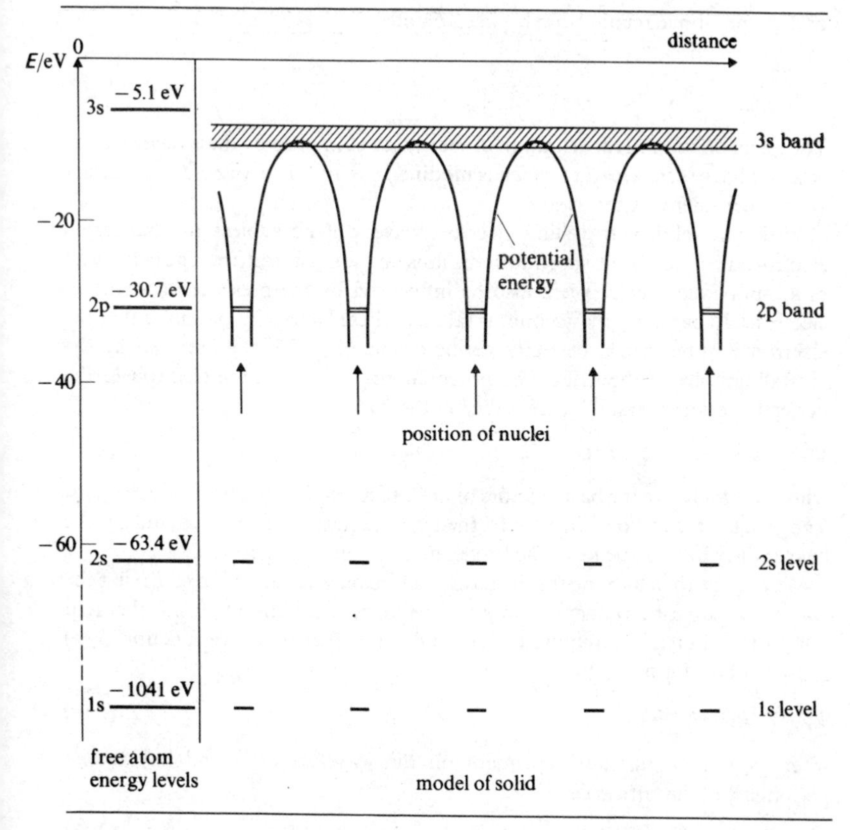

Fig. 7.8 A comparison of the free atom levels in sodium with the bands predicted on a model for the solid.

7.3 The free electron approach to energy bands

In the free electron theory a plane wave of the form

$$\psi = A\exp(\mathrm{i}\boldsymbol{k}\cdot\boldsymbol{r}) \qquad \ldots(5.46)$$

was used as a solution of the wave equation. The wave carries momentum $\hbar \boldsymbol{k}$. Two important results of the free electron theory may be recalled, namely the dependence of the density of states on energy through

$$S(E) = \frac{1}{2\pi^2}\left(\frac{2m_e}{\hbar^2}\right)^{\frac{3}{2}} VE^{\frac{1}{2}} \qquad \ldots(5.41)$$

and the parabolic relationship between E and k

$$E = \frac{\hbar^2 k^2}{2m_e}. \qquad \ldots(5.49)$$

The quasi-continuous distribution of energy states implied by these relationships is, as we have seen, a feature which is modified in band theory and it is necessary now to consider why this is so.

Within a solid, a travelling electron wave will be subject to changes in electrostatic potential arising due to the ions on the crystal lattice. The behaviour of an individual electron will also be influenced by the presence of the other electrons. In band theory, account is taken of these factors by assuming that the electron is subject to a perfectly periodic potential which is imposed by the crystallography of the solid. The potential energy $V(\boldsymbol{r})$ of the electron in the perfectly periodic crystal lattice is then of the form

$$V(\boldsymbol{r}) = V(\boldsymbol{r} + l\boldsymbol{a} + m\boldsymbol{b} + n\boldsymbol{c}), \qquad \ldots(7.15)$$

where $\boldsymbol{a}$, $\boldsymbol{b}$ and $\boldsymbol{c}$ are the basic vectors of the lattice and l, m and n are any integers. The exact form of $V(\boldsymbol{r})$ will be governed by the nature of the ions and by the charge distribution due to all the travelling electrons in the crystal.

We require to determine the form of the electron wave-function satisfying the wave equation for a potential energy in the form of equation (7.15). A theorem due to Bloch gives the required result and states that the wave-function $\psi_{\boldsymbol{k}}(\boldsymbol{r})$ must be of the form

$$\psi_{\boldsymbol{k}}(\boldsymbol{r}) = u_{\boldsymbol{k}}(\boldsymbol{r}) \exp(\mathrm{i}\boldsymbol{k}\cdot\boldsymbol{r}) \qquad \ldots(7.16)$$

where $u_{\boldsymbol{k}}(\boldsymbol{r})$ is a function dependent on the wave-vector $\boldsymbol{k}$ and having the periodicity of the lattice, i.e.

$$u_{\boldsymbol{k}}(\boldsymbol{r}) = u_{\boldsymbol{k}}(\boldsymbol{r} + l\boldsymbol{a} + m\boldsymbol{b} + n\boldsymbol{c}), \qquad \ldots(7.17)$$

where l, m and n are any integers. The solution $\psi_{\boldsymbol{k}}(\boldsymbol{r})$ is known as a Bloch function. The function $u_{\boldsymbol{k}}(\boldsymbol{r})$ is closely related to the free atom wave-function and Fig. 4.17 illustrates, for a one-dimensional array, a simple example of this modulation.

The theorem may be illustrated for a one-dimensional array as follows. For such an array the Bloch function will be

$$\psi_k(x) = u_k(x) \exp(\mathrm{i}kx). \qquad \ldots(7.18)$$

Applying the method of periodic boundary conditions as in Section 6.2.1 we note that the solution obtained must be periodic over a sufficiently large number N of the atomic spacing a, i.e.

$$\psi_k(x) = \psi_k(x + Na). \qquad \ldots(7.19)$$

Since in addition the function $u_k(x)$ is the same for each atom we must have

$$u_k(x) = u_k(x + Na). \qquad \ldots(7.20)$$

These two conditions will be satisfied by a solution of the Bloch form as in equation (7.18), provided that

$$\exp(ikNa) = 1, \qquad \ldots(7.21)$$

which gives

$$k = \pm\frac{2\pi}{a}\frac{n}{N} \qquad (n = 0, 1, 2, 3 \ldots). \qquad \ldots(7.22)$$

This condition will be studied further in Section 7.8 and we note at this point that it defines the number of allowed wave-functions for any range of values of k.

7.4 The Krönig–Penney Model

The Bloch theorem gives the general form of the wave-functions in a periodic potential and thus allows the development and study of models representing the behaviour of electrons in crystals. An extension of the delta well model of Section 7.2 to an array of delta wells has been used by Smith (1959). An alternative model, based more closely on the Krönig–Penney approach, will be utilised here. In this model attention is focused on an array of regularly spaced barriers in the form of potential humps. The mathematics of the model may be simplified considerably by taking the barriers to be in the form of delta functions. This is illustrated in Fig. 7.9.

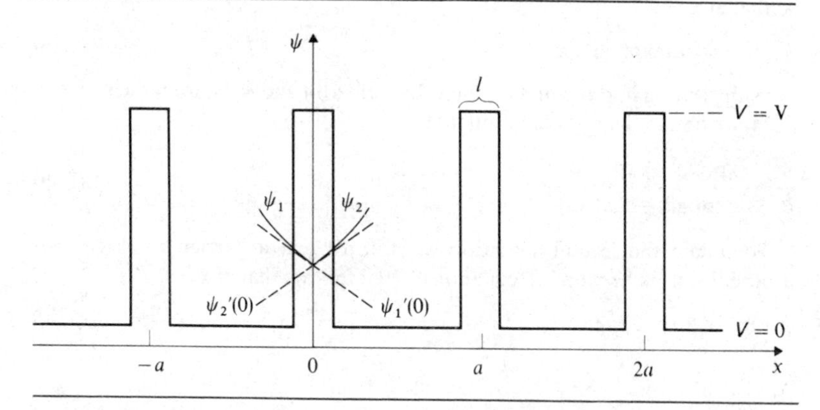

Fig. 7.9 The Krönig–Penney model. The one-dimensional potential energy barriers are of height V, width l and have periodicity a. Note that the wave-function is shown for $l \to 0$.

Solutions of the wave equation (1.104) for one dimension are looked for. These are required to be in the Bloch form and must conform with the boundary conditions of the model. For the regions between the barriers the electron potential energy $V = 0$ so that solutions to

$$\frac{\hbar^2}{2m_e}\frac{d^2\psi}{dx^2}+E\psi = 0 \qquad \ldots(7.23)$$

for positive values of E are required. Such solutions are of the form

$$\psi = A\exp(i\alpha x)+B\exp(-i\alpha x), \qquad \ldots(7.24)$$

where

$$\alpha = \frac{(2m_e E)^{\frac{1}{2}}}{\hbar}. \qquad \ldots(7.25)$$

In order also to be Bloch functions as in equation (7.18) it will be seen that the necessary condition is that

$$u_k(x) = A\exp[i(\alpha-k)x]+B\exp[-i(\alpha+k)x]. \qquad \ldots(7.26)$$

The procedure, which will now be followed, is similar to that adopted for the previous model, i.e. boundary conditions are applied so that the constants A and B are eliminated and a relationship found between the constants k, α and the barrier strength γ defined below. The wave-function to the left and to the right of the spike at $x = 0$ is designated ψ_1 and ψ_2, respectively.

First, since ψ is a Bloch function of the form of equation (7.18), it may be verified that

$$\psi(x) = \psi(x+a)\exp(-ika) \qquad \ldots(7.27)$$

so that at $x = 0$

$$\psi_1(0) = \psi_2(a)\exp(-ika). \qquad \ldots(7.28)$$

On substituting in this equation the values of $\psi_1(0)$ and $\psi_2(a)$ from the solution (7.24) we obtain the first condition that

$$\frac{A}{B} = \frac{\exp(ika)-\exp(-i\alpha a)}{\exp(i\alpha a)-\exp(ika)}. \qquad \ldots(7.29)$$

To obtain the second condition we note that at a barrier $E \ll V$ so that proceeding as in Section 7.2 (equation (7.7)) it follows that at $x = 0$

$$\psi_2'(0)-\psi_1'(0) = 2\gamma\psi(0) \qquad \ldots(7.30)$$

where

$$\gamma = \frac{m_e}{\hbar^2}Vl. \qquad \ldots(7.31)$$

The parameter γ is a measure of the strength of the barrier. It follows from equations (7.28) and (7.30) that

$$\psi_2'(0)-\psi_2'(a)\exp(-ika) = 2\gamma\psi(0). \qquad \ldots(7.32)$$

On differentiating (7.24) and substituting in this equation the second condition for A/B is obtained in the form

$$\frac{A}{B}=\frac{1-\exp(-\mathrm{i}\alpha a)\exp(-\mathrm{i}ka)-2\mathrm{i}(\gamma/\alpha)}{1-\exp(\mathrm{i}\alpha a)\exp(-\mathrm{i}ka)+2\mathrm{i}(\gamma/\alpha)}. \qquad \ldots(7.33)$$

The conditions in (7.29) and (7.33) may be combined and simplified to give the relation

$$\cos ka = \cos \alpha a+\gamma a\left(\frac{\sin \alpha a}{\alpha a}\right). \qquad \ldots(7.34)$$

This equation is central to the Krönig–Penney model. For any particular barrier strength γ it provides through the variable α given by

$$\alpha = \frac{(2m_e E)^{\frac{1}{2}}}{\hbar} \qquad \ldots(7.25)$$

a relationship between the electron energy E and the propagation constant k of the electron waves.

The significance of the interrelationship may best be understood by reference to Fig. 7.10. This gives a plot of the quantity $\cos \alpha a+(\gamma a/\alpha a)\sin \alpha a$ versus αa for

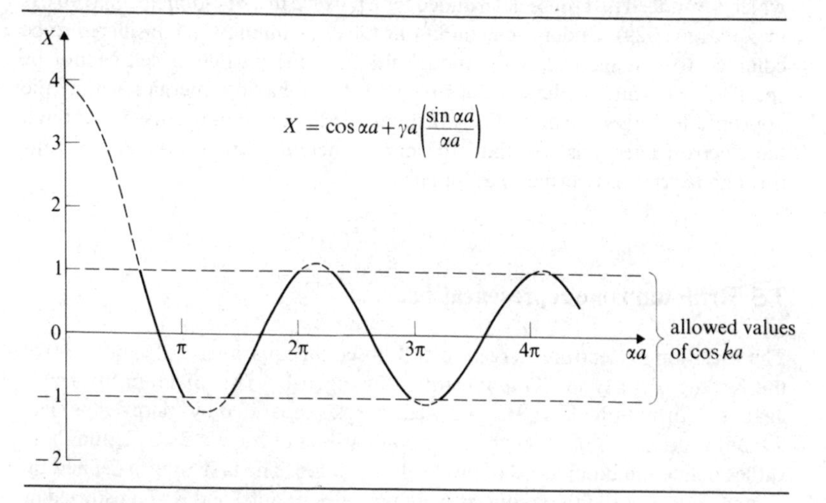

Fig. 7.10 The Krönig–Penney model. Allowed values of the parameter αa are restricted to those which yield values of the function $X = \cos \alpha a+(\gamma a/\alpha a)\sin\alpha a$ lying between $+1$ and -1. In this figure the parameter $\gamma a = \pi$.

one specific value of γa ($\gamma a = \pi$). Since values of $\cos ka$ may lie only between -1 and 1, only certain ranges of values of αa will satisfy equation (7.34). These ranges are indicated on the figure. In accordance with equation (7.25) we find therefore that there are certain allowed ranges within which the energy of the wave-function must lie, i.e. the model predicts the existence of energy bands separated by energy gaps.

The width of the energy bands is governed by the parameter γ. When γ is very small, i.e. for a very weak barrier, equation (7.34) gives $\alpha = k$; so that

$$E = \frac{\hbar^2 k^2}{2m_e}, \qquad \ldots(5.49)$$

which is the energy relationship in the free electron theory. When γ is large, solutions become restricted to those for which

$$\sin \alpha a = 0 \qquad \ldots(7.35)$$

i.e.

$$\alpha a = \pm n\pi \qquad (n = 1, 2, 3 \ldots) \quad \ldots(7.36)$$

so that in accordance with equation (7.25)

$$E = \frac{\hbar^2}{2m_e}\frac{\pi^2}{a^2}n^2 \qquad (n = 1, 2, 3 \ldots). \quad \ldots(7.37)$$

Thus for a tightly bound electron the model predicts a series of discrete levels which are indeed just those determined for a particle in a one-dimensional box as in equation (5.29). Under such conditions the electron may be considered to be confined to a single cell of the model although the particular cell cannot be specified. For intermediate values of γ electrons having energies within the appropriate ranges can move freely through the crystal. This occurs even though the electron energy is less than the barrier height. Penetration of electrons through barriers is referred to as tunnelling.

7.5 Brillouin zone representation

The variation of electron energy E with k based on equations (7.25) and (7.34) of the Krönig–Penney model is illustrated in Fig. 7.11. This diagram illustrates again the formation of bands of allowed energies separated by energy gaps. The discontinuities in the E vs. k curves occur at k values of $\pm\pi/a, \pm 2\pi/a, \ldots$ and these values define the boundaries of the Brillouin zones. The first zone is defined by $-\pi/a < k \leqslant \pi/a$ and the second and higher zones lie outside this as illustrated in the figure (see also Section 6.2). This type of representation is known as the *extended zone* representation and it has the bands of higher energy lying in the higher zones. The method of deriving zones in the three-dimensional reciprocal lattice will be discussed in Section 7.7.

Other types of representation may also be used and they follow from the properties of wave-vectors in the three-dimensional reciprocal lattice. Thus, two

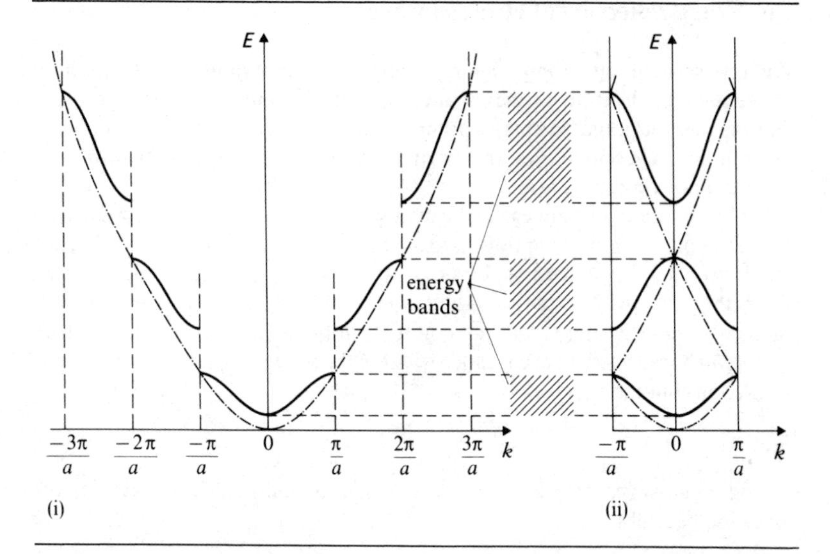

Fig. 7.11 The variation of electron energy E with k on the Krönig–Penney model showing the formation of bands of allowed energies. This is shown in (i) in the extended zone representation and in (ii) in the reduced zone representation. The parabolic variation of E with k on the free electron model is shown dotted.

wave-vectors $\boldsymbol{k}$ and $\boldsymbol{k}'$ are equivalent if they are related by

$$\boldsymbol{k}' = \boldsymbol{k} + \boldsymbol{G}, \qquad \ldots(7.38)$$

where $\boldsymbol{G}$ is a reciprocal lattice vector. We have already come across this relationship in Section 3.10 where $\boldsymbol{k}$ and $\boldsymbol{k}'$ then denoted the wave-vectors of incident and elastically reflected waves. The relationship denotes the equivalence of corresponding points in $\boldsymbol{k}$ space. It follows that, if the wave-vector of the Bloch wave-function lies outside the first Brillouin zone, it can always be brought within it by choice of a suitable lattice vector $\boldsymbol{G}$. The wave-function can then be shown to remain unchanged (see e.g. Kittel, 1973, p. 307). By this means, therefore, we can have a representation in which all the bands lie within the first zone. It is termed a *reduced zone* representation and is illustrated in Fig. 7.11 for the Krönig–Penney model. Each branch corresponds to an energy band. Different energy bands correspond to different modulating functions $u_{\boldsymbol{k}}(\boldsymbol{r})$ which describe different degrees of binding of the electron. The greater the degree of binding, the narrower the band is in energy so that in the limit a band may degenerate into a single level. This occurs when there is no appreciable interaction between corresponding energy levels in adjacent atoms. Finally we note that the properties of the reciprocal lattice vectors imply that every band may be represented in every zone. This is termed a *periodic zone* representation.

7.6 Bragg reflection of electron waves

We have seen, in the Krönig–Penney model, that the assumption of a periodic lattice potential leads to a series of alternate energy bands and energy gaps and that the discontinuities in energy occur at values of wave-vector corresponding to Brillouin zone boundaries. It is pertinent to enquire more closely how these discontinuities in energy arise.

Let us again assume a linear array of atoms spaced a apart and now consider the behaviour of an electron travelling along this array. For any given value of k the electron may be represented by a travelling wave which will, in general, travel along the array without any modification. However, for certain values of k, namely those satisfying the Bragg condition for the linear array, the electron waves will be reflected. In accordance with the theory developed in Chapter 3 the diffraction condition is given by

$$2a = n\lambda. \qquad \ldots(7.39)$$

Noting that waves travelling in either direction may suffer reflection, the condition becomes

$$k = \pm\frac{n\pi}{a} \qquad (n = 1, 2, 3, \ldots) \qquad \ldots(7.40)$$

i.e. electron waves having k values corresponding to the Brillouin zone boundaries suffer Bragg reflection. Such waves undergo multiple reflection and the travelling wave is replaced by equal components of waves in each direction. They become standing waves.

We may take as a specific example the reflection of an electron wave at the first Brillouin zone boundary corresponding to $k = \pm\pi/a$. A travelling wave in the crystal must be a Bloch function and the modulating factor $u_k(x)$ will have the periodicity of the lattice. Let us assume that this has the simple form $A\cos\left[(2\pi/a)x\right]$. In accordance with equation (1.22), two waves travelling in opposite directions in the crystal might thus be represented by

$$\begin{aligned} \psi_1 &= A\cos\frac{2\pi}{a}x\exp\mathrm{i}\left(\frac{\pi}{a}x-\omega t\right) \\ \psi_2 &= A\cos\frac{2\pi}{a}x\exp\mathrm{i}\left(-\frac{\pi}{a}x-\omega t\right). \end{aligned} \qquad \ldots(7.41)$$

These waves may be combined to give the standing wave

$$\psi_S = \psi_1 + \psi_2 \qquad \ldots(7.42)$$

$$= A\cos\frac{2\pi}{a}x\exp(-\mathrm{i}\omega t)\left[\exp\left(\frac{\mathrm{i}\pi}{a}x\right)+\exp\left(-\frac{\mathrm{i}\pi}{a}x\right)\right]$$

which, on taking the real part and omitting the time factor becomes

$$\psi_S = 2A\cos\frac{2\pi}{a}x\cos\frac{\pi}{a}x. \qquad \ldots(7.43)$$

This is a standing wave of symmetrical form.

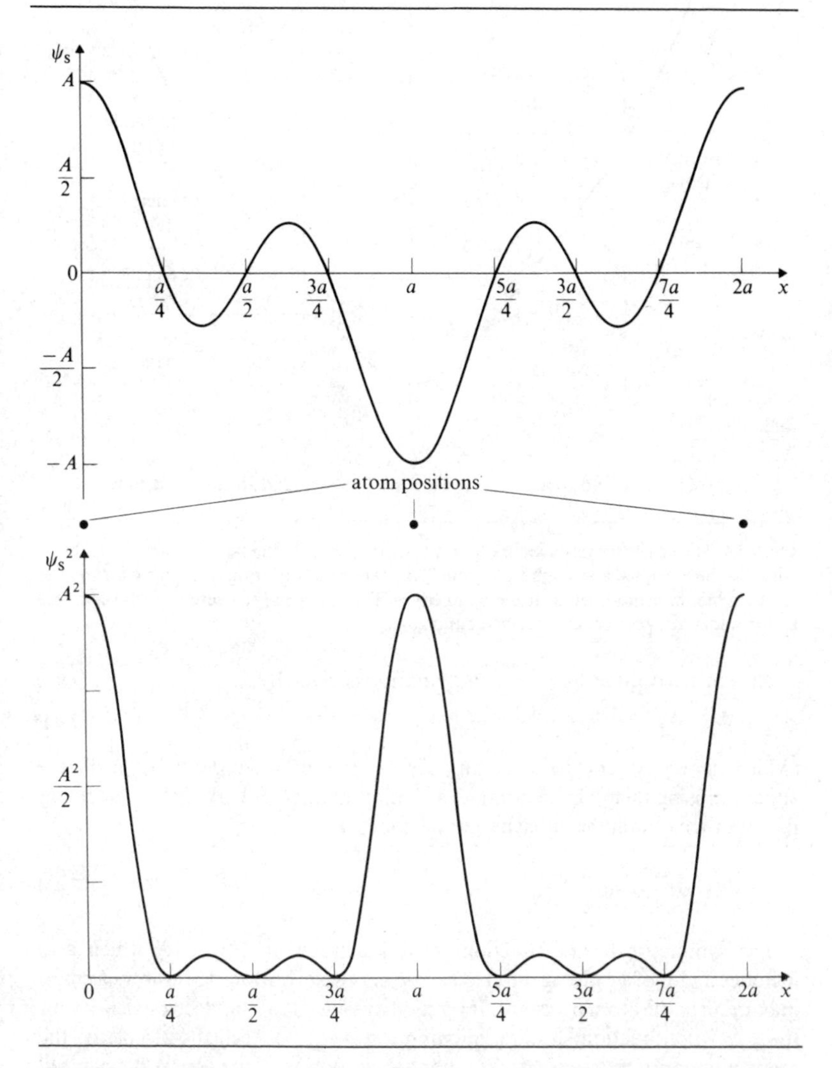

Fig. 7.12 The symmetrical wave-function ψ_S of equation (7.43) and the corresponding electron density ψ_S^2 for a wave reflected at the first Brillouin zone boundary for a linear lattice of spacing a.

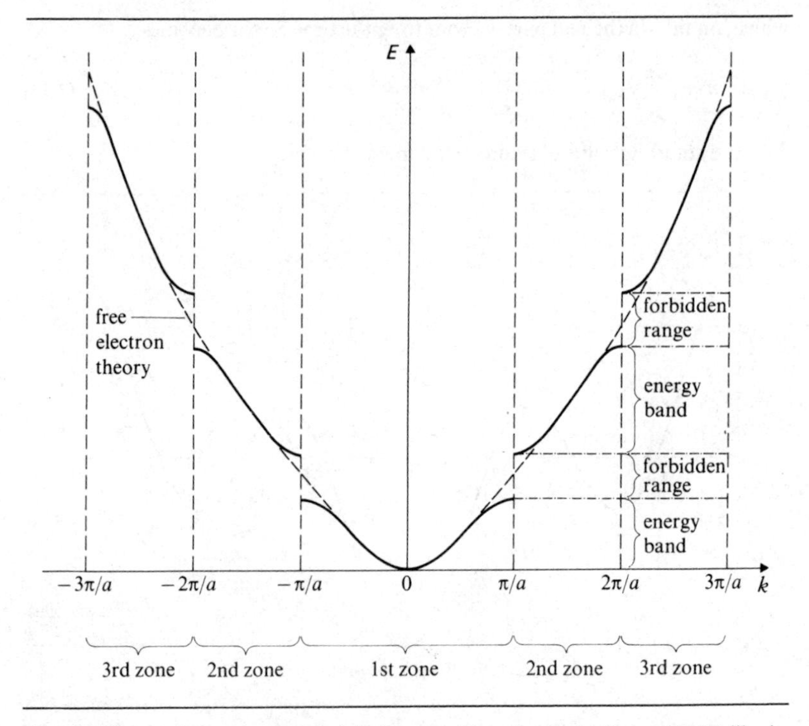

Fig. 7.13 The nearly free electron model is illustrated here for the 1st, 2nd and 3rd Brillouin zones in the extended zone representation. The parabolic variation of E with k for the free electron theory is modified near the band edges. This gives rise to energy bands separated by forbidden ranges of energy termed band gaps.

An alternative standing wave combination is given by

$$\psi_{AS} = \psi_1 - \psi_2 \qquad \ldots(7.44)$$

(which is equivalent to changing the phase of one wave by π before superimposing them). Proceeding in a similar manner and taking the imaginary part we then obtain the antisymmetrical function

$$\psi_{AS} = 2A \cos\frac{2\pi}{a}x \sin\frac{\pi}{a}x. \qquad \ldots(7.45)$$

The symmetrical wave-function ψ_S is illustrated in Fig. 7.12, which also includes a plot of ψ_S^2 giving the electron density distribution. This shows a sharp maximum at the atom positions. It is readily verified that the electron density for the ψ_{AS} wave-function shows a maximum *midway* between the atoms. At the atom positions the electrostatic potential is high and the electron potential energy is correspondingly low. The ψ_S wave-function which is making the maximum use of the lower potential energy thus has a much lower average energy than the ψ_{AS} wave-functions. This accounts for the splitting of the energy at the zone boundary. The difference in energies of the ψ_S and ψ_{AS} wave-functions is a measure of the energy gap at the boundary.

If the periodic potential is weak the effect of the modulating function is much reduced and the solution then approaches that of the free electron model. This is the nearly free electron model illustrated in Fig. 7.13. If E_g is the width of the band gap the energies of the two wave-functions at the boundary are given by

$$E = \frac{\hbar^2 k^2}{2m_e} \pm \frac{E_g}{2}, \qquad \ldots(7.46)$$

where the + and − signs refer to the ψ_{AS} and ψ_S wave-functions, respectively.

7.7 Brillouin zones and crystal structures

We have seen in the previous section that discontinuities in energy, arising because of the periodic lattice potential, can be identified with Bragg reflection of electron waves at zone boundaries. In a real solid the symmetry of the crystal lattice will govern the position of the zone boundaries in the different directions in $\boldsymbol{k}$ space. The form and periodicity of the lattice potential will also be dependent on crystal direction. The principal features of the simple model are, however, retained, namely that electron waves in the crystal are reflected at zone boundaries and that such boundaries correspond to discontinuities in energy for electrons having the appropriate wave-vectors.

The form of the Brillouin zones in a solid therefore strongly influences its band structure and the method of determining the zone structure will now be considered in some detail.

Let us first look at the reflection of electron waves by a one-dimensional array. From the standpoint of the reciprocal lattice the diffraction condition may be written as

$$2\boldsymbol{k}\cdot\boldsymbol{G} + G^2 = 0. \qquad \ldots(3.31)$$

We recall also that both $\boldsymbol{G}$ and $-\boldsymbol{G}$ are reciprocal lattice vectors so that for the linear one-dimensional array the condition is simply

$$k = \pm\frac{G}{2} \qquad \ldots(7.47)$$

i.e. a wave-vector which bisects a reciprocal lattice vector will be diffracted. If the linear spacing is a the primitive translation vector $\boldsymbol{a}$ of the array is $a\,\hat{\boldsymbol{x}}$, where $\hat{\boldsymbol{x}}$ is the unit vector along the array. In accordance with Section 3.10 the primitive translation vector $\boldsymbol{a}^*$ of the reciprocal array will be given by

$$\boldsymbol{a}^* = \frac{2\pi}{a}\hat{\boldsymbol{x}} \qquad \ldots(7.48)$$

and the general reciprocal vector $\boldsymbol{G}$ by

$$\boldsymbol{G} = n\boldsymbol{a}^*$$

$$= \frac{2\pi}{a} n\hat{\boldsymbol{x}}. \quad \ldots(7.49)$$

It follows, therefore, that the diffraction condition (7.47) for the array can be put in the form

$$k = \pm\frac{n\pi}{a} \qquad (n = 1, 2, 3, \ldots) \quad \ldots(7.40)$$

which, as we have seen, defines the boundaries of the Brillouin zones for this array. This is shown in Fig. 7.13 for the nearly free electron model.

The Brillouin zone concept can now be extended to two and three dimensions. The simplest two-dimensional array is that of a square lattice and following the geometrical construction in Section 3.9 it is readily verified that the reciprocal lattice is also a square lattice. The same result can be obtained by use of the transformation equations in Section 3.10 and it is then convenient to assume a third perpendicular c axis and to put $\boldsymbol{c}$ as a unit vector. Figure 7.14 illustrates the

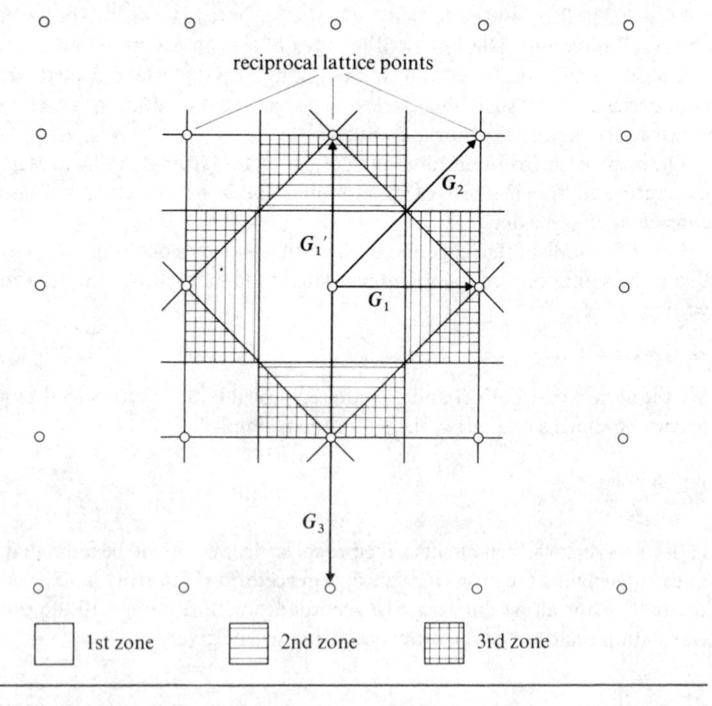

Fig. 7.14 The first three Brillouin zones for a square lattice.

construction of the first three Brillouin zones in this reciprocal lattice. We note firstly that any vector $\boldsymbol{k}$ from the origin terminating on the perpendicular bisector of the vector $\boldsymbol{G}_1$ must satisfy the diffraction condition (3.31). Moreover, since $\boldsymbol{G}_1$ is the *shortest* vector its bisector must form a boundary to the first Brillouin zone. Similarly, the perpendicular bisectors of $-\boldsymbol{G}_1$, $\boldsymbol{G}_1'$ and $-\boldsymbol{G}_1'$ form the other boundaries of the square Brillouin zone. The second zone consists of a number of separate parts which lie between the first zone and the perpendicular bisectors of vectors such as $\boldsymbol{G}_2$. Higher zones are constructed in a similar manner but may be transferred into the first zone in a reduced zone representation. It may be verified, in the example chosen, that all zones are of equal area. A similar result holds for three dimensions where all zones are of equal volume in any given reciprocal lattice.

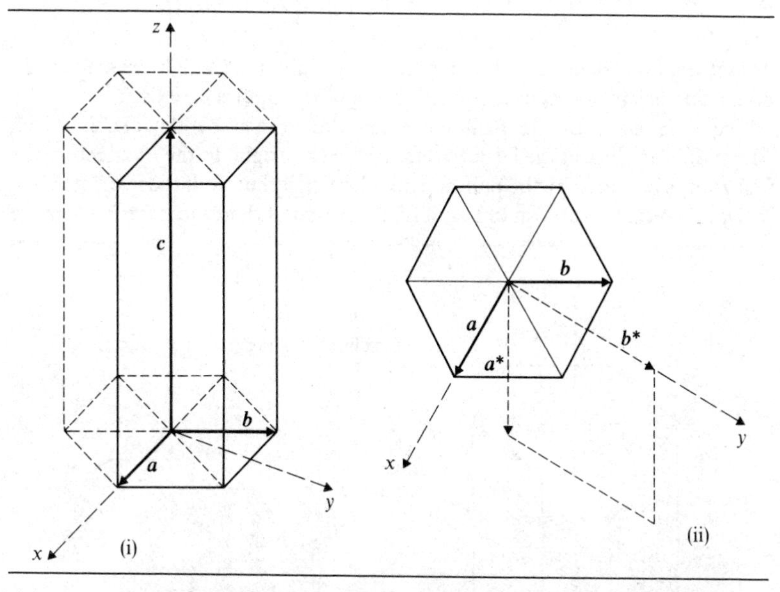

Fig. 7.15 (i) The primitive cell of the simple hexagonal lattice and the three translation vectors $\boldsymbol{a}$, $\boldsymbol{b}$ and $\boldsymbol{c}$. (ii) The relationship between the vectors $\boldsymbol{a}$ and $\boldsymbol{b}$ and the vectors $\boldsymbol{a}^*$ and $\boldsymbol{b}^*$ of the reciprocal lattice.

The procedure for deriving the Brillouin zone for a three-dimensional lattice may be illustrated by using as an example the simple hexagonal lattice ($a = b \neq c$, $\gamma = 120°$) shown in Fig. 7.15. The lattice may be defined by the primitive translation vectors

$$\begin{aligned} \boldsymbol{a} &= a\hat{\boldsymbol{x}} \\ \boldsymbol{b} &= -\frac{a}{2}\hat{\boldsymbol{x}} + \frac{\sqrt{3}}{2}a\hat{\boldsymbol{y}} \\ \boldsymbol{c} &= c\hat{\boldsymbol{z}}, \end{aligned} \qquad \ldots(7.50)$$

where $\hat{x}$, $\hat{y}$ and $\hat{z}$ are three perpendicular unit vectors. The translation vectors are shown in Fig. 7.15 although we should note that other alternative descriptions of the lattice could have been adopted with a different choice of unit vectors.

The primitive translation vectors of the reciprocal lattice are determined by use of the transformation equations (3.22)–(3.24) which yield

$$\begin{aligned} \boldsymbol{a}^* &= \frac{2\pi}{a}\left(\hat{x} + \frac{\hat{y}}{\sqrt{3}}\right) \\ \boldsymbol{b}^* &= \frac{4\pi}{\sqrt{3}a}\hat{y} \\ \boldsymbol{c}^* &= \frac{2\pi}{c}\hat{z}. \end{aligned} \qquad \ldots(7.51)$$

These equations define another simple hexagonal lattice which, relative to the direct axis, has two of its axes rotated through 30° about the c axis.

The boundaries of the Brillouin zones are formed by the planes which perpendicularly bisect the lattice vectors from the origin. In the k_z direction the first zone is bounded by the pair of planes bisecting the two lattice vectors $\boldsymbol{G} = \pm(2\pi/c)\hat{z}$ so that, as shown in Fig. 7.16, the height of the zone is $2\pi/c$. The zone

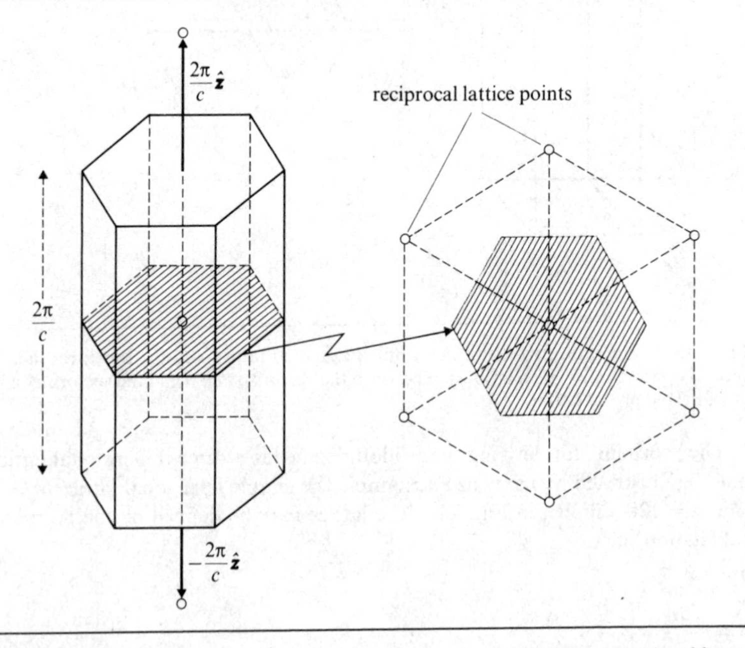

Fig. 7.16 The first Brillouin zone for a simple hexagonal lattice. The zone boundary bisects the reciprocal lattice vectors.

has one lattice point at its centre so that the volume of the zone is the same as that of a primitive cell of the reciprocal lattice. It may be confirmed that the second Brillouin zone for the structure consists of two pyramids having six faces and situated on the top and bottom faces of the first zone, together with pyramids (having four faces) on the other six faces of the first zone. The volume of the second zone is the same as that of the first zone.

Similar procedures are adopted for the determination of zone structures for other lattices. Thus one finds that a b.c.c. lattice has a f.c.c. reciprocal lattice. The first Brillouin zone in this case is a regular rhombic dodecahedron formed by the planes bisecting the 12 shortest vectors from a reciprocal lattice point (Fig. 7.17).

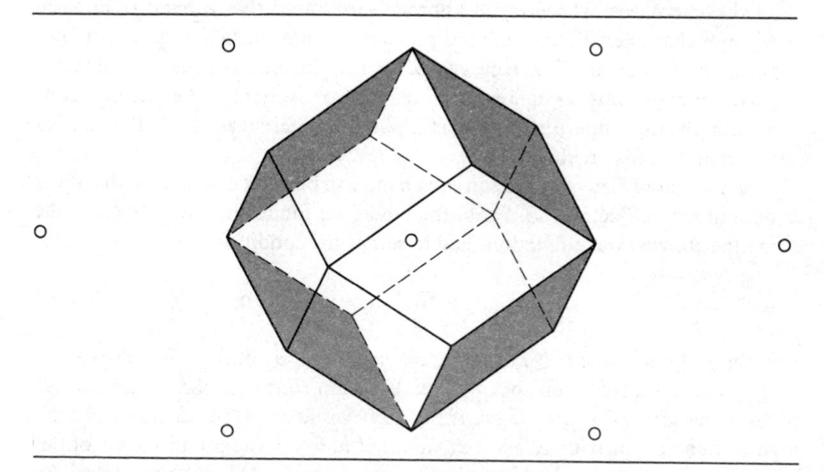

Fig. 7.17 The first Brillouin zone for a b.c.c. lattice. The reciprocal lattice is f.c.c. so that each lattice point has twelve nearest neighbours. Taking the face-centred lattice position as origin the four planes shown shaded are perpendicular bisectors of the four lattice vectors from the origin and are perpendicular to the plane of the diagram. The zone boundaries form a regular dodecahedron, the twelve planes being of the $\{110\}$ type. Each boundary plane therefore lies in one of three mutually perpendicular directions.

For a f.c.c. lattice the reciprocal lattice is b.c.c. and the first Brillouin zone is a truncated octahedron (Fig. 7.22). The shapes of Brillouin zones have been determined for many structures and are known in considerable detail. Many of the higher zones are complex in form and difficult to represent pictorially in a satisfactory manner.

7.8 Number of states in an energy band

It is useful at this point to bring together some of the results of the free atom and the free electron approach to band theory. In the free atom approach we noted that each energy level in the isolated atom is broadened out in the solid into a band of energies. Each energy band in a solid is therefore traceable to a specific level, although as we shall see at the end of this section some interactions may occur which can modify this simple picture. A central result of the model is that when a total of N atoms are brought together to form a solid then the number of quantum states in any band is equal to the total number of states for the corresponding level in the N atoms. Thus for a 3p atomic level there are six states per atom and in a solid of N atoms a total of $6N$ states are thus traceable to this 3p level. The energy distribution of states within the 3p band will not be uniform and the band may also overlap the adjacent 3s and 3d bands.

In the alternative free electron approach we found that a band of allowed energies is characterised in a reduced zone representation as in Fig. 7.11 (ii) by a separate branch of the $\boldsymbol{E}$ vs. $\boldsymbol{k}$ curve and that in the limit of tight binding each band degenerated into a single energy level. Alternatively, as in the extended zone scheme in the same figure each band is drawn in a different zone and the number of zones and bands are then equal.

The number of electron states in each band can be determined from the linear crystal model of Section 7.3. From the model we found that for N cells in the array the allowed wave-functions had to satisfy the condition

$$k = \pm\frac{2\pi}{a}\frac{n}{N} \qquad (n = 0, 1, 2, 3, \ldots). \qquad \ldots(7.22)$$

For the range $-\pi/a < k \leqslant \pi/a$ there are, therefore, a total of N values of k. Taking account of electron spin, the first Brillouin zone thus contributes a total of $2N$ states for *each* energy band. It is readily verified that the second and each higher zone also provide N allowed values of k. From the point of view of the extended zone scheme each band will thus contain $2N$ electron states. As indicated previously the equality of states in each zone is also true in three dimensions and for which we found the volumes of all zones were equal. It should be noted that electron wave-functions having the same wave-vector but lying in different bands are independent of each other. The wave-functions differ in the $u_{\boldsymbol{k}}(\boldsymbol{r})$ factor and the Pauli exclusion principle is not contravened since the bands essentially arise from different atomic levels.

We have seen, from the free atom approach, that a direct relationship exists between the atomic levels of isolated atoms and the band structure of a solid. In view of the results derived from equation (7.22) the correspondence must imply that s states give rise to one zone, p states to three zones, d states to five zones, etc. Thus the band arising from a 3d atomic level, as in a transition metal, will consist of five sub-bands overlapping in energy. Each is capable of accommodating $2N$ electrons and thereby give a total of $10N$ states for the 3d band. The density of states, plotted as a function of electron energy, is found as a composite variation for the band as a whole and it will show for these metals a number of maxima and minima throughout the band. As illustrated in Fig. 7.18 the band as a whole is overlapped by the 4s band and this has to be taken into account in determining, for example, the position of the Fermi level.

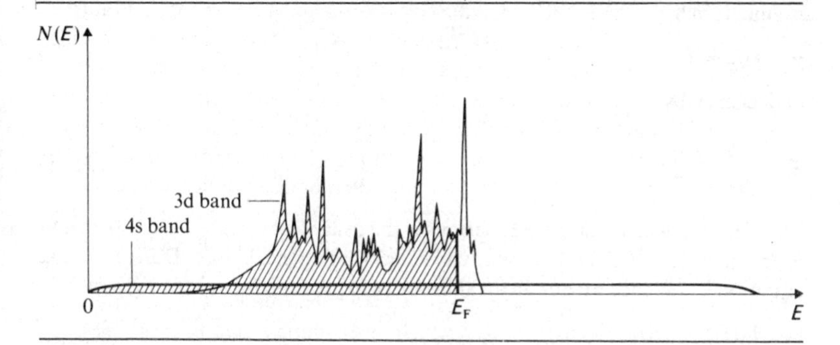

Fig. 7.18 The 3d band for nickel showing the complex variation in density of states. The 4s band overlaps the 3d band and the properties of the transition metals are strongly influenced by this type of band structure.

The availability of electron states in the energy bands in semiconductors may be briefly introduced here. In semiconductors such as germanium and silicon the s and p orbitals of the isolated atoms combine in the solid to form hybrid sp^3 orbitals (see Section 4.4.3). This gives a band structure consisting of a lower, or valence band, and an upper conduction band separated by an energy gap. The total of $8N$ states of the isolated atoms, $2N$ from s and $6N$ from p levels, combine in the solid to give $4N$ states in each of the two bands. Since the elements are quadrivalent the electrons participating in the bonding of the solid fill all available states in the valence band and the conduction bands is unoccupied at $T = 0$ K. The consequences of this type of band structure will be developed in subsequent sections and an illustration of the form of the bands is given in Fig. 7.27.

7.9 Electron velocity and crystal momentum

In developing the electrical properties of solids it is necessary to explain and be able to predict the behaviour of electrons in the solid under the action of applied electric and magnetic fields. One is therefore interested in quantities such as electron velocity, momentum, energy and acceleration. Of particular importance also is the form of the variation of E with k within the energy bands and the associated density of states.

Let us initially look at some of the quantities as they were defined and used in the free electron theory in Section 1.6.3. The electron velocity was then given by

$$u = \frac{d\omega}{dk} = \frac{\hbar k}{m_e} \quad \ldots(1.79)$$

momentum by

$$p = m_e u = \hbar k \qquad \ldots(1.73)$$

and energy by

$$E = \frac{\hbar^2 k^2}{2m_e}. \qquad \ldots(1.78)$$

The electron mass m_e in these equations is the usual rest mass of the electron and is independent of E and k, relativity effects being negligible. Differentiating equation (1.78) once we obtain an alternative expression for electron velocity

$$u = \frac{1}{\hbar}\frac{dE}{dk} \qquad \ldots(7.52)$$

and differentiating again we can obtain

$$m_e = \frac{\hbar^2}{\dfrac{d^2E}{dk^2}}. \qquad \ldots(7.53)$$

This constant mass will govern the behaviour of an electron in an applied electric field $\boldsymbol{E}$ irrespective of the value of k. Its acceleration in the field will be

$$\begin{aligned} \frac{du}{dt} &= -\frac{eE}{m_e} \\ &= -\frac{eE}{\hbar^2}\frac{d^2E}{dk^2}. \end{aligned} \qquad \ldots(7.54)$$

These quantities may now be reconsidered in the context of band theory. In a solid the electron energy E will be greatly influenced by the form of the periodic potential, and except for certain restricted values of the wave-vector $\boldsymbol{k}$, the energy is no longer a simple parabolic function of $\boldsymbol{k}$. Implicit also in the reduced zone representation of different energy bands is that energy is not a single-valued function of $\boldsymbol{k}$.

Due to the form of the $E:\boldsymbol{k}$ relationship it is not valid in band theory to equate electron velocity given by

$$u = \frac{\partial \omega}{\partial k} \qquad \ldots(7.55)$$

to the quantity $h\boldsymbol{k}/m_e$. A quantity

$$p = \hbar \boldsymbol{k} \qquad \ldots(7.56)$$

termed crystal momentum is, however, defined and this is conserved in electron collision processes. We note, however, from equation (7.55) that the electron velocity can be put in the alternative form

$$u = \frac{1}{\hbar}\frac{\partial E}{\partial \boldsymbol{k}}$$

$$= \frac{1}{\hbar}\nabla_k E \qquad \ldots(7.57)$$

where

$$\nabla_k = \boldsymbol{i}\frac{\partial}{\partial k_x} + \boldsymbol{j}\frac{\partial}{\partial k_y} + \boldsymbol{k}\frac{\partial}{\partial k_z}.$$

Let us consider the implications of this relationship. Firstly, the magnitude and direction of the electron group velocity is governed by $\partial E/\partial \boldsymbol{k}$ rather than by k as on the free electron theory. Close to the bottom of a band the relationship between E and $\boldsymbol{k}$ may be parabolic for each crystal direction in which case the constant energy surfaces are spherical as on the free electron model. This is illustrated in Fig. 7.19(i), point A, and it is noted that the velocity $\boldsymbol{u}$, governed by

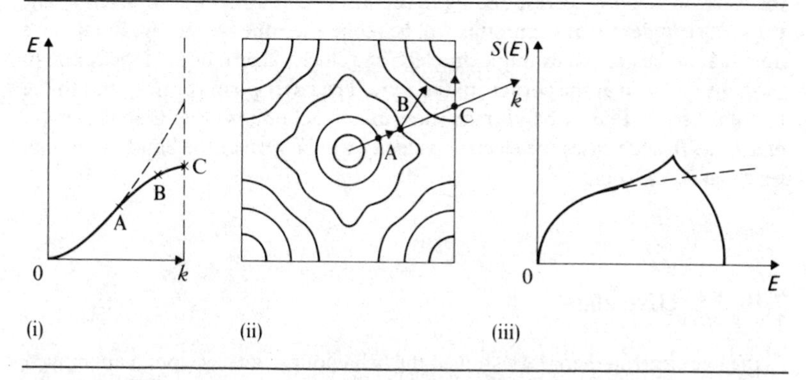

Fig. 7.19 (i) The variation of E with k on the band model. This is drawn for a specific direction in $\boldsymbol{k}$-space and illustrates the deviation from the free electron model towards the edge of a band (points B and C). (ii) The constant energy surfaces for different values of $\boldsymbol{k}$. For small values of $\boldsymbol{k}$ the surfaces are spherical as on the free electron model but become distorted for larger values. (iii) The density of states as a function of energy. A rapid fall in the density occurs as the corners of the zone become filled.

$\partial E/\partial \boldsymbol{k}$, will then lie in the direction of $\boldsymbol{k}$. Under these conditions an electron in a state defined by a particular value of the wave-vector $\boldsymbol{k}$ will move in the corresponding direction in real space at a constant velocity defined by equation (7.57). Its wave-vector $\boldsymbol{k}$ will remain unchanged until eventually the electron is scattered by an impurity, a lattice imperfection, a crystal boundary or a lattice phonon. If the momentum of the lattice phonon is denoted by $\hbar\boldsymbol{K}$ the conservation of crystal momentum will give, on absorption of the phonon a modified electron wave-vector $\boldsymbol{k}'$ where

$$\boldsymbol{k}' = \boldsymbol{k}+\boldsymbol{G}+\boldsymbol{K}. \quad \ldots(7.58)$$

For values of $\boldsymbol{k}$ away from the band minimum the constant energy surfaces are no longer spherical. This is manifested by a warping of the constant energy surfaces and the formation of protuberances towards the zone boundary (see Fig. 7.19(ii)). The density of states shown in (iii), thus increases above the free electron value but subsequently falls off to zero as the corners of the zone are progressively filled. Since the electron velocity is perpendicular to the constant energy surface, the electron velocity at points such as B will not lie in the direction of the wave-vector.

We may also query the value of electron velocity at a zone boundary. It was shown in Section 7.6, by consideration of a one-dimensional model, that at zone boundaries the appropriate wave-functions representing electrons are stationary waves. This implies that for three dimensions the component of electron velocity perpendicular to the zone boundary is zero and that the constant energy surfaces must intersect the boundary perpendicularly. In the example shown in the figure the electron velocity corresponding to the wave-vector at C will be given by the value of $(1/\hbar)(\partial E/\partial \boldsymbol{k})$ at C and this will lie along the zone boundary. Finally, we note that in the present example the electron velocity is zero both at the centre of the zone and also at the corners of the zone. The latter points correspond to the top of the band. Energy bands may show minima at points other than the centre of a zone. In such cases the electron velocity will be zero at the band minimum, i.e. when $\partial E/\partial \boldsymbol{k} = 0$.

7.10 Effective mass†

In the free electron theory we studied the behaviour of an electron in an applied electric field and we found that a direct relationship existed between the curvature of the E vs. k curve and the electron mass (equation (7.53)). Let us now look at the corresponding behaviour on the band model, i.e. the behaviour of an electron in the presence of both an applied external field and the periodic lattice potential. An applied external electric field $\boldsymbol{E}$ will exert a force $\boldsymbol{F}$ of magnitude $-e\boldsymbol{E}$ on the electron and will produce a rate of change of electron energy given by

$$\frac{\mathrm{d}E}{\mathrm{d}t} = \boldsymbol{u}\cdot\boldsymbol{F}. \quad \ldots(7.59)$$

† The detailed mathematics in this Section may be omitted on a first reading.

The electron acceleration in accordance with equation (7.57) is then

$$\frac{d\boldsymbol{u}}{dt} = \frac{1}{\hbar}\nabla_k \frac{dE}{dt}$$

$$= \frac{1}{\hbar}\nabla_k(\boldsymbol{u}\cdot\boldsymbol{F}) \qquad \ldots(7.60)$$

$$= \frac{1}{\hbar^2}\nabla_k[(\nabla_k E)\cdot\boldsymbol{F}].$$

The vector acceleration is thus obtained as the scalar product $\nabla_k E\cdot\boldsymbol{F}$ acted upon by the operator ∇_k. This result should be compared with the corresponding equations of the free electron model. It is seen that the quantity $(1/\hbar^2)\nabla_k(\nabla_k E)$ has the dimensions of mass^{-1} and it is known as the effective mass tensor. Its components are seen to be of the form

$$\frac{1}{m^*_{rs}} = \frac{1}{m^*_{sr}} = \frac{1}{\hbar^2}\frac{\partial^2 E}{\partial k_r \partial k_s}, \qquad \ldots(7.61)$$

where r and s stand for any of x, y and z. The components of acceleration in equation (7.60) are thus

$$\frac{du_x}{dt} = \frac{1}{\hbar^2}\left(\frac{\partial^2 E}{\partial k_x^2}F_x + \frac{\partial^2 E}{\partial k_x \partial k_y}F_y + \frac{\partial^2 E}{\partial k_x \partial k_z}F_z\right) \qquad \ldots(7.62)$$

with symmetrical expressions for

du_y/dt and du_z/dt.

For a solid with a complex band structure all nine components of the tensor may be different and one may also note from equation (7.62) that components such as

$(1/\hbar^2)(\partial^2 E/\partial k_x \partial k_y)$

which are not zero will produce an electron acceleration in a different direction from that of the applied field. For simpler band structures such components are zero and three effective masses m_1^*, m_2^* and m_3^* then suffice to describe the electron acceleration, e.g. equation (7.62) will then reduce to

$$m_1^*\frac{du_x}{dt} = F_x \qquad \ldots(7.63)$$

with corresponding expressions for F_y and F_z. Assuming that the band minimum occurs at $\boldsymbol{k} = 0$, the variation of E near the band minimum is then of the form

$$E = E_0 + \frac{\hbar^2}{2}\left(\frac{k_x^2}{m_1^*} + \frac{k_y^2}{m_2^*} + \frac{k_z^2}{m_3^*}\right). \qquad \ldots(7.64)$$

E_0 determines the zero energy of the band and the constant energy surfaces described by the equation are ellipsoids.

For the simplest band a single scalar effective mass m_e^* may be used to describe the behaviour of the electron. The energy relationship is then

$$E = E_0 + \frac{\hbar^2}{2m_e^*}(k_x^2 + k_y^2 + k_z^2), \qquad \ldots(7.65)$$

where m_e^* is given by

$$m_e^* = \hbar^2 \bigg/ \frac{d^2E}{dk^2} \qquad \ldots(7.66)$$

i.e. by an expression of the same type as that obtained in equation (7.53) on the free electron theory. The constant energy surfaces are then spherical and such energy bands are often referred to as being spherical.

In practice energy bands are more complex than might be inferred from these approximations. Some of the more important features of energy bands in metals and semiconductors are described in Sections 7.12 and 7.13.

7.11 Electrons and holes

In the simple band structure referred to in the previous section the similarity in the expression for effective mass to the one in the free electron model was pointed out. Let us briefly consider the significance of this in terms of the nearly free electron model mentioned at the end of Section 7.6 and illustrated in Fig. 7.13. Clearly the two models differ most significantly at points close to the zone boundary. We note also that whereas the free electron mass is a constant, the effective mass m_e^* on the n.f.e. model is a function of $\boldsymbol{k}$. Both masses are, of course, given by $\hbar^2/(d^2E/dk^2)$ in accordance with equations (7.53) and (7.66). Referring therefore to the E vs. $\boldsymbol{k}$ curve on the n.f.e. model the effective mass at the bottom of the band is positive and equal to m_e. It then gradually increases with $\boldsymbol{k}$ becoming infinite at the point of inflection of the curve. At the top part of the band $d^2E/d\boldsymbol{k}^2$ is negative so that the effective mass is then negative. Whereas an electron in a state near the zone boundary has both E and $\boldsymbol{k}$ increased by the field, the component of electron velocity perpendicular to the boundary is decreased. This is in accord with the development of a stationary wave near the band edge. The increase in total energy E is due to an increase in potential energy which is such as to outweigh the loss of kinetic energy.

The use of an effective mass thus allows the behaviour of an electron in an applied external field to be derived as if the electron were a classical particle possessing translational inertial mass. The electrons are, however, interacting with the periodic lattice potential and this accounts for the complex behaviour of the effective mass parameter.

For nearly filled bands it is simpler to deal with the behaviour of the vacant states rather than that of the electrons. In order to achieve this the electron effective mass tensor defined by equation (7.61), whose components are normally all negative near the top of a band, is replaced by a hole mass tensor having positive components. For electrons with scalar effective mass m_e^* one simply puts

$$m_h^* = -m_e^* \qquad \ldots(7.67)$$

where m_h^* is the hole effective mass. Under the action of an applied electric field $\boldsymbol{E}$ we therefore have

$$m_e^* \frac{du}{dt} = -eE \qquad \ldots(7.68)$$

replaced by

$$m_h^* \frac{du}{dt} = eE \qquad \ldots(7.69)$$

i.e. the hole behaves as if it has both positive mass and positive charge. Under the action of an applied field the electrons in a nearly filled band will have their wave-vectors changed systematically, and those electrons reaching a zone boundary will be reflected back to the opposite boundary. If there is an electron missing from a given state then it is termed to be occupied by a positive hole. The vacant state moves in **k**-space in the *same* direction as that of the electrons and the motion of the hole is thus governed by the response to the applied field of all the electrons in the occupied states. It is clear that holes lower in a band have the greater energy and it follows therefore that representation of hole energy in a band must be opposite to that of electron energy.

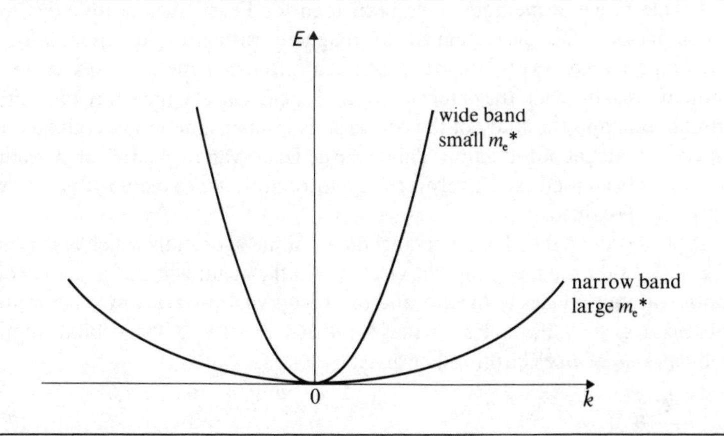

Fig. 7.20 The dependence of effective mass m_e^* on the curvature and width of an energy band.

A point regarding the wave-vectors of holes is of interest. In accordance with the discussion in Section 3.3 a Brillouin zone must possess inversion symmetry and it follows that for a filled band the sum of all the wave-vectors is zero. If an electron of wave-vector $\boldsymbol{k}_e$ is removed from the band then the total wave-vector for the band is $-\boldsymbol{k}_e$ and this gives the wave-vector of the hole. A more complete discussion of this point is given by Kittel (1971, 4th edn, p. 328).

The width of an energy band for a particular solid depends on the form of the lattice potential for that band of the solid. As illustrated in Fig. 7.20 a wide band will have a larger value of $\mathrm{d}^2E/\mathrm{d}k^2$ than will a narrow band. Consequently, small effective masses are associated with wide bands and large effective masses with narrow bands. An energy band will normally consist of a number of sub-bands, each of which may exhibit different effective masses. The occurrence of light and heavy holes in a semiconductor such as germanium is one such example.

7.12 Energy bands in metals

The constant energy surface in $\boldsymbol{k}$-space corresponding to the Fermi level is termed the Fermi surface and its form is a major factor determining the conduction properties of a solid. Considerable effort has been put into elucidating the form of this surface for different solids and a number of different experimental techniques have been evolved. Although for a particular solid the band may be spherical for small $\boldsymbol{k}$, if the band is half filled the effective mass at the Fermi level may well be rather different and it is the behaviour of electrons at this level which governs the conduction properties. When a band is nearly filled the solid will exhibit hole conduction and give a changed sign of the Hall constant.

A wide range of methods have been used for Fermi surface investigations. These include the measurement of magnetoresistance, the attenuation of ultrasonic sound waves, the absorption of electromagnetic waves as in the anomalous skin effect, the determination of specific heats, the study of positron annihilation and the study of the oscillations in diamagnetic susceptibility as a function of magnetic induction. This is the de Haas–van Alphen effect. A method which has been used extensively is the cyclotron resonance method and this will be briefly described.

Typical is the Azbel–Kaner arrangement for metal crystals which is shown in Fig. 7.21. In the magnetic flux the electrons in the metal will perform an orbital motion at right-angles to the flux and this is superimposed on any other motion the electron may have. For a magnetic flux density $\boldsymbol{B}$ the orbital angular frequency ω_c of an electron will be given by

$$\omega_c = \frac{e}{m_e^*}B \qquad \ldots(7.70)$$

which is the cyclotron frequency. When the angular frequency ω of the applied

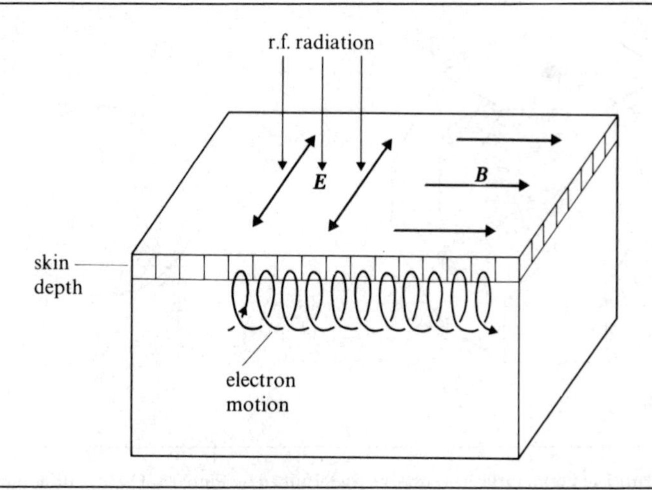

Fig. 7.21 The Azbel–Kaner arrangement for observation of cyclotron resonance.

r.f. electric field is such that it coincides with ω_c the electrons will extract energy from the oscillator. This will occur during that part of the helical motion of an electron when it lies within the skin depth of the surface. It is termed the Azbel–Kaner resonance. Measurement of the surface impedance as a function of the r.f. frequency and the magnetic flux density will thus provide the data yielding the effective mass m_e^*. The electrons, whose behaviour is observed, are those at the Fermi level and to attain the long mean free paths, which are necessary to obtain extended orbits, measurements are carried out on very pure crystals at low temperatures.

The simplest example is that of a metal which exhibits a nearly spherical Fermi surface. The electron orbits in $\boldsymbol{k}$-space lie on the surface of this sphere and the orbit of maximum diameter, termed an extremal orbit, will correspond to circular electron paths in real space, i.e. those orbits which do not have a superimposed velocity along the direction of the magnetic flux. For non-spherical Fermi surfaces the topography is determined by taking sets of measurements with samples of different crystal orientation. Extremal orbits, i.e. those which show maximum or minimum resonant periods, can be identified since the absorption of the r.f. field is dominant for such orbits. This allows the Fermi surface to be mapped and the type of result obtained is illustrated in Fig. 7.22 for copper.

We may now consider the band structure of specific elements and deal first with the alkali metals. The energy bands in solid sodium have been compared with the atomic levels in the isolated atom in Figs. 7.7 and 7.8 and it is noted that, of the occupied atomic levels, only the 3s level is significantly broadened into a band in the solid. The form of the band has been deduced from X-ray spectra, the intensities of the soft X-ray photons emitted as a result of electron transitions

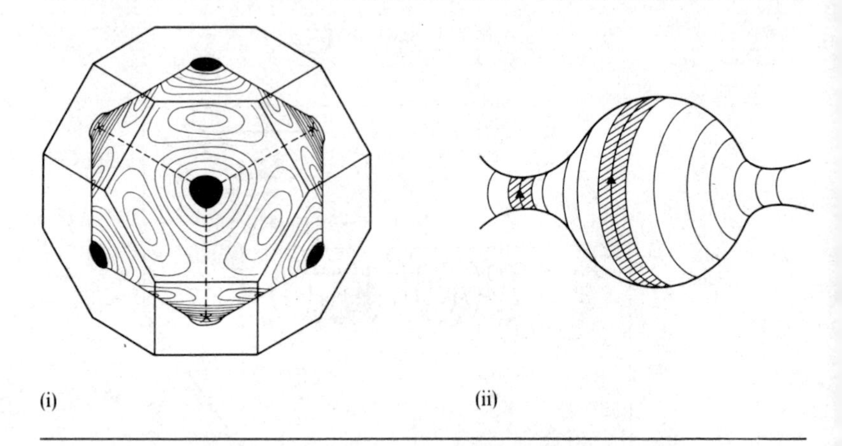

Fig. 7.22 (i) The Fermi surface of copper as determined by Pippard. The first Brillouin zone of the f.c.c. structure is a truncated octahedron and the Fermi surface touches the hexagonal faces of the zone in the eight ⟨111⟩ directions in *k*-space. (ii) Extremal orbits.

from the 3s band into the 2p level giving a measure of the density of occupied states for each value of wavelength or energy. A parabolic density of states relationship has been confirmed for the s band in sodium in this way. In addition X-ray absorption measurements have been used to determine the width of the unoccupied states.

One therefore differentiates between core electrons such as 2p which are completely localised and which do not overlap between adjacent atoms and the 3s valence electrons which extend throughout the metal. The success of the Sommerfeld model in explaining the main features of electrical conductivity and the heat capacity of metals has been described in previous chapters. As might therefore be expected the nearly free electron model has been successfully applied to the alkali metals. In the model, the wave-function approximates to that of an atomic orbital near an ion core whereas elsewhere it is in the form of a plane wave. Although the variation in lattice potential near an ion core may be rather large the scattering experienced by a conduction electron in a metal is small. This has led to the use of pseudo-potential methods in which the actual core potentials are replaced by much smaller pseudo-potentials in order to derive the wave-function. Account is taken in the method of the requirement that the valence electrons cannot, in accordance with the Pauli exclusion principle, occupy the core atomic orbitals. As a consequence the theory shows that the effective, or pseudo-potential, for the core region, which is determining the form of the wave-function, is much less than the true potential. The corresponding pseudo-function is then similar to a plane wave and this property explains why the valence electrons in metals behave as free particles as originally postulated on the free electron theory.

As was discussed in Section 7.6 and illustrated in Fig. 7.13 the nearly free electron model gives a parabolic relationship between E and k for most of $\boldsymbol{k}$-space. For an alkali metal the variation over a substantial part of the band can be well represented by

$$E = E_0 + \frac{\hbar^2 k^2}{2m_e^*}, \qquad \ldots(7.71)$$

where E_0 is the minimum energy of the band and m_e^* is an effective electron mass. The density of states $S(E)$ is similarly given by the corresponding relationship

$$S(E) = \frac{1}{2\pi^2}\left(\frac{2m_e^*}{\hbar^2}\right)^{\frac{3}{2}} V(E-E_0)^{\frac{1}{2}} \qquad \ldots(7.72)$$

which should be compared with equation (5.41) of the free electron model. Some variation of m_e^* throughout the band is found to occur for the alkali metals. A mean value over the band of $m_e^* = 1.22m_e$ is found for sodium whereas the electrons at the Fermi level involved in conduction processes have $m_e^* \simeq m_e$.

At the top of the band the density of states must finally decrease to zero and for this region the energy variation forms an inverted parabola. As is clear from the previous discussion this region of the band is not occupied. In addition an incompletely filled band gives a situation analogous to that on the free electron theory and, as described in Section 5.10, there will be electrical conduction typical of a metal.

The form of the Fermi surface for the alkali metals can be inferred by comparing the diameter of the free electron Fermi sphere with the dimensions of the corresponding Brillouin zone. In terms of the volume V of the unit cell in the direct lattice the volume V^* of a zone is given by

$$V^* = \frac{(2\pi)^3}{V}. \qquad \ldots(7.73)$$

A Brillouin zone may contain two electrons per atom, so that in order to accommodate x valence electrons per atom a volume of $\boldsymbol{k}$-space equal to $\frac{1}{2}x(2\pi)^3/V$ per atom is required. This assumes that there is one atom per unit cell in the direct lattice. Putting k_F as the radius of the Fermi sphere we have

$$\tfrac{4}{3}\pi k_F^3 = \frac{x}{2}\frac{(2\pi)^3}{V}. \qquad \ldots(7.74)$$

The value of V depends on the crystal structure, and is $\frac{1}{4}a^3$ for a f.c.c., $\frac{1}{2}a^3$ for a b.c.c. and $\frac{1}{2}\sqrt{3}a^2c$ for a h.c.p. structure.

The alkali metals are monovalent and have a b.c.c. structure so that we find

$$k_F = \frac{3.9}{a}. \qquad \ldots(7.75)$$

As noted in Section 7.7 the Brillouin zone for this structure consists of a regular rhombic dodecahedron and it may be confirmed that the shortest distance k_A from the centre of the zone to a boundary face is

$$k_A = \frac{\sqrt{2}\pi}{a} = \frac{4.4}{a}. \qquad \ldots(7.76)$$

The Fermi sphere does not therefore intersect the zone boundary in an alkali metal. Experimental evidence confirms this result and indicates very little distortion of the actual Fermi surface from the free electron surface. The periodic lattice potential is having little effect on the conductivity of the alkali metals.

The Fermi surface for the noble metals, copper, silver and gold is of interest. These metals are all univalent and crystallise in a f.c.c. structure. The corresponding Brillouin zone is a truncated octahedron (see Fig. 7.22) and it may be confirmed, by following the same procedures as for the alkali metals, that the radius of the Fermi sphere is rather less than the shortest distance from the centre of the zone to the hexagonal shaped faces of the zone. As discussed in Section 7.9 and illustrated in Fig. 7.25 warping of the constant energy surfaces occurs near a zone face and in these metals is such as to make the Fermi surface overlap the hexagonal faces of the zone. The constant energy surface is illustrated in Fig. 7.22(ii). This is hyperbolic in form near the zone boundary and it is termed a neck. Also shown on the figure are extremal orbits which confirm experimentally the form of the Fermi surface for these metals as first proposed by Pippard.

The Group IIA alkaline earth metals beryllium and magnesium are divalent so that the number of valence electrons per atom would be sufficient to fill completely their respective s bands. A completely filled band cannot provide conduction since an applied electric field cannot raise electrons to adjacent states of higher energy. Since these elements are metallic we conclude, therefore, that the band structure is such as to provide overlap between the s and p bands with a measure of electron occupation of the higher lying p band.

The nature of zone overlaps is illustrated in Fig. 7.23 for a two-dimensional square lattice. In the first figure the lattice potential is assumed to be vanishingly small and the constant energy contours in $\boldsymbol{k}$-space are all circular. The Fermi surface separates the filled regions, which are shown shaded, from the adjacent unfilled regions. The reduced zone representation is obtained, in accordance with Section 7.5, by translating the second zone into the first by use of appropriate lattice vectors. This representation shows clearly the unfilled states, or holes, in the $\langle 11 \rangle$ directions at the corners of the first zone and the overlap of electrons in the $\langle 10 \rangle$ direction across the $\{10\}$ faces. The nature of the orbits at the Fermi surface is perhaps more easily appreciated from the periodic zone representation in Fig. 7.24. Hole orbits are enclosing unfilled states whilst electron orbits enclose filled states. The possibility of open orbits, as obtained for copper, is shown in Fig. 7.25 for a rectangular lattice.

The effect of a periodic potential is illustrated in Figs. 7.19 and 7.26, from which we may infer the situation in a real metal. First, we find that the constant energy surfaces are warped so that they bulge towards the zone boundary faces.

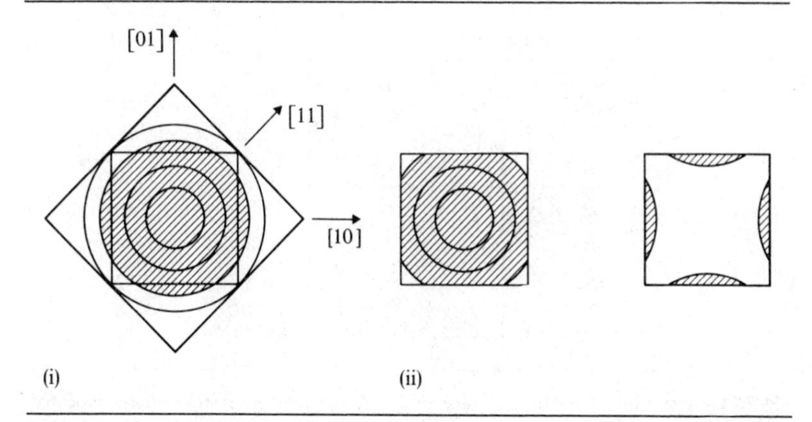

Fig. 7.23 (i) The first and second Brillouin zones of a square lattice are shown together with a series of constant energy surfaces on a free electron model. The Fermi surface encloses the shaded area and overlaps into the second zone are occurring over the planes of the {10} type. (ii) These diagrams show the first and second zones in the reduced representations.

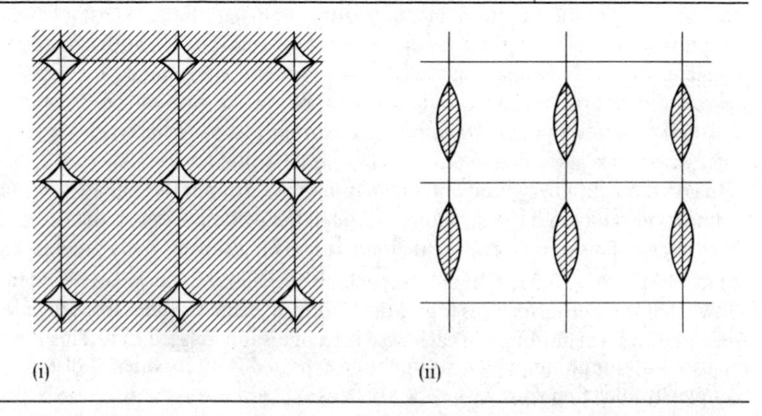

Fig. 7.24 The Fermi surface of the previous diagram is drawn here in the periodic zone scheme. (i) The Fermi surface lying within the first zone encloses unoccupied states. (ii) The Fermi surface lying within the second zone encloses occupied states.

Intersections occur at right-angles to the zone faces. In a metal the distortions of the constant energy surfaces would be such as to give a density of states above that expected on the free electron model. Secondly, there are now energy discontinuities across the zone boundaries. In a real structure, zone boundaries in different directions will be reached at different values of energy and sharp discontinuities in the total density of states occur at such points. These are referred to as van Hove discontinuities.

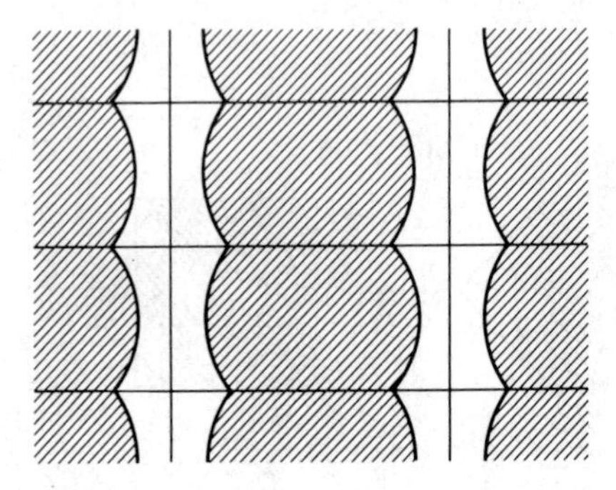

Fig. 7.25 Open orbits. The Fermi surface for a rectangular lattice is shown in the repeated zone representation for free electrons.

The example chosen illustrates a further point, namely that states in the ⟨10⟩ directions in the second zone are being filled before all the states in the ⟨11⟩ directions in the first zone are occupied. As shown in Fig. 7.26 there is therefore, for this model, no range of forbidden values of energy, i.e. there is no band gap. This is the type of situation in magnesium and beryllium. These elements crystallise in a h.c.p. structure and the s bands are overlapped in both cases by the adjacent p band. The low conductivity of beryllium is explained by the small degree of overlap. Similar considerations apply in the case of the three other Group IIA metals, calcium, strontium and barium, although the types of overlap differ, strontium and calcium having a f.c.c. and barium a b.c.c. structure.

In common with beryllium and magnesium the Group IIB elements, zinc and cadmium also have a h.c.p. structure although the Brillouin zone is distorted for these metals. This is as a result of their c/a ratio being substantially greater than the ideal value of $\sqrt{8/3}$ for the closest packing of spheres. Experimental evidence shows that the Fermi surfaces for all the Group II elements differ from the ideal free electron form although in each case is recognisably related to it. This is also the case for aluminium which, as might be expected from its valency of 3, has a completely filled first zone with zone overlaps extending as far as the fourth zone.

The transition metals, iron, cobalt and nickel differ from copper in having an incompletely filled 3d band overlapping the 4s band, i.e. the Fermi surface separates filled and unfilled states in both bands for these metals whereas for copper the 3d states are all filled and the Fermi level lies within the higher lying 4s band. This has been illustrated in Fig. 7.18 for nickel and it is seen that for a transition metal such as nickel the 3d band is quite narrow compared with the 4s band. The electron core overlap is small, indicating low interaction between atoms, and the tight binding approach has been successful in the study of this band. Since the band arises from d atomic levels there are 10 electron states per atom in the band and combined with its narrow width gives a high density of states. This form of band structure giving high density of states and small

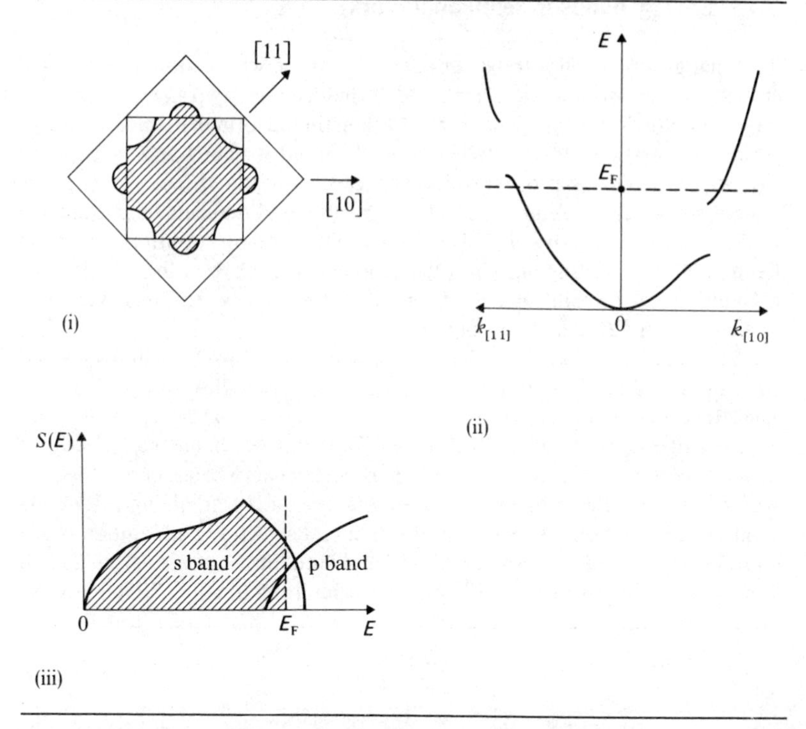

Fig. 7.26 The nearly free electron model. (i) The Fermi surface is shown for the first and second zones. The periodic potential gives rise to energy discontinuities at the boundary of the zone. Note that low energy states in direction [10] in the second zone are being filled before higher energy states in direction [11] in the first zone. (ii) The E vs. k curves for directions [10] and [11] illustrate the energy discontinuities at the zone boundary. (iii) The density of states $S(E)$ for the lower s band and the upper p band. The occupied states are shown shaded.

electron core overlap has been used to explain the magnetic properties of the transition metals. The ferromagnetic metals have low conductivity compared with copper. This has been explained in terms of the trapping of electrons from the 4s band in vacant 3d states which are at a high density at the Fermi level.

The elements discussed so far have all been metals and we have seen how this comes about in terms of their band structure. In view of their position in Group V it might have been expected that the elements arsenic, antimony and bismuth would have been metallic with a large overlap into the third zone. These elements, however, crystallise with 2 atoms in their unit cell so that the 10 electrons in the unit cell are just sufficient to fill 5 bands completely. In practice the fifth band is incompletely filled with a small overlap into the sixth band. The elements therefore exhibit both electron and hole conduction. Since the overlap is maintained at all temperatures down to 0 K they are classified as semimetals.

7.13 Energy bands in semiconductors

The final group of solids to be considered are those whose atomic and crystal structures lead to the occurrence of forbidden energy ranges. The most important are the Group IV elements. Carbon (in the form of diamond), silicon, germanium and grey tin, crystallise in the diamond structure and have a band gap which decreases progressively from carbon to tin. Lead, which is the remaining Group IV element, has a f.c.c. structure and is a metal. Some of these elements have alternative crystal structures. Thus grey tin transforms at room temperature to a tetragonal form which is metallic. Carbon in its hexagonal graphite form exhibits anisotropy in its conductivity and behaves as a semiconductor with zero band gap.

The diamond structure, as we have seen in Chapter 3, is a tetrahedrally bonded structure based on the hybrid sp^3 orbitals and has been illustrated in Fig. 4.15. The lattice is f.c.c. with a basis of atoms at 000 and $\frac{1}{4}\frac{1}{4}\frac{1}{4}$ so that each atom has four nearest neighbours tetrahedrally disposed. The similarity in the outer electronic structure of the isolated atoms of the Group IV elements is clear; in each case the atoms have two electrons in the s level and two in the adjoining p level. As mentioned in Section 7.8 there is in the diamond structure a separation of the levels into two bands, the lower valence band with four states per atom and an upper conduction band also with four states per atom (see Fig. 7.27). Since the

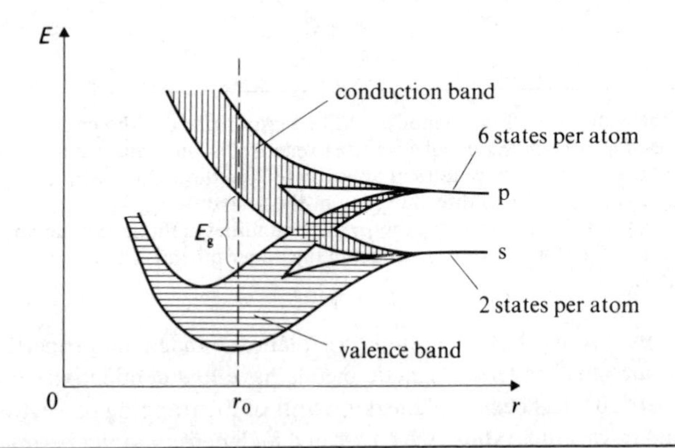

Fig. 7.27 The formation of energy bands in a semiconductor such as silicon or germanium. At the equilibrium lattice spacing r_0 the hybrid sp^3 bonding orbitals provide four states per atom in the valence band. At absolute zero the antibonding conduction band, also with four states per atom, is empty.

elements are quadrivalent the valence band is able to accommodate all the electrons. This is the situation at absolute zero and the conduction bands are then empty. At this temperature the elements behave as insulators.

In semiconductors such as germanium and silicon thermal energy allows excitation of a small proportion of the electrons from the valence band into the conduction band. At ordinary temperatures, therefore, these solids exhibit conduction, this arising both from the conduction band electrons and the holes in the valence band. In diamond, the band gap is sufficiently wide to prevent any appreciable thermal excitation and diamond therefore remains an insulator.

We have noted that there are two atoms in each unit cell in the diamond structure. From the standpoint of zone theory the valence band consists of 4 sub-bands each accommodating 2 electrons per unit cell. Overall the 8 electrons in the unit cell may be accommodated within this band. The upper conduction band will similarly consist of 4 sub-bands. The form taken by the various valence and conduction bands will now be discussed. Of interest are the positions of the various maxima and minima; in particular those which lie close to the Fermi surface and which are significant in conduction phenomena.

Some of the important features of energy band structures can be developed directly from the empty lattice approach. In this method a lattice is assumed which has the translational periodicity of the real lattice but with a vanishingly small lattice potential. Energy relationships are derived for different directions in the Brillouin zone for the assumed structure and are then modified to take account of symmetry elements of the actual crystal. A band structure developed in this manner is shown in Fig. 7.28 for the diamond structure. The similarity to the band structures of germanium and silicon given in Figs. 7.29 and 7.30 is apparent. Details of the method are given by Long (1968).

The valence bands of germanium and silicon are qualitatively similar and may therefore be discussed together. In each case the top of the valence band is defined by two maxima coincident at $\boldsymbol{k} = 0$. The bands involved, designated V1 and V2, exhibit distortion of the constant energy surfaces as is illustrated in Fig. 7.31 for silicon. This distortion is a result of the degeneracy of the bands at $\boldsymbol{k} = 0$. For the region close to $\boldsymbol{k} = 0$ the relationship between E and $\boldsymbol{k}$ for both bands may be represented by the single relationship

$$E = -\frac{\hbar^2}{2m_e}[Ak^2 \pm \{B^2k^4 + C^2(k_x^2k_y^2 + k_y^2k_z^2 + k_z^2k_x^2)\}^{\frac{1}{2}}], \qquad \ldots(7.77)$$

where the zero of energy is taken at the top of the band and where A, B and C are constants (see Kittel, 1963). These may be accurately determined from cyclotron resonance experiments. The positive sign in the relationship gives the band of smaller curvature, i.e. the heavy hole V1 band; the negative sign gives the light hole V2 band.

Anisotropy in the hole masses is governed by the value of the constant C and is found to be most marked for the heavy hole. An average effective mass can, however, be determined for each band in terms of the constants A, B and C and the values obtained for the different hole effective masses are given in Table 7.1. Their adoption is equivalent to the replacement of the warped energy surfaces by spherical surfaces and the use of a parabolic relationship between E and k. The values can be utilised in the density of states function in the form of equation

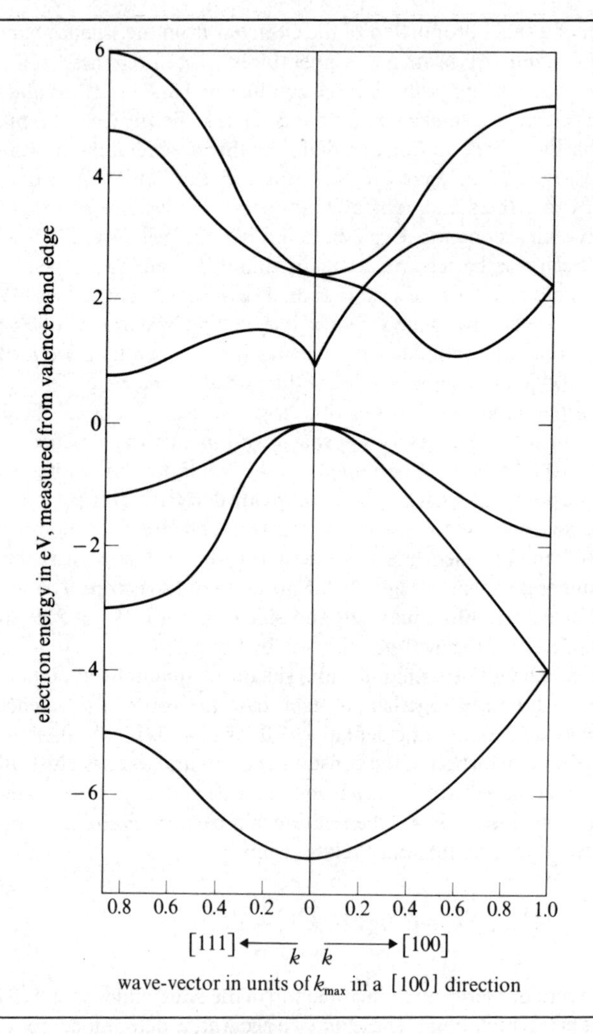

Fig. 7.28 Energy band structure of a diamond type crystal.

(7.72) each band being treated separately. This is equivalent to using a combined density of states effective mass m_h^* given by

$$m_h^* = (m_{V1}^{*3/2} + m_{V2}^{*3/2})^{2/3} \qquad \ldots(7.78)$$

for the two bands.

Away from the valence band edge at $\boldsymbol{k} = 0$ the parabolic relationship between E and k breaks down. The valence bands are, however, very nearly full so that the relationship given applies, with the conduction dominated by the heavy hole

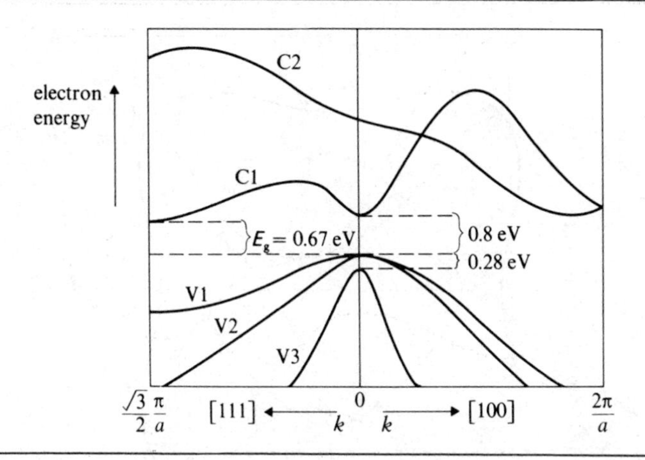

Fig. 7.29 The energy bands of germanium adjacent to the band gap.

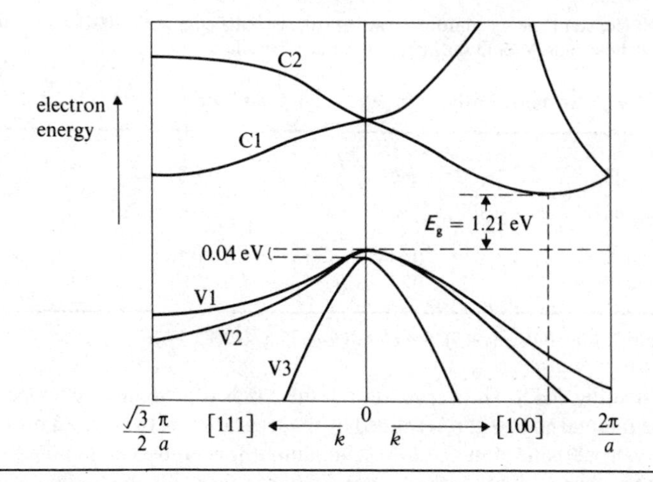

Fig. 7.30 The energy bands of silicon adjacent to the band gap.

band V1. There is, in addition, a further band V3 split off from V1 and V2 at $\boldsymbol{k} = 0$. This splitting is due to a spin-orbit interaction and it amounts to 0.04 eV for silicon and 0.28 eV for germanium. Since there is no degeneracy at $\boldsymbol{k} = 0$ for this band the energy surfaces are spherical. Values of the scalar effective mass m^*_{V3} are included in Table 7.1.

The conduction bands of both silicon and germanium are more complex in form than their valence bands and in each case a number of band minima are

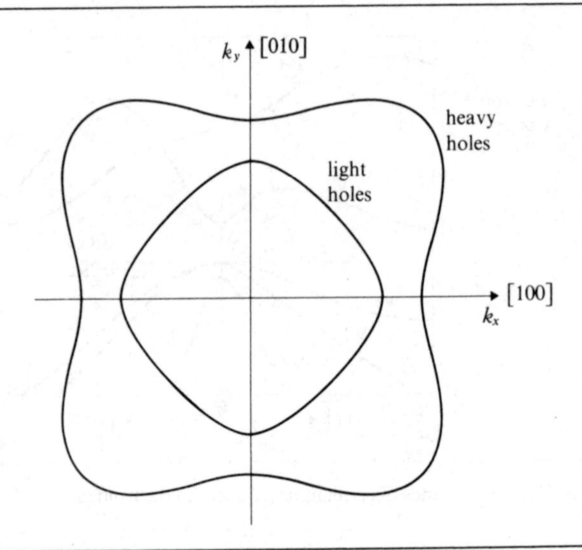

Fig. 7.31 Constant energy contours near the valence band edge in the (001) plane showing the heavy hole band V1 and the light hole band V2 in silicon.

Table 7.1 Effective masses of holes in germanium and silicon†

	$\dfrac{m^*_{V1}}{m_e}$	$\dfrac{m^*_{V2}}{m_e}$	$\dfrac{m^*_{V3}}{m_e}$	$\dfrac{m^*_h}{m_e}$
Ge	0.36	0.043	0.075	0.37
Si	0.50	0.15	0.23	0.55

† *Data for Ge from Putley* (1960). *Data for Si from Long* (1968).

found (see Fig. 7.32). The occupation of the bands is governed by temperature and the thermal energy gap is defined by the separation of the highest maximum of the valence band and the lowest minimum (or equivalent minima) of the conduction band. Since these occur at different values of $\boldsymbol{k}$ the thermal gaps for these elements are indirect gaps. Their values are given in Table 7.2.

As illustrated in Fig. 7.32 the lowest conduction band minima in germanium lie at the surface of the zone in the $\langle 111 \rangle$ directions. The constant energy surfaces near the band minima are spheroids with their symmetry axes along the $\langle 111 \rangle$ directions. These minima are termed valleys and from the figure they can be seen to be equivalent to four complete spheroids. The conduction band has four equivalent minima.

For silicon the lowest conduction band minima do not lie at the zone boundary but at a value of k of $0.85k_{\max}$ in the $\langle 100 \rangle$ directions. There are,

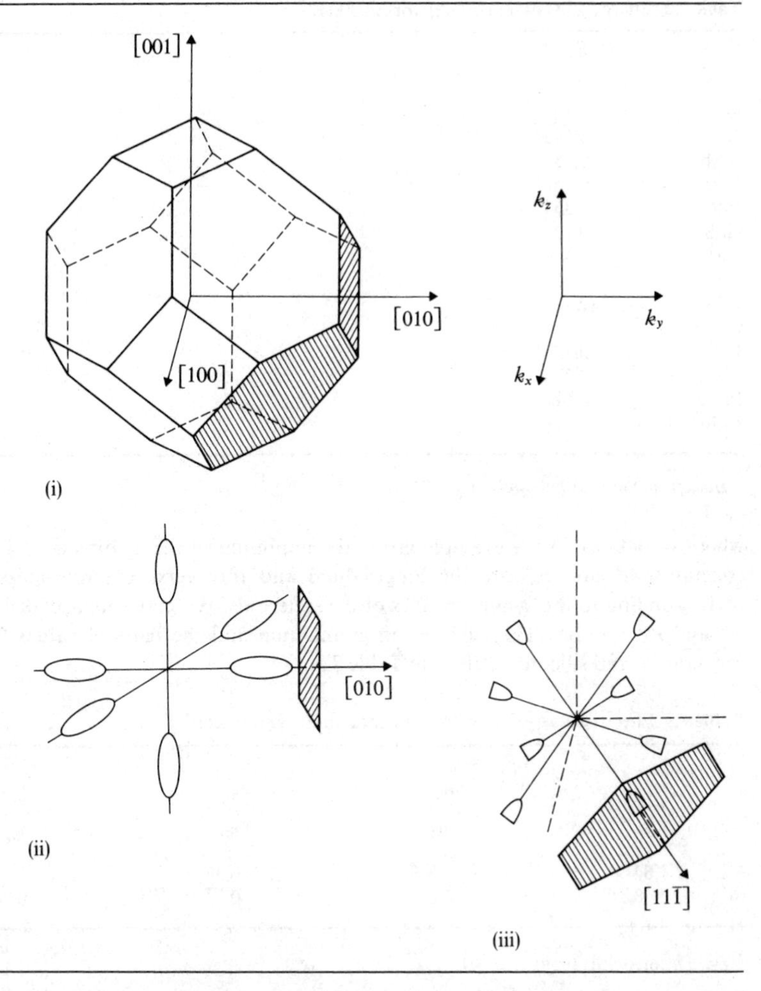

Fig. 7.32 (i) The first Brillouin zone for a diamond lattice. The zone boundaries consist of the six planes of the {100} type and eight planes of the {111} type. (ii) The surfaces of constant energy near the conduction band edge in silicon consist of ellipsoids having their axes along the six ⟨100⟩ directions. (iii) The surfaces of constant energy near the conduction band edge in germanium consist of half ellipsoids having axes along the eight ⟨111⟩ directions.

therefore, in this case 6 equivalent valleys and the variation of E with k for each will be given by an expression of the type

$$E = \frac{\hbar^2}{2}\left[\frac{(k_x - k_0)^2}{m_l^*} + \frac{k_y^2 + k_z^2}{m_t^*}\right] \qquad \ldots(7.79)$$

Table 7.2 Energy gaps of semiconductors at 300 K[†]

	E_g/eV
Si	1.12
Ge	0.665
InSb	0.18
InAs	0.35
InP	1.35
GaSb	0.72
GaAs	1.4
GaP	2.26
CdS	2.4
CdSe	1.7
PbS	0.32
PbSe	0.27
PbTe	0.22
AlSb	1.52

[†] *Data from Smith* (1959) *and Long* (1968).

where k_0 defines, in the example given, the minimum in the k_x direction. The constants m_l^* and m_t^* are the longitudinal and transverse effective masses corresponding to the symmetry axes of the spheroids. We may similarly define m_l^* and m_t^* for the $\langle 111 \rangle$ minima in germanium and the pairs of values for germanium and silicon are given in Table 7.3.

Table 7.3 Effective masses of electrons in germanium and silicon[†]

	$\frac{m_l^*}{m_e}$	$\frac{m_t^*}{m_e}$	$\frac{m_e^*}{m_e}$	$\frac{m_c^*}{m_e}$
Ge	1.64	0.008	0.22	0.12
Si	0.98	0.19	0.33	0.26

[†] *Data from Smith* (1959)

For the calculation of the density of states in the conduction band an effective mass m_e^* defined by

$$m_e^* = (m_l m_t^2)^{\frac{1}{3}} \qquad \ldots(7.80)$$

is used. This is equivalent to reducing equation (7.79) to the parabolic form and the density of states will then be given by

$$S(E) = \frac{M_c}{2\pi^2}\left(\frac{2m_e^*}{\hbar^2}\right)^{\frac{3}{2}} VE^{\frac{1}{2}} \qquad \ldots(7.81)$$

where M_c is the number of equivalent minima in the conduction band. For germanium and silicon the values of M_c are, respectively, 4 and 6 in accordance with Fig. 7.32.

In the free electron theory electrical conductivity is directly proportional to the electron mobility defined by the expression

$$\mu = \frac{e\tau}{m_e}. \qquad \ldots(5.9)$$

In band theory an average conductivity mobility

$$\mu_c = \frac{e\tau}{m_c^*} \qquad \ldots(7.82)$$

is defined where the conductivity effective mass m_c^* is given by

$$\frac{1}{m_c^*} = \frac{1}{3}\left(\frac{1}{m_l} + \frac{2}{m_t}\right). \qquad \ldots(7.83)$$

Further information regarding density of states and conductivity effective masses (equations (7.78 to (7.83)) is given by Smith (1959).

The energy band structures of many semiconductors are known in considerable detail. Of interest are studies on grey tin (α-Sn) which indicate that, contrary to earlier measurements placing grey tin as a semiconductor with a narrow gap ~ 0.1 eV, it is now considered to be a semimetal with a zero band gap at $\boldsymbol{k} = 0$.

Closely related to the diamond structure is the zinc-blende type. The two structures differ only in that the two interpenetrating f.c.c. sub-lattices in the zinc-blende structure have atoms of different elements associated with each lattice. Binary compounds between Group III and Group V elements, and some between Group II and Group VI, crystallise in this structure. These are semiconductors closely allied to silicon and germanium and a typical band structure for the III–V compound GaAs is shown in Fig. 7.33. There is evidence that the absence of inversion symmetry in the structure gives a more complex valence band edge than is indicated in the figure. The energy gaps of some III–V compounds are included in Table 7.2. Apart from AlSb and GaP, which have indirect gaps similar to silicon, the III–V compounds are direct gap semiconductors.

Another closely related structure is the hexagonal wurtzite type which differs from the zinc-blende only in the arrangement of the second nearest neighbours. As a consequence the energy band structure obtained is very similar to that found for the zinc-blende structure. The II–VI compounds CdS and CdSe crystallise in this hexagonal wurtzite form. They are of interest because of their use as phosphors and the very marked photoconductive effects shown by CdS.

Finally one may note that, although the semiconducting elements and compounds discussed have covalent bonding, the fairly highly polar II–VI compounds such as PbS, PbSe and PbTe are also semiconductors. These

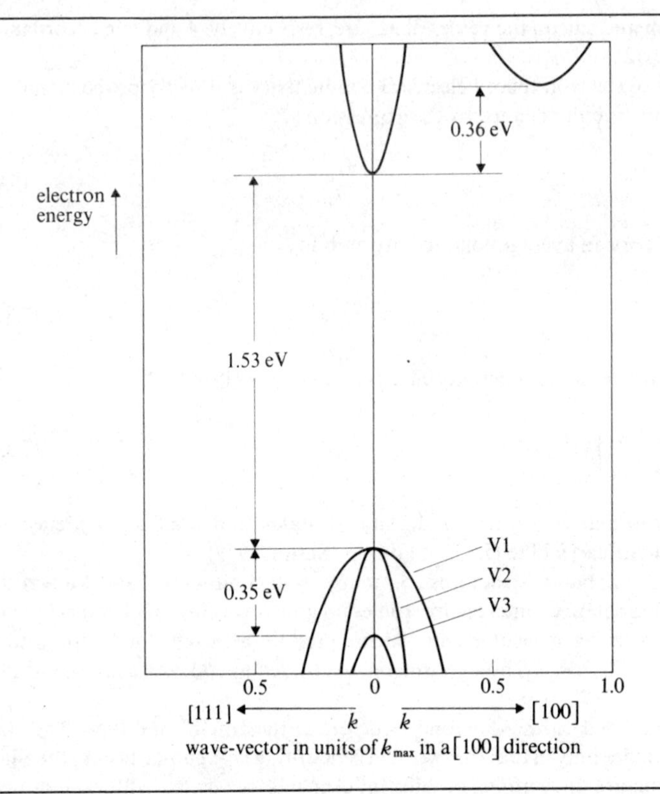

Fig. 7.33 The energy bands for GaAs in the [100] and [111] directions. The energy differences indicated are those determined experimentally at 0 K.

compounds crystallise in the rock-salt structure which has the two f.c.c. sub-lattices displaced by half a diagonal vector with respect to each other. They have a fairly narrow direct band gap, which is located at the Brillouin zone boundary in the ⟨111⟩ directions, and have been rather widely studied because of their photoconductivity in the infrared region.

References and further reading

Altmann, S. L. (1970) *Band Theory of Metals*, Pergammon Press.
Blakemore, J. S. (1969) *Solid State Physics*, Saunders.
Dekker, A. J. (1967) *Solid State Physics*, Macmillan.
Hart-Davis, A. (1975) *Solids, an Introduction*, McGraw-Hill.
Holden, A. (1971) *Bonds between Atoms*, Oxford University Press.

Kittel, C. (1963) *Quantum Theory of Solids*, Wiley.
Long, D. (1968) *Energy Bands in Semiconductors*, Interscience.
McKelvey, J. P. (1966) *Solid State and Semiconductor Physics*, Harper and Row.
Omar, M. A. (1975) *Elementary Solid State Physics*, Addison-Wesley.
Pincherle, L. (1971) *Electronic Energy Bands in Solids*, Macdonald.
Putley, E. H. (1960) *The Hall Effect and Semiconductor Physics*, Dover Publications.
Rosenberg, H. M. (1974) *The Solid State*, Oxford University Press.
Smith, R. A. (1959) *Semiconductors*, Cambridge University Press.
Smith, R. A. (1963) *Wave Mechanics of Crystalline Solids*, Chapman and Hall.
Wert, C. A. and **Thomson, R. M.** (1970) *Physics of Solids*, McGraw-Hill.
Wilkes, P. (1973) *Solid State Theory in Metallurgy*, Cambridge University Press.
Ziman, J. M. (1964) *Electrons in Metals*, Taylor and Francis.
Ziman, J. M. (Ed.) (1971) *The Physics of Metals I*, Cambridge University Press.
Ziman, J. M. (1972) *Principles of the Theory of Solids*, Cambridge University Press.

Problems

7.1 A simple model illustrating Bragg reflection of electron waves is given in Section 7.6. Sketch the form of the antisymmetric wave-function (equation (7.45)) and show that it yields an electron density exhibiting maxima midway between the atom positions.

7.2 Figure 7.13 illustrates the nearly free electron model in the extended zone representation. Redraw this figure in the reduced zone and in the periodic zone representations.

7.3 Sketch the symmetric and antisymmetric wave-functions for a linear array of 6 atoms.

7.4 The first, second and third zones for a square lattice are shown in Fig. 7.14. Consider which lattice vectors would be required in order to transfer each separate portion of the second and third zones into a reduced zone diagram. Use suitable construction lines to develop the fourth zone of the lattice.

7.5 One type of zone overlap for a square lattice is illustrated in Fig. 7.23. Consider the form of the overlaps into the second and third zones which occur as the Fermi surface radius is increased and illustrate on a reduced zone representation.

7.6 The X-ray emission spectra corresponding to the 2p to 1s transitions are sharp for both gaseous and solid sodium whereas the 3s to 2p transition for solid sodium gives an emission band of about 3 nm width. Given that the 2p and 3s atomic levels in sodium are at -38.7 eV and -5.0 eV, respectively, determine the wavelength of the corresponding X-ray photons. Estimate the width of the occupied region of the 3s band in solid sodium.

7.7 If $\boldsymbol{a}$, $\boldsymbol{b}$, $\boldsymbol{c}$ and $\boldsymbol{a}^*$, $\boldsymbol{b}^*$, $\boldsymbol{c}^*$ are the fundamental vectors of the direct and reciprocal lattices respectively show that

$$\boldsymbol{a}^*\cdot\boldsymbol{a} = \boldsymbol{b}^*\cdot\boldsymbol{b} = \boldsymbol{c}^*\cdot\boldsymbol{c} = 2\pi.$$

Show that the volume V^* of the unit cell of the reciprocal lattice is given by

$$V^* = \frac{8\pi^3}{V},$$

where V is the volume of the unit cell of the direct lattice. (See also problem 3.1.)

7.8 The primitive cell of the f.c.c. structure is a rhombohedron with the three translation vectors defined by the lines drawn from the origin to the atom positions $\frac{1}{2}\frac{1}{2}0$, $0\frac{1}{2}\frac{1}{2}$, and $\frac{1}{2}0\frac{1}{2}$. Write down the three translation vectors $\boldsymbol{a}$, $\boldsymbol{b}$, $\boldsymbol{c}$ in terms of the unit vectors

$\hat{x}$, $\hat{y}$, $\hat{z}$ and derive the three vectors $\boldsymbol{a}^*$, $\boldsymbol{b}^*$, $\boldsymbol{c}^*$ of the reciprocal lattice. Hence show that the first Brillouin zone is a truncated octahedron.

7.9 The primitive cell of the b.c.c. structure is a rhombohedron with the three translation vectors defined by the lines drawn from the origin to the atom position $\frac{1}{2}\frac{1}{2}-\frac{1}{2}$, $-\frac{1}{2}\frac{1}{2}\frac{1}{2}$, and $\frac{1}{2}-\frac{1}{2}\frac{1}{2}$. Write down the three vectors $\boldsymbol{a}$, $\boldsymbol{b}$, $\boldsymbol{c}$ in terms of the unit vectors $\hat{x}$, $\hat{y}$, $\hat{z}$ and derive the three vectors $\boldsymbol{a}^*$, $\boldsymbol{b}^*$, $\boldsymbol{c}^*$ of the reciprocal lattice. Hence show that the first Brillouin zone is a regular rhombic dodecahedron.

7.10 An alternative description of the simple hexagonal lattice illustrated in Fig. 7.15 is that in which the unit cell is rotated through 60° about the c-axis. Write down the primitive translation vectors $\boldsymbol{a}$, $\boldsymbol{b}$ and $\boldsymbol{c}$ with respect to the new system of axes and derive $\boldsymbol{a}^*$, $\boldsymbol{b}^*$ and $\boldsymbol{c}^*$.

7.11 Show that the free electron Fermi sphere for sodium corresponds to energy 3.2 eV whilst the minimum energy corresponding to a boundary face of the first Brillouin zone is 4.1 eV. Derive the corresponding values for copper and comment on the significance of the values.

7.12 Show that for the first Brillouin zone of the f.c.c. lattice the shortest distance k_A from the centre of the zone to a boundary face is given by

$$k_A = \frac{\sqrt{3}\pi}{a}.$$

Hence show that the free electron sphere will touch the zone boundary at an electron concentration of 1.36 electrons per atom.

8. Properties of semiconductors

The free electron theory forms a useful basis for the understanding of many of the properties of metals. In band theory account is taken of the interaction between the electrons and the periodic lattice potentials and this has been shown, in the case of metals, to lead to incompletely filled bands and overlapping bands. A most significant result is that the behaviour of electrons in such bands is not dissimilar to that expected from the free electron theory. This accounts for the success of the free electron theory in explaining metallic properties.

In this chapter we shall be concerned with the properties of semiconductors and the way these properties are related to the distinctive band structure. A summary of the properties of germanium and silicon is given in Table 8.1. It will be convenient first to give a qualitative picture of the behaviour of pure semiconductors, the so-called intrinsic semiconductors, and then to describe how this behaviour is modified in extrinsic semiconductors by the presence of very small amounts of specific impurities. This will be followed by a more quantitative discussion of the factors governing the concentrations of charge carriers in semiconductors.

8.1 Intrinsic semiconductors

A semiconductor only exhibits its intrinsic properties provided that it is sufficiently free of impurity. The necessary level of purity depends upon the width of its thermal band gap and on the temperature. Thus the semiconductor germanium, which has a band gap of 0.665 eV at room temperature, is intrinsic provided that impurities are present at a level of less than about 1 part in 10^{10}. On the other hand, at the same temperature, intrinsic silicon with the larger band gap of 1.12 eV would require a purity level of better than 1 in 10^{13}. Purification methods, such as the zone refining techniques, have been extensively used for preparing semiconductor crystals and intrinsic germanium is available by these methods.

The conductivity of a semiconductor is governed by the occupancy both of states in the lower, or valence band, and also of the states in the upper, or conduction, band. As already indicated it is the conductivity exhibited by a

Table 8.1 Properties of germanium and silicon†

		Germanium	Silicon
Density	$\rho/\mathrm{kg\ m^{-3}}$	5.32×10^3	2.33×10^3
Atomic density	$N/\mathrm{m^{-3}}$	4.42×10^{28}	5.00×10^{28}
Electron mobility	$\mu_e/\mathrm{m^2\ V^{-1}\ s^{-1}}$	0.39	0.13
Hole mobility	$\mu_h/\mathrm{m^2\ V^{-1}\ s^{-1}}$	0.19	0.048
Intrinsic carrier density	$n_i/\mathrm{m^{-3}}$	2.0×10^{19}	1.4×10^{16}
Intrinsic conductivity	$\sigma_i/\Omega^{-1}\ \mathrm{m^{-1}}$	1.95	4.1×10^{-4}
Energy gap	E_g/eV	0.665	1.12
Energy gap temperature coefficient	$\alpha/\mathrm{eV\ K^{-1}}$	-3.9×10^{-4}	-2.4×10^{-4}
Relative permittivity	ε	16	11.5
Melting point	$/K$	1210	1693
Electron effective mass	m_e^*/m_e	0.22	0.33
Hole effective mass	m_h^*/m_e	0.37	0.55
Donor (arsenic) ionisation energy	$(E_c - E_d)/\mathrm{eV}$	0.013	0.050
Acceptor (boron) ionisation energy	$(E_a - E_v)/\mathrm{eV}$	0.010	0.045

† *The data in the first part of the table refer to a temperature of 300 K. The effective masses are mean density of states values.*

semiconductor at ordinary temperatures which distinguishes a semiconductor from an insulator. We are therefore particularly interested in the reasons underlying the variation in the occupancy of states in both bands with a change of temperature.

As we have seen in the previous chapter the band structure of semiconductors is such that the number of states available in the valence band is exactly sufficient to accommodate all the valence, or bonding, electrons. At absolute zero these valence electrons, in the ideal intrinsic semiconductor, will be occupying all the available states in the valence band and the conduction band is therefore unoccupied (Figs. 7.27 and 8.1). The reason why the crystal is then non-conducting is as follows. Under the influence of an applied electric field there can be no change in the number of states in the valence band since the density of states is determined only by the form of the Brillouin zones and the electronic structure of the solid. Neither can there be a change in the occupation of the states since these are all already filled; interchange of electrons between states has no physical significance since electrons are indistinguishable particles. Whereas in the free electron theory an electric field leads to an overall momentum change, in the band model this is no longer possible. This is illustrated in Fig. 8.2 for electrons accelerated in the direction k_x. We note that

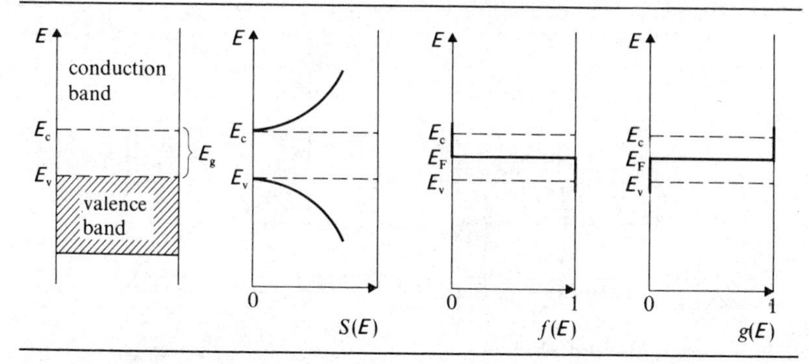

Fig. 8.1 Occupation of bands at 0 K. The valence band is completely filled, and the conduction band is empty. The density of states in each band is related to the corresponding effective masses. $f(E)$ and $g(E)$ give respectively the probability of occupation of states by electrons and holes.

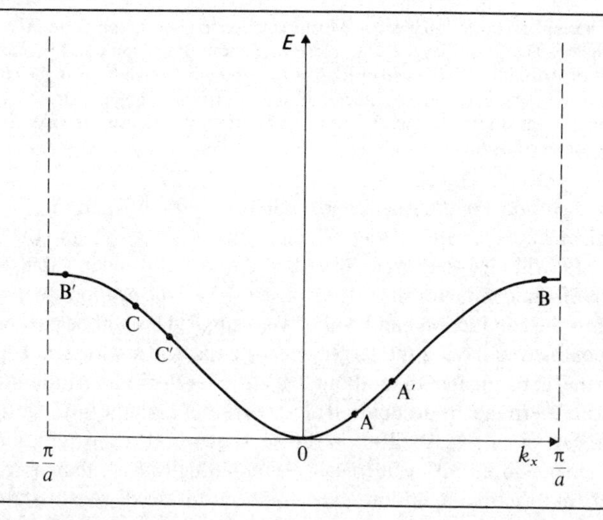

Fig. 8.2 Effect of an electric field on the distribution of electrons in a filled band. Electrons occupying states such as those at A and C are accelerated to energy states A′ and C′. Those in states such as B suffer Bragg reflection and then occupy states such as B′. Overall there is no change in electron momenta or energies.

electrons accelerated near the zone boundary are reflected to the opposite boundary so that the momentum distribution remains unaffected.

If the semiconductor is at ordinary temperatures thermal excitation occurs of a proportion of electrons from the top of the valence band across the band gap to the bottom of the conduction band (Fig. 8.3). This is an equivalent description to

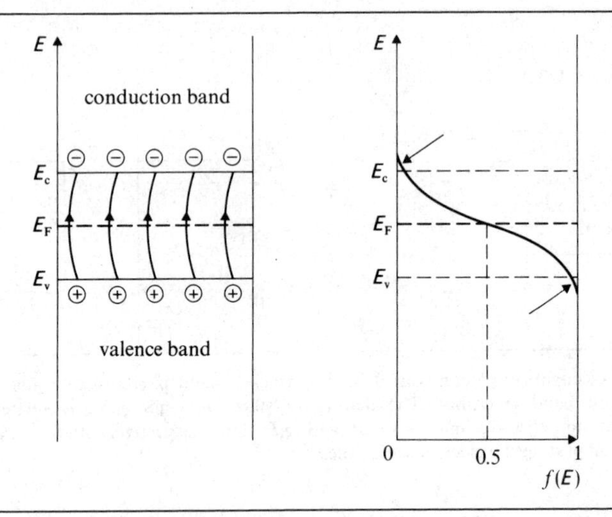

Fig. 8.3 Thermal excitation of electrons from valence to conduction band in an intrinsic semiconductor. The Fermi level E_F lies close to the middle of the band gap and at this energy the probability $f(E)$ of occupation of states by electrons is $\frac{1}{2}$. For the conduction band, $f(E)$ is very small, but finite, indicating that at ordinary temperatures there is some occupation of electron states. A corresponding number of states in the valence band are empty, i.e. occupied by holes.

the breaking of some of the valence bonds in the crystal by thermal agitation. At room temperature the value of thermal energy kT is ~ 0.025 eV and the electrons in the crystal will receive energies distributed about this mean value. Since the band gaps of semiconductors lie in the range 0.1–2 eV the proportion of electrons excited into the conduction band will be very small. This will be particularly so for semiconductors having the larger energy gaps. Indeed for semiconductors having band gaps greater than about 1 eV the level of impurity unavoidably present will determine the occupation of states rather than the intrinsic thermally generated carriers. As we shall see in the next section thermal agitation gives a charge carrier concentration increasing exponentially with temperature. It is also clear that thermal agitation gives a concentration of vacant states in the valence band, i.e. hole concentrations, equal to the electron concentration in the conduction band.

The excitation of electrons into the conduction band provides the necessary condition for conductivity to occur in the presence of an applied electric field. Taking first the electrons in the conduction band. Their situation resembles those in a metal, in that the conduction electrons are occupying states adjacent to higher level energy states which are empty. As described in Section 5.10 there can thus be a systematic change in electron momenta due to the applied field. The magnitude of the associated drift velocity is governed by the value of the electric field and by the mobility of the electrons, the latter quantity being determined by the effective mass for the energy band. The behaviour differs from that in a metal in that the electron concentration is very low so that one is concerned only with states close to the bottom of the conduction band, i.e. those near the band edge E_c in Fig. 8.3. Secondly, the low level of occupation means that the statistics applicable are the Maxwell–Boltzmann rather than the Fermi–Dirac statistics used in the free electron theory of metals.

In addition to the conduction due to electrons in the conduction band, there will also be conduction arising from the presence of empty states at the top of the valence band. The hole effective mass will differ from the electron effective mass and both will vary from one energy band to another. Since the electron and hole mobilities differ, the contributions of each band to the intrinsic conductivity will also be different.

Putting n and p as the electron and hole concentrations in a semiconductor, the conductivity will be given by a relationship analogous to equation (5.5) for a metal, i.e.

$$\sigma = ne\mu_e + pe\mu_h \qquad \ldots(8.1)$$

where μ_e and μ_h are the electron and hole mobilities. Denoting the intrinsic electron and hole concentrations as n_i and p_i, respectively, we therefore have for the intrinsic semiconductor that

$$n_i = p_i \qquad \ldots(8.2)$$

and

$$\begin{aligned} \sigma &= n_i e(\mu_e + \mu_h) \\ &= p_i e(\mu_e + \mu_h) \end{aligned} \qquad \ldots(8.3)$$

8.2 Extrinsic semiconductors

Whereas the conductivity of an intrinsic semiconductor is a function only of temperature, that of an extrinsic semiconductor is governed by the impurity content. In an extrinsic semiconductor, therefore, a specific impurity element is added in controlled amounts, and its concentration will determine the conductivity of the sample and the type of current carriers present. When these are electrons, i.e. *negative* carriers, the sample is termed n-type; when these are holes, i.e. *positive* carriers, the sample is p-type. The feasibility of control of conductivity and of carrier type, underlies the use of semiconductor materials in devices.

A semiconductor such as germanium has the typical tetrahedral covalent bonding in which each atom shares its four valence electrons with each of its four nearest neighbours. Let us consider the effect of incorporating a very small amount of a Group V element such as phosphorus into a germanium crystal. As illustrated in Fig. 8.4 the phosphorus impurity atom is substituted for a

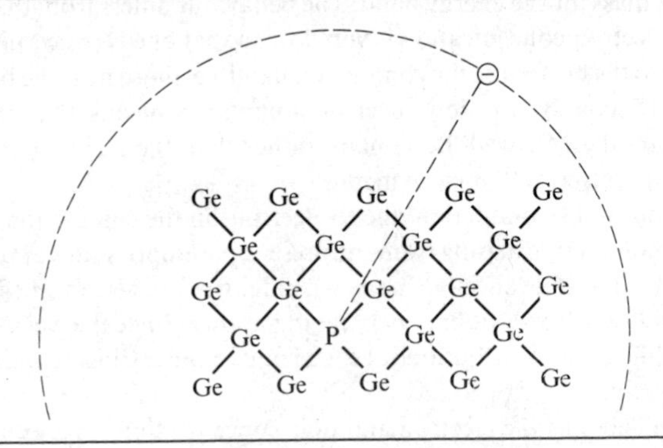

Fig. 8.4 A schematic representation of the tetrahedral bonding in germanium showing a substitutional phosphorus atom. The electron is shown weakly bound to the parent donor atom and the orbital radius is ~4 nm. This is the situation at very low temperatures whereas at ordinary temperatures the electron is essentially free of the donor atom.

germanium atom on the lattice so that it is tetrahedrally bound to its nearest neighbours. In the free atom state phosphorus has two 3s electrons and three 3p electrons so that in the substitutional alloy the atom has one surplus electron over and above that necessary to form the covalent bonds. At very low temperatures, close to absolute zero, this electron is loosely bound with its parent atom due to the coulombic attraction of the nucleus. However, if the temperature of the semiconductor is raised to about 20 K or above, this electron is released from the attraction of the parent impurity atom by thermal energy, and is then

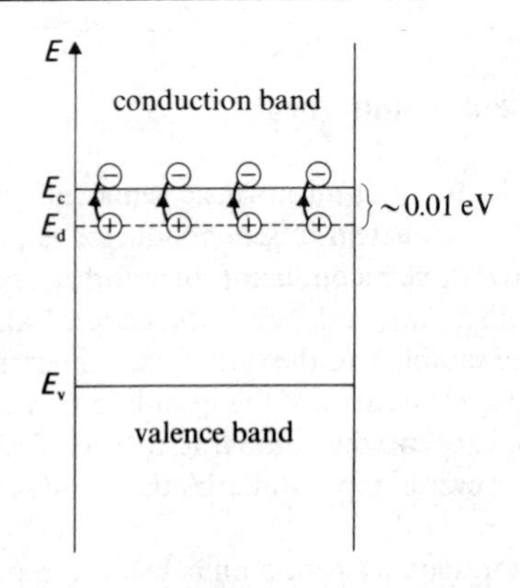

Fig. 8.5 A schematic representation of electrons from the donor level E_d in n-type germanium occupying states in the conduction band. Although of no significance in the present example the horizontal scale can be used to represent distance when dealing with samples showing inhomogeneous impurity distribution.

free to move throughout the crystal. Each impurity atom thus gives rise to an electron in the conduction band and such atoms are termed *donors.* The concentration of donors in the crystal governs the electron concentration, and therefore the conductivity of a sample. The semiconductor is then termed extrinsic. It is clear that with a suitable donor concentration in a sample the electron concentration may be orders of magnitude greater than that of an intrinsic sample. As will be shown later the condition $p \ll n$ then obtains so that the electrical conductivity is given by

$$\sigma = ne\mu_e. \qquad \ldots(8.4)$$

The band representation of an n-type semiconductor is shown in Fig. 8.5. At room temperature each donor atom is ionised and gives rise to an electron in the conduction band. As is evidenced by the low temperature at which ionisation occurs, as verified for example in conductivity measurements, it is apparent that the corresponding ionisation energy must be small. Thus for the example chosen of phosphorus in germanium, the value is 0.012 eV. As is illustrated, the donor atoms give rise to a donor level E_d lying within the forbidden energy gap at a distance of 0.012 eV below the level of the bottom of the conduction band E_c.

In addition to phosphorus the other Group V elements, arsenic, antimony and bismuth also behave as donor impurities in both germanium and silicon. All three elements have a closely similar outer electronic configuration. The ionisation energies for the donors in germanium lie within about 20 per cent of 0.011 eV whereas in silicon the energies are rather larger being $\sim$0.05 eV. These values may be determined experimentally by measuring the variation of the Hall constant for the range of temperature over which the ionisation of the impurities is occurring.

It is possible to estimate the ionisation energy of an impurity atom in a semiconductor from a simple model. Whereas the appropriate wave-functions of electrons within the bands are travelling waves of the Bloch form we require in the model the solutions of wave-functions representing localised or bound states associated with each impurity atom. In the bound state the electron is under the influence of the central coulombic force of the nucleus of the donor atom shielded by the electrons of the impurity atom. Anticipating the solution, one finds that the wave-function of the bound electron extends over very many lattice spacings. Consequently the potential energy term in the wave equation can be put in the form

$$V = -\frac{e^2}{4\pi\varepsilon_r\varepsilon_0 r} \qquad \ldots(8.5)$$

so that the influence of adjoining atoms is averaged through the inclusion of the relative permittivity factor ε_r. This equation should be compared with equation (2.3) used in the solution of the wave equation for the hydrogen atom. In order also to take into account the periodic potential within the crystal the electron mass m_e in the wave equation may be replaced by the effective mass m_e^*

appropriate to the adjacent band. The wave equation is now of the same general form as that used in the solution for the hydrogen atom and may be solved in a similar manner to yield

$$E = -\frac{m_e^* e^4}{32\pi^2 n^2 \varepsilon_r^2 \varepsilon_0^2 \hbar^2} \qquad \text{...(8.6)}$$

For the ground state of the bound donor electron we therefore have

$$E = -\frac{13.6}{\varepsilon_r^2}\left(\frac{m_e^*}{m_e}\right)\text{eV}. \qquad \text{...(8.7)}$$

Putting $m_e^* = 0.2m_e$ and $\varepsilon_r = 16$ for germanium the ionisation energy of the donor electron is 0.01 eV, which is in agreement with the values already quoted.

The validity of the assumption regarding the spatial extension of the bound electron wave-function can be tested. The maximum of the radial probability function for the bound electron is obtained as in Section 2.3.2 and the first Bohr radius a_0 is replaced by

$$a = a_0 \varepsilon_r \left(\frac{m_e}{m_e^*}\right). \qquad \text{...(8.8)}$$

Inserting numerical values one finds that the orbital for the bound electron has its maximum value at about 4 nm. This value should, however, only be regarded as approximate in view of the assumptions made regarding the use of ε_r and m_e^* in the theory. A similar approach is valid for the study of the so-called acceptor states adjacent to the valence band but the assumptions of the theory cease to be valid for states lying more deeply within the energy gap.

We have up to now considered only the incorporation of Group V impurities into silicon and germanium. The isolated atoms of the Group III elements boron, aluminium, gallium and indium have outermost electronic shells of two electrons in s states and one electron in a p state. Consequently when one of these elements is alloyed into silicon or germanium there is a deficiency of one electron for tetrahedral bonding of the impurity atom into the crystal lattice. At very low temperatures this missing bond is localised in the coulombic field of the impurity atom. At ordinary temperatures, however, the impurity atom, which is termed an acceptor, receives an electron from another atom in the crystal and the missing bonding electron, or hole, becomes mobile. On the band representation the acceptor atom becomes ionised on receiving an electron from the valence band, with the simultaneous creation of a mobile hole in the valence band (see Fig. 8.6). Alternatively ionisation may be regarded as the release of a bound hole from the acceptor atom to the valence band. The ionisation energies of the acceptor atoms are ~0.01 eV in germanium and ~0.05–0.16 eV in silicon.

The conductivity exhibited by a sample containing acceptor type impurity is due to the mobile positively charged holes and the semiconductor is p-type. Provided that there is complete ionisation, the mobile hole concentration p is

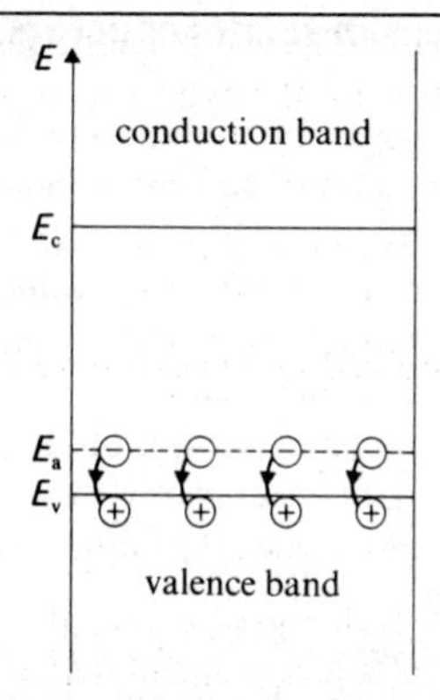

Fig. 8.6 Occupation of hole states in the valence band in a p-type crystal. The process may be regarded as one in which the acceptor atoms receive electrons from the valence band or alternatively the release of bound holes from the acceptor atom to the band.

equal to the concentration of acceptor atoms and the conductivity will be given by

$$\sigma = pe\mu_h. \quad \ldots(8.9)$$

The use of this simplified equation is valid provided that the impurity concentration is not small enough for the material to be near intrinsic and in which case the complete form of equation (8.1) is used. It should also be noted that when the impurity concentration is sufficiently high, in either an n-type or in a p-type semiconductor, there will be incomplete ionisation of the impurity atoms. In such cases the mobile carrier concentrations are less than the impurity atom.concentrations.

A semiconductor may also contain simultaneously both donor and acceptor type impurity atoms. In such a case the conductivity is governed by the net excess impurity of one type, and the semiconductor is then termed to be *compensated.* When an impurity element from a group other than III or V is present it may give rise to one or more separate levels lying within the forbidden energy gap and many such levels lie deep within the gap. Their state of ionisation is governed by the conductivity of the semiconductor sample and such impurities are of importance in that the behaviour of semiconductor devices is markedly affected by their presence even at very low concentration levels.

Although the foregoing discussion has been given in terms of the Group IV semiconductors, similar considerations apply to the presence of donor and acceptor impurities in other semiconductors. Of some importance are the Group III–V compounds such as gallium arsenide crystallising in the zinc-blende structure. In such compounds a departure from a strict stoichiometric composition leads to either an n-type or a p-type sample depending on which element is present in the greater concentration.

8.3 Carrier concentrations in semiconductors

As we have seen the conductivity of a semiconductor is governed by the density of charge carriers and on their mobility. Carrier mobilities in semiconductors are substantially larger than those in many metals, e.g. electron mobility in germanium is 0.39 $m^2 V^{-1} s^{-1}$ compared with a value of 0.0035 $m^2 V^{-1} s^{-1}$ in copper. There is, however, considerable variation in values between different semiconductors and a value of as high as 7.7 $m^2 V^{-1} s^{-1}$ is obtained for the electron mobility in indium antimonide. Hole mobilities are generally lower than electron mobilities and all mobilities show some variation with carrier concentration. This is most marked when the impurity concentration is high, e.g. in degenerate semiconductors.

Carrier concentration is a function of temperature for an intrinsic semiconductor and of donor or acceptor concentration in an extrinsic sample. Variations in carrier concentration over many orders of magnitude are obtained and these produce proportional changes in conductivity (in accordance with equation (8.1)). These changes in impurity concentration and of temperature are reflected in changes in the occupation of states in the valence and conduction bands.

Before deriving the relationships for carrier concentrations in semiconductors we may look again briefly at the concept of the Fermi level. As we have seen in Section 5.9, states at the Fermi level have a 50 per cent probability of being occupied. Since electron states in a semiconductor will normally have a high occupation in the valence band and a correspondingly low occupation in the conduction band, the Fermi level will almost always lie within the forbidden gap. As well as determining the occupation of levels within the bands the Fermi level will also govern that of the discrete levels within the band gap due to donors, acceptors or other impurity atoms, i.e. the position of the Fermi level governs the state of ionisation of these impurities.

Conductivity may arise due to the presence of mobile carriers in either or both bands and we may deal with the two bands separately. Let us consider first the carrier concentration in the conduction band. This is governed by the number of available states and their mode of occupation. Proceeding as in Section 5.8 the number of occupied states $N(E)\,dE$ in the energy range E to $E+dE$ is

$$N(E)\,dE = S(E)f(E)\,dE, \qquad \ldots(5.53)$$

where $S(E)$ is the density of states in the conduction band and $f(E)$ is the appropriate distribution function governing the occupation of the states. The total electron concentration n in the conduction band is thus obtained as

$$n = \frac{1}{V}\int_{E_c}^{E_c'} N(E)\,dE = \frac{1}{V}\int_{E_c}^{E_c'} S(E)f(E)\,dE, \qquad \ldots(8.10)$$

where E_c' is the energy corresponding to the top of the conduction band.

The form of the density of states function $S(E)$ for the conduction bands of semiconductors has been discussed in Section 7.13. If we assume a spherical band then following the procedure in Section 7.13 the density of states will be given by

$$S(E) = \frac{M_c}{2\pi^2}\left(\frac{2m_e^*}{\hbar^2}\right)^{\frac{3}{2}} V(E-E_c)^{\frac{1}{2}}, \qquad \ldots(8.11)$$

where the density of states effective electron mass is m_e^* and $(E-E_c)$ is the energy of a state measured from the bottom of the conduction band. If we use the Fermi–Dirac distribution function

$$f(E) = \frac{1}{\exp[(E-E_F)/kT]+1} \qquad \ldots(5.54)$$

for the occupancy of the states in the conduction band we then have

$$n = \int_{E_c}^{\infty} \frac{\dfrac{M_c}{2\pi^2}\left(\dfrac{2m_e^*}{\hbar^2}\right)^{\frac{3}{2}}(E-E_c)^{\frac{1}{2}}}{1+\exp\left(\dfrac{E-E_F}{kT}\right)} \mathrm{d}E. \qquad \ldots(8.12)$$

We note here that it is permissible to set the upper limit of the integral as ∞ since the function $f(E)$ rapidly approaches zero for $E > E_c$.

For a wide range of impurity concentration the integral in equation (8.12) can be solved by making the simplifying assumption that

$$1+\exp\left(\frac{E-E_F}{kT}\right) \approx \exp\left(\frac{E-E_F}{kT}\right). \qquad \ldots(8.13)$$

This is equivalent to using Maxwell–Boltzmann instead of Fermi–Dirac statistics and is a valid assumption provided that

$$E-E_F > 2kT \qquad \ldots(8.14)$$

i.e. the Fermi level must lie more than $2kT$ below the conduction band edge. This will be the case provided that the impurity concentration is not very high or alternatively the semiconductor is not at a low temperature. The conduction band electrons are then behaving as a non-degenerate or classical gas and the Pauli exclusion principle is not a significant restriction on the mode of occupation of the levels. Using the approximation in equation (8.13) and changing the variable $(E-E_c)/kT$ to x we have

$$n = \frac{M_c}{2\pi^2}\left(\frac{2m_e^* kT}{\hbar^2}\right)^{\frac{3}{2}} \exp\left(\frac{E_F-E_c}{kT}\right)\int_0^{\infty} x^{\frac{1}{2}}\exp(-x)\,\mathrm{d}x. \qquad \ldots(8.15)$$

The integral is standard and has a value of $\frac{1}{2}\sqrt{\pi}$ so that

$$n = N_c \exp\left(\frac{E_F - E_c}{kT}\right), \qquad \ldots(8.16)$$

where

$$N_c = 2M_c\left(\frac{m_e^* kT}{2\pi\hbar^2}\right)^{\frac{3}{2}}. \qquad \ldots(8.17)$$

As we have seen in Section 7.13 there may be a number of equivalent minima in the conduction band and the additional constant M_c is included here as was the case in equation (7.81) for the total density of states.

Expression (8.16) relates the electron concentration in the conduction band with the position of the Fermi level. The exponential factor in the expression is of the same form as the simplified form of the Fermi–Dirac probability factor for the occupation of a state of energy E_c. This feature allows the conduction band to be treated as if it were composed of states localised in energy at a value E_c and having a density of states N_c. N_c is termed the *effective density of states* in the conduction band. Subsituting numerical values for the constants this yields the value

$$N_c = 4.83 \times 10^{21} M_c \left(\frac{m_e^*}{m_e}\right)^{\frac{3}{2}} T^{\frac{3}{2}}\ \mathrm{m}^{-3}. \qquad \ldots(8.18)$$

At a sufficiently high impurity concentration the condition (8.14) ceases to be valid and numerical solutions of the Fermi–Dirac integrals are utilised. The Fermi level may then move into the conduction band and when the condition

$$E_F - E_c \gg kT \qquad \ldots(8.19)$$

holds the electron gas becomes degenerate as in a metal.

The hole concentration in the valence band may be derived by a similar procedure. The probability $g(E)$ of a level at energy E being occupied by a hole (i.e. not being occupied by an electron) is

$$g(E) = 1 - f(E) \qquad \ldots(8.20)$$

which using equation (5.55) becomes

$$g(E) = \frac{1}{1 + \exp\left(\frac{E_F - E}{kT}\right)}. \qquad \ldots(8.21)$$

Making similar assumptions to those used for the conduction band, i.e. assuming a spherical band and a Fermi level displaced from the band edge by more than $2kT$, we obtain

$$p = N_v \exp\left(\frac{E_v - E_F}{kT}\right) \qquad \ldots(8.22)$$

where N_v, the effective density of states in the valence band is given by

$$N_v = 2\left(\frac{m_h^* kT}{2\pi\hbar^2}\right)^{\frac{3}{2}}. \qquad \ldots(8.23)$$

It should be noted that m_h^* here is the density of states effective mass which takes into account both the light and heavy holes.

The two expressions (8.16) and (8.22) relate the carrier concentrations in a semiconductor to the position of the Fermi level. They are valid for both extrinsic and intrinsic semiconductors provided that the Fermi level does not lie closer than about $2kT$ to either band edge.

8.3.1 *The intrinsic semiconductor*

In an intrinsic semiconductor electrons are thermally excited across the band gap so that the carrier concentrations are dependent upon the temperature and the width of the energy gap of the semiconductor. The form of the relationship can be derived from the results of the previous subsection. Denoting the position of the Fermi level in an intrinsic semiconductor by E_i and the intrinsic electron and hole concentrations by n_i and p_i respectively it follows from equations (8.16) and (8.22) that

$$n_i = N_c \exp\left(\frac{E_i - E_c}{kT}\right) \qquad \ldots(8.24)$$

and

$$p_i = N_v \exp\left(\frac{E_v - E_i}{kT}\right). \qquad \ldots(8.25)$$

The carriers are thermally generated in the form of electron hole pairs so that we may put $n_i = p_i$ and thus obtain

$$n_i = p_i = (N_c N_v)^{\frac{1}{2}} \exp\left(-\frac{E_g}{2kT}\right), \qquad \ldots(8.26)$$

where

$$E_g = E_c - E_v \qquad \ldots(8.27)$$

is the energy gap of the semiconductor.

The value of the intrinsic carrier concentration can be derived from equation (8.26) and as an example of the calculation we may consider intrinsic germanium. The density of states effective masses for germanium are $m_e^* = 0.22m_e$ and $m_h^* = 0.37m_e$ and we note that in accordance with the discussion in Section 7.13 the conduction band has four equivalent minima. This yields the values for N_c and N_v at 300 K of

$$N_c = 1.04 \times 10^{25}\ \text{m}^{-3}$$

$$N_v = 5.64 \times 10^{24}\ \text{m}^{-3}.$$

Taking the value of the band gap E_g as 0.665 eV this yields

$n_i = p_i = 2.0 \times 10^{19}\ m^{-3}$.

The value of n_i derived here is extremely sensitive to the exact value used in equation (8.26) for the energy gap E_g. Accurate values of E_g can be obtained from infrared absorption measurements, a marked increase in absorption occurring at the threshold optical frequency ν_g for which $h\nu_g$ corresponds to E_g. The method gives information as to whether the band gap is direct or indirect and can be used to study the temperature variation of E_g. A plot of ln n_i versus $1/T$ for equation (8.26), as in Fig. 8.7 for germanium and silicon is nearly linear and will provide an approximate value of E_g.

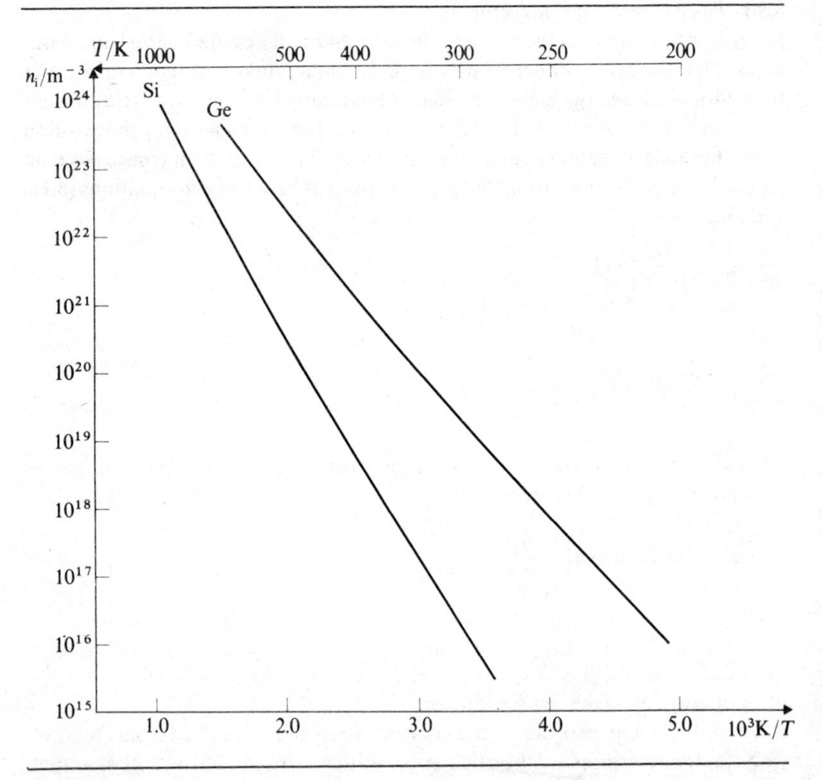

Fig. 8.7 Temperature variation of carrier concentration in intrinsic germanium and silicon.

In order to obtain an accurate value of the band gap its temperature dependence must be taken into account. For temperatures above 200 K germanium and silicon show a linear decrease of energy gap with increasing temperature so that for this range we may put

$$E_g = E_0 + \alpha T \qquad \ldots(8.28)$$

where the temperature coefficient α is negative. E_0 is the extrapolated value of E_g at 0 K from this linear range and will differ from the experimental E_g value at 0 K. This equation allows equation (8.26) to be expressed in the form

$$n_i = AT^{\frac{3}{2}} \exp\left(-\frac{E_0}{2kT}\right) \qquad \ldots(8.29)$$

so that a plot of $\ln n_i/T^{\frac{3}{2}}$ against $1/T$ will be linear. Its exact form will be governed by the values of E_0 and α for the particular semiconductor. The values of E_0 for germanium and silicon are respectively 0.78 and 1.21 eV and the temperature coefficients are given in Table 8.1. Problems (8.7)–(8.10) at the end of this chapter deal with the use of equation (8.29).

An independent approach to equation (8.29) is that of determining the temperature variation of the intrinsic carrier concentration n_i from conductivity and mobility measurements using equation (8.1). The values of electron and hole mobilities for the intrinsic semiconductors show a temperature dependence (see Section 8.4) and for an accurate comparison this dependence needs to be taken into account. The experimental measurements on germanium and silicon (Morin and Maita, 1954) show agreement with equation (8.29) which is particularly close for germanium.

The position of the Fermi level in an intrinsic semiconductor is close to the middle of the band gap. This may be seen from equations (8.24) and (8.25) which yield

$$\begin{aligned} E_i &= \tfrac{1}{2}(E_v + E_c) + \tfrac{1}{2}kT \ln\left(\frac{N_v}{N_c}\right) \\ &= \tfrac{1}{2}(E_v + E_c) + \tfrac{3}{4}kT \ln\left(\frac{m_h^*}{m_e^*}\right) \end{aligned} \qquad \ldots(8.30)$$

from equations (8.17) and (8.23).

The displacement from the middle of the gap described by the last term is normally very small but can be substantial in a semiconductor such as indium antimonide which has an electron effective mass very much smaller than the hole effective mass.

8.3.2 *The extrinsic semiconductor*

When considering an extrinsic semiconductor one is interested in the relationships between the concentrations of the mobile current carriers and the donor or acceptor atom concentrations giving rise to the carriers. Over a substantial range of impurity concentration this may take on a very simple form, e.g. in an n-type semiconductor the electron concentration n will be equal to the concentration N_d of donor atoms. This simple picture will, however, cease to be valid in the near intrinsic region where intrinsic generation of carriers is significant; neither will it be valid when N_d is rather large and when a proportion of the donor atoms are not ionised. As in the previous section it will be convenient to describe the carrier concentrations in terms of the Fermi level whose position will be found to vary according to the concentration of impurity element contained in the semiconductor.

We first note that an important relationship is inherent in equations (8.16), (8.22) and (8.26), namely that

$$np = N_c N_v \exp\left(-\frac{E_g}{kT}\right) = n_i^2. \qquad \ldots(8.31)$$

Thus, if in an n-type semiconductor the electron concentration n is increased by addition of donor impurity, then the hole concentration p must decrease in the same proportion. It is usual to refer, therefore, to the majority and minority carriers in an extrinsic semiconductor. Either can of course be electrons or holes. The equilibrium carrier concentrations are obtained as a balance between equal generation and recombination rates of electron–hole pairs. The generation rate is governed by thermal energy and therefore temperature, whilst the recombination rate is determined by the number of electrons and the number of empty states available for them in the valence band, i.e. recombination rate is determined by the product np. Carrier concentration in excess of those given by equation (8.31) can, however, occur in non-equilibrium conditions and the recombination mechanisms will be discussed further in this connection in Section 8.5.

The interdependence of carrier concentration and the position of the Fermi level is given by equations (8.16) and (8.22). By the use of equations (8.24) and (8.25) these relationships can be written in the modified form

$$n = N_c \exp\left(\frac{E_F - E_c}{kT}\right) = n_i \exp\left(\frac{E_F - E_i}{kT}\right) \qquad \ldots(8.32)$$

and

$$p = N_v \exp\left(\frac{E_v - E_F}{kT}\right) = p_i \exp\left(\frac{E_i - E_F}{kT}\right). \qquad \ldots(8.33)$$

These equations are valid for both extrinsic and intrinsic semiconductors provided that the carrier concentrations are not high enough for the semiconductors to become degenerate. They are termed Boltzmann's equilibrium equations.

In an n-type sample, provided that there is complete ionisation of the donor atoms, the electron concentration n will equal the donor impurity concentration N_d. In accordance with equation (8.32) the Fermi level will move towards the conduction band edge E_c as N_d is increased. Similarly for a p-type sample it will move towards E_v as the acceptor concentration N_a is increased. For compensated samples the position of the Fermi level will be determined by the net donor or acceptor concentration. The movement of the Fermi level with donor and acceptor concentration is shown in Fig. 8.8 and the occupancy of states in n- and p-type semiconductors in Fig. 8.9.

We may take as a specific example the position of the Fermi level in a sample of n-type germanium at room temperature. Putting $n = N_d$ we have from equation (8.32) that

$$E_c - E_F = kT \ln \frac{N_c}{N_d}. \quad \ldots(8.34)$$

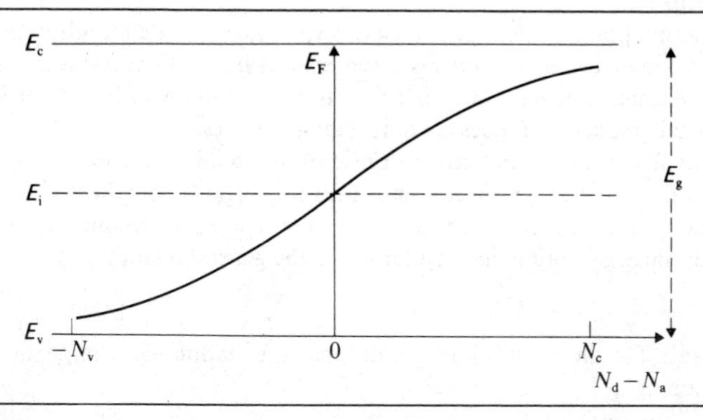

Fig. 8.8 The variation in the position of the Fermi level E_F with net donor or acceptor concentration. For $N_d = N_a$ the position of the Fermi level E_i lies close to the middle of the band gap E_g.

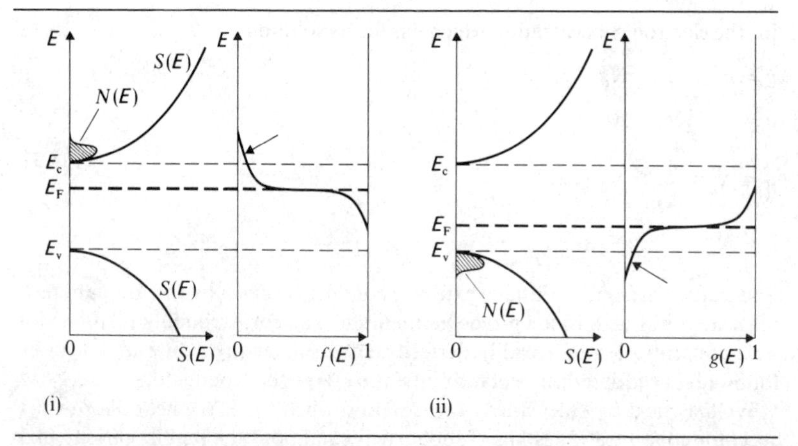

Fig. 8.9 Illustration of the occupation of states in a semiconductor. In (i) the impurity concentration is such as to make the semiconductor n-type. The Fermi level E_F has moved towards E_c and the shaded area bounded by $N(E)$ represents electron states occupied in the conduction band. In (ii) the semiconductor is p-type and the Fermi level has moved towards E_v. The probability of occupation of hole states in the valence band is $g(E)$ and the occupied states are shown bounded by $N(E)$.

Since $N_c \approx 10^{25}\ m^{-3}$ the Fermi level will lie at more than $2kT$ from the band edge provided that $N_d < 10^{24}\ m^{-3}$ and the semiconductor will remain non-degenerate. The intrinsic concentration n_i is $2 \times 10^{19}\ m^{-3}$ and for $N_d > 10^{20}\ m^{-3}$ the donor impurity will control the carrier concentration. Throughout this range, therefore, the carrier concentration is governed only by the donor atom concentration and is nearly independent of temperature. A similar type of situation will apply in p-type germanium and also in other semiconductors such as silicon.

The assumption that $n = N_d$ breaks down at low donor concentrations when the semiconductor is near intrinsic and also at high concentrations when the donors are incompletely ionised. This latter condition is of particular interest when semiconductor samples are at low temperatures.

Let us first consider the near intrinsic n-type semiconductor. There will be contributions to the electron concentration arising both from the ionised donors and also from the intrinsic generation process. The electron concentration may be determined by noting that in addition to the general relationship

$$np = n_i^2 \qquad \ldots(8.31)$$

applicable for all equilibrium conditions, the additional charge neutrality condition

$$n = N_d + p \qquad \ldots(8.35)$$

must also hold. This gives the quadratic expression

$$n^2 - nN_d - n_i^2 = 0 \qquad \ldots(8.36)$$

for the electron concentration which has for its solution

$$\begin{aligned} n &= n_i &&;\ N_d \ll n_i \\ n &= n_i + \tfrac{1}{2}N_d &&;\ N_d < n_i \\ n &= N_d + \frac{n_i^2}{N_d} &&;\ N_d > n_i \\ n &= N_d &&;\ N_d \gg n_i \end{aligned} \qquad \ldots(8.37)$$

The transition from intrinsic to extrinsic conduction given by these equations is illustrated in Fig. 8.10 for n-type germanium. The corresponding relations for hole concentration are readily derived and a similar procedure may also be followed for the description of near intrinsic p-type semiconductors.

We may next consider briefly the situation when either the net donor or net acceptor concentration is high. Under such conditions the mobile carriers in a band will tend to fill the adjoining impurity level which is equivalent to the impurity atoms being incompletely ionised. In an n-type semiconductor the net donor concentration N_d will then approach the value N_c and the Fermi level will lie close to the donor level E_d. Although the statistics of occupation of impurity levels differ from those of the band states, the position of the Fermi level will still

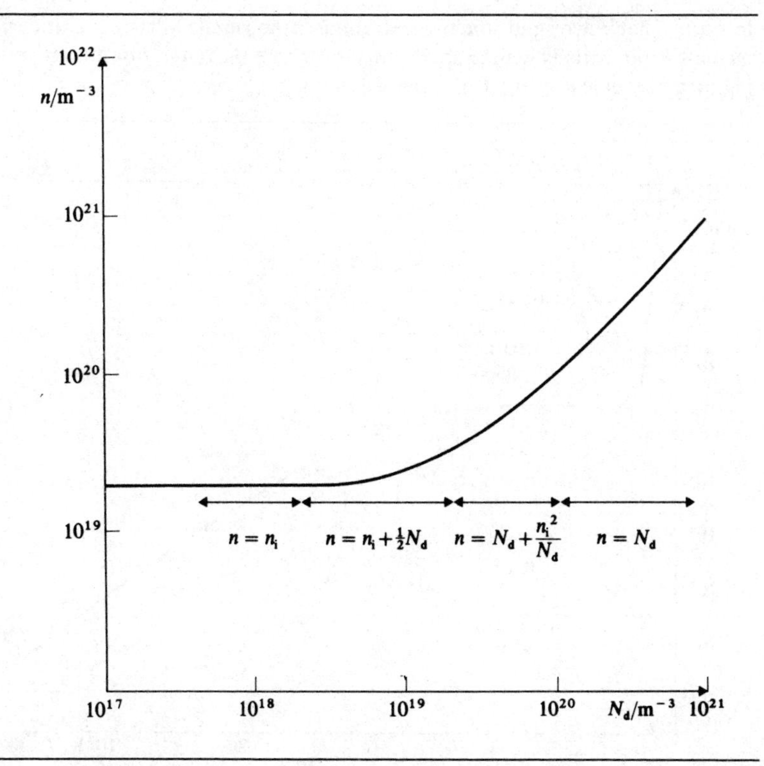

Fig. 8.10 The transition from intrinsic to extrinsic conduction with increasing concentration of donor impurity N_d.

describe the occupancy of the impurity levels, i.e. their degree of ionisation. For very high impurity concentrations E_F can rise within the band and the impurity levels themselves broaden out and overlap into the band. There is no longer an activation energy associated with the impurity levels and the carrier concentration becomes nearly independent of temperature for a wide temperature range. The occupancy of states is governed by Fermi–Dirac statistics and, as in metals, only carriers near the top of the distribution take part in the conduction process.

Carrier concentration changes in a semiconductor can be usefully summarised by considering the changes which occur as its temperature is varied over a wide range of values. As a specific example we may take an n-type silicon sample containing arsenic at a concentration level N_d of 2×10^{21} m^{-3} and assume that the acceptor impurity concentration is negligible. In practice some level of compensation is difficult to avoid.

At a temperature of 600 K the intrinsic carrier concentration in silicon is 4.8×10^{21} m^{-3} so that at this, and at any higher temperature, the sample is intrinsic. The electron and hole concentrations in the sample will be equal and will increase exponentially with temperature throughout the range. For this intrinsic region a plot of $\ln n$ versus $1/T$ as shown in Fig. 8.11 has a slope of $-E_g/2k$. We

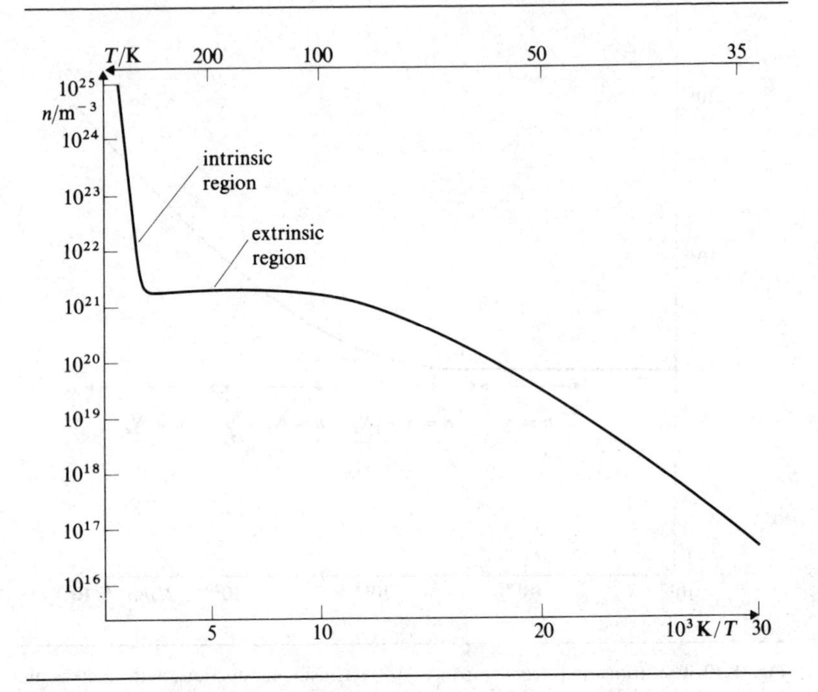

Fig. 8.11 The variation of carrier concentration with temperature for silicon. At high temperatures the semiconductor is intrinsic and the slope of $\ln n$ versus $1/T$ is governed by E_g. Below 600 K the semiconductor is extrinsic with a constant carrier concentration determined by the impurity level. At low temperatures the impurities become deionised and the slope of $\ln n$ vs. $1/T$ is governed by $E_c - E_d$.

neglect here the small departure from linearity due to the $T^{\frac{3}{2}}$ factor in equation (8.26) and also the decrease in the width of the band gap with increasing temperature. The steepness of the slope of $\ln n$ in the intrinsic region results in quite a sharp transition from intrinsic to extrinsic behaviour below 600 K.

For a wide temperature range, down to about 150 K the electron concentration is then determined only by the donor concentration and since all the donors are ionised the carrier concentrations remain constant throughout. As the temperature is lowered within this range the Fermi level gradually moves from the centre of the band gap towards the conduction band edge in accordance with equation (8.32).

If the temperature is further lowered the concentration of ionised donors decreases so that $n < N_d$. The form of the carrier concentration variation with temperature within this lower temperature range may be obtained as follows. The carrier concentration is governed by the state of ionisation of the donor atoms and a donor level is able to accept one electron only, this being of either spin. It may be shown that the probability $P(E)$ of a donor level being occupied, i.e. unionised, is given by

$$P(E) = \frac{1}{1+\frac{1}{2}\exp\left(\frac{E_d - E_F}{kT}\right)} \qquad \ldots(8.38)$$

The concentration n of electrons in the conduction band is thus

$$n = N_d - \frac{N_d}{1+\frac{1}{2}\exp\left(\frac{E_d - E_F}{kT}\right)}$$

$$= \frac{N_d}{1+2\exp\left(\frac{E_F - E_d}{kT}\right)}. \qquad \ldots(8.39)$$

Assuming that the semiconductor is not degenerate, we also have

$$n = N_c \exp\left(\frac{E_F - E_c}{kT}\right) \qquad \ldots(8.32)$$

so that eliminating E_F between these two equations we obtain

$$n^2 + \frac{nN_c}{2\exp\left(\frac{E_c - E_d}{kT}\right)} - \frac{N_c N_d}{2\exp\left(\frac{E_c - E_d}{kT}\right)} = 0. \qquad \ldots(8.40)$$

At a sufficiently low temperature the second term in the expression may be neglected and the electron concentration n is then given by

$$n = \left(\frac{N_c N_d}{2}\right)^{\frac{1}{2}} \exp - \left(\frac{E_c - E_d}{2kT}\right). \qquad \ldots(8.41)$$

The slope of $\ln n$ versus $1/T$ for this temperature region is therefore $-(E_c - E_d)/2k$ and its evaluation enables a value of the ionisation energy of the donors to be determined. It may be shown that if the semiconductor is compensated the presence of the acceptor impurity leads instead to a slope of $-(E_c - E_d)/k$. Further details concerning the temperature variation of carrier concentration have been given by Smith (1959).

8.4 Electrical conductivity and carrier mobility

The concept of electron mobility and its application to the theory of metals has been discussed in Chapters 5 and 6. The observed decrease in the conductivity of metals with temperature was then explained in terms of a decrease in electron mobility arising from the increased lattice scattering occurring at the higher temperatures.

For a semiconductor, equation (8.1) relates conductivity with carrier concentrations and electron and hole mobilities. The type of conductivity variation with temperature, which is experimentally observed in semiconductors, is illustrated in Fig. 8.12 for some samples of n-type germanium. As is

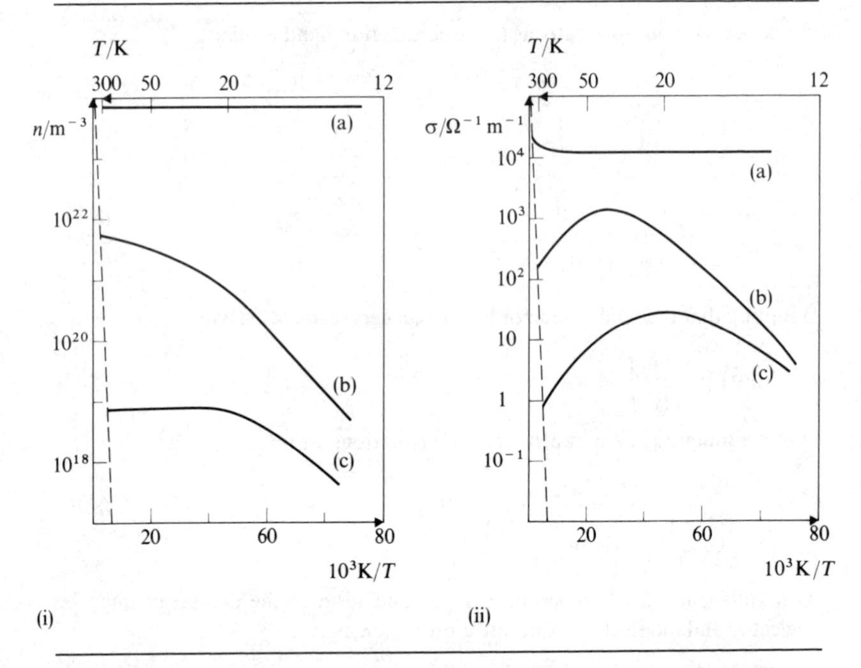

Fig. 8.12 Temperature dependence of (i) carrier concentration and (ii) conductivity in n-type germanium samples. Corresponding pairs of curves are marked (a), (b) and (c) and refer to three separate samples possessing substantially different impurity content. (Data from Debye P P, Conwell E M, Phys Rev 93 No. 4, 693, 1954.)

clear from the figure, conductivity changes in the intrinsic region are dominated by the exponential increase of carrier concentration with temperature. The influence of mobility changes is, however apparent in the extrinsic range where a fall in mobility with a rise in temperature, coupled with an essentially constant carrier concentration, accounts for the observed conductivity decrease with temperature.

There are a number of different scattering mechanisms which influence the observed electron and hole mobilities of semiconductors. Detailed explanation of the observed mobility variations with both temperature and impurity content of samples has proved difficult. The dominant mechanism in fairly pure samples is the scattering by lattice phonons and, assuming spherical constant energy surfaces for the semiconductor, a mobility varying according to $T^{-\frac{3}{2}}$ is then predicted. This dependence is observed for near intrinsic samples of germanium (as in Fig. 8.13), but departures from this relationship are found in most

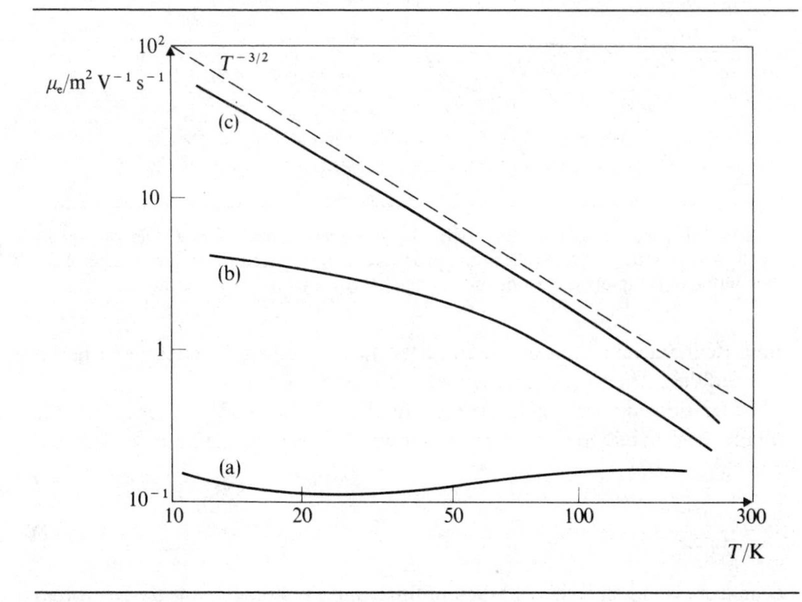

Fig. 8.13 Temperature dependence of the mobility of n-type germanium samples. The three samples correspond to those compared in Fig. 8.12.

semiconductors. This has been attributed to departures from spherical energy bands, and the scattering by optical branch phonons.

Scattering due to ionised donors or acceptors is most pronounced when the velocity of the free carriers is low, so that it is observed at low temperatures when lattice scattering is also less important. Ionised impurity scattering has been investigated by Conwell and Weisskopf using the classical Rutherford scattering formula and a dependence of scattering according to $T^{\frac{3}{2}}$ is then predicted. At very low temperatures, however, the donors and acceptors are not ionised and scattering by neutral impurity atoms may then be important.

8.4.1 *Conductivity and mobility measurements*

The conductivity of bulk samples of semiconductors is normally determined by the four-point probe method. This was first described in detail by Valdes and is illustrated in Fig. 8.14. The outer pair of probes A and D are current carrying whilst the inner pair B and C allow the potential drop to be measured. In order to

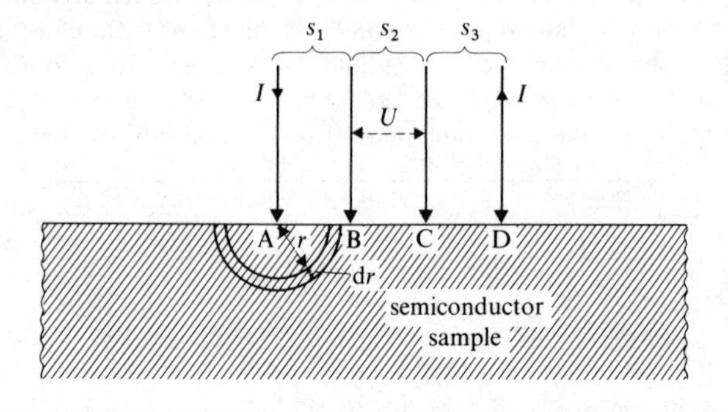

Fig. 8.14 Measurement of conductivity by the four-point probe method. The four spring loaded probes, mounted together in the probe head are pressed on to the surface of the semiconductor sample, the current and voltage being measured in a separate unit.

minimise the effect of non-ohmic contacts, an abraded semiconductor surface is used and the voltage drop is measured potentiometrically.

The relationship between sample conductivity σ, the sample current I, and the voltage drop U may be obtained as follows. Referring to the figure a potential drop

$$dU = \frac{I\,dr}{2\pi\sigma r^2} \qquad \ldots(8.42)$$

occurs acrosss a hemispherical shell radius r and thickness dr due to the current probe at A. This gives

$$U_B - U_C = \int_{s_1}^{s_1+s_2} \frac{I\,dr}{2\pi\sigma r^2}$$

$$= \frac{I}{2\pi\sigma}\left(\frac{1}{s_1} - \frac{1}{s_1+s_2}\right). \qquad \ldots(8.43)$$

A potential drop from B to C will also occur due to the second current probe at D so that the total measured potential drop U will be

$$U = \frac{I}{2\pi\sigma}\left(\frac{1}{s_1} + \frac{1}{s_3} - \frac{1}{s_1+s_2} - \frac{1}{s_2+s_3}\right). \qquad \ldots(8.44)$$

Normally $s_1 = s_2 = s_3 = s$, where s may have a value $\sim$0.5 to 1 mm and the conductivity is then

$$\sigma = \frac{I}{2\pi s U}. \qquad \ldots(8.45)$$

The formula applies only to samples whose dimensions are large compared with the probe spacing, although correction methods are available for smaller samples. Other similar techniques are available for determining the conductivity of thin sheets and of surface layers.

The mobility of current carriers may be determined by injecting excess carriers, via a voltage pulse and a metal contact, into a rod sample of an extrinsic semiconductor. An electric field sweeps the minority carriers along the rod which are detected as a pulse and the time of transit determines the drift velocity. The method essentially measures the drift velocity of minority carriers in the presence of a large excess of majority carriers. A drift mobility measured in this way gives the conductivity mobility (either μ_h or μ_e) provided that the semiconductor is not near intrinsic. In the latter case the measured drift mobility is a function of both μ_e and μ_h. (See Section (9.3.3).

An entirely independent method of determining the carrier concentration in a semiconductor sample is provided by a measurement of the Hall effect. In this method, a longitudinal electric field and a transverse magnetic field are applied simultaneously to a semiconductor sample, which is in the form of a rectangular slab. The current carriers experience a Lorentz force perpendicular to both the electric and magnetic fields as is illustrated in Fig. 8.15. An equilibrium condition

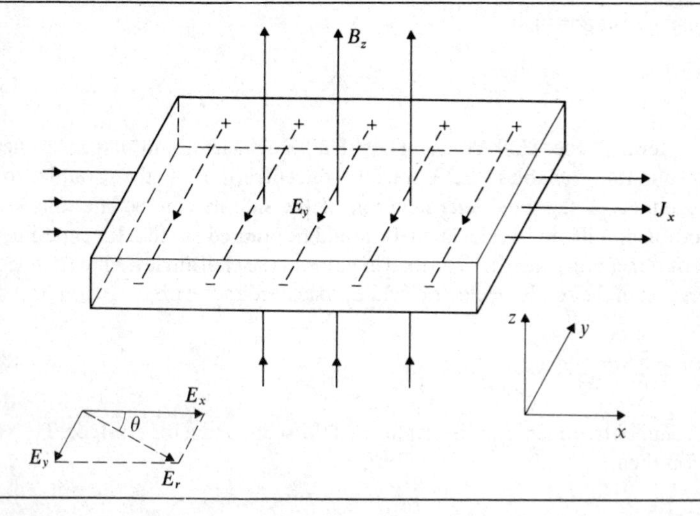

Fig. 8.15 Measurement of the Hall effect. The magnetic induction B_z produces a net force on the electron in the – y direction and the charge build up shown produces the balancing field E_y. In the absence of B_z the equipotentials are transverse to E_x and are rotated through angle θ by the induction.

is attained with a transverse electric field opposing transverse carrier flow. The observed voltage is referred to as the Hall voltage. Having noted that the direction of the force on the carriers is governed by the direction of the *current*, it is clear that the sign of the Hall voltage enables the current carriers to be identified as either electrons or holes.

Let us assume in the first instance that the semiconductor sample is extrinsic n-type. Referring to the diagram, a magnetic induction B_z at right-angles to the direction of current flow produces a transverse force of $B_z ev_x$ on an electron, where v_x is the electron drift velocity. The Hall field E_y set up gives a balancing force eE_y where

$$eE_y = -B_z ev_x. \qquad \ldots(8.46)$$

The current density J_x in the sample is given by

$$J_x = E_x\sigma = nev_x, \qquad \ldots(8.47)$$

where σ is the sample conductivity. We thus have

$$E_y = -RB_zJ_x = -RB_zE_x\sigma, \qquad \ldots(8.48)$$

where

$$R = -\frac{1}{ne}. \qquad \ldots(8.49)$$

or using equation (8.4)

$$R = -\frac{\mu_e}{\sigma}. \qquad \ldots(8.50)$$

R is termed the Hall constant or Hall coefficient and its experimental determination enables the electron concentration in the sample to be determined. A separate determination of the sample conductivity allows the electron mobility to be calculated. It should be pointed out that the experimental measurements can also be described in terms of the Hall angle θ which defines the direction of the resultant electric field E_r in accordance with

$$\tan\theta = \frac{E_y}{E_x} = -B_z\mu_e. \qquad \ldots(8.51)$$

A similar treatment can be developed if the semiconductor is p-type. The value of R is then

$$R = \frac{1}{pe} \qquad \ldots(8.52)$$

and a transverse electric field, opposite to that observed with the n-type sample, is obtained.

The relationships (8.50) and (8.52) are, however, not exact since implicit in their derivation is the assumption that the carrier relaxation time τ is a constant. τ is in practice a function of carrier velocity and the experimentally determined values of R differ from those given by a numerical factor. However, the quantity

$$R\sigma = \mu_H \qquad \text{...(8.53)}$$

may be used to define a Hall mobility μ_H which differs from the conductivity mobility μ_e. The ratio μ_H/μ_e has a value between 1 and 2 depending on the scattering mechanism predominant in the sample. For lattice scattering and for ionised impurity scattering the factors are respectively $\frac{3}{8}\pi$ and 1.98 whilst for degenerate semiconductors or in the presence of large applied magnetic fields the factor tends to unity.

Finally we note that for a near intrinsic semiconductor the Hall coefficient may be shown to be given (Smith, 1st edition, 1959, p. 103) by

$$R = \frac{1}{e}\frac{(p\mu_h^2 - n\mu_e^2)}{(p\mu_h + n\mu_e)^2}. \qquad \text{...(8.54)}$$

Since the mobility of electrons is normally greater than that of holes the Hall constant of an intrinsic semiconductor is negative. This is in accordance with the greater contribution of electrons to the intrinsic conductivity.

8.5 Minority carrier lifetime

The equilibrium carrier densities in a semiconductor are governed by the relationship

$$np = n_i^2 = \text{constant}. \qquad \text{...(8.31)}$$

The numerical value of the constant varies from one semiconductor to another and, as shown previously, increases exponentially with temperature. This relationship was obtained, using the results of statistical mechanics, without reference to the nature of the process leading to the equilibrium state. This state is achieved as a dynamic balance between a thermal generation rate of electron hole pairs and a recombination rate determined by the product np. In semiconductor devices one is often concerned with carrier concentrations in excess of the equilibrium values and the rate of recombination of excess carriers is an important factor in devices, e.g. it determines the distance injected excess carriers may travel before being dissipated by recombination.

A simple method of producing excess carriers is that of optical irradiation of the semiconductor. Provided that the photon energy is greater than the energy gap, electrons are excited from the valence band into the conduction band, i.e. electron-hole pairs are formed. The corresponding optical absorption edge lies in the near or intermediate infrared, its exact position depending on the value of the band gap. Semiconductors are thus opaque in the visible range. For a large energy gap insulator such as diamond the absorption edge is in the ultraviolet and diamond is therefore transparent.

Let us consider the irradiation of a semiconductor by a single pulse of light. This will produce an excess of electron-hole pairs and the cloud of excess carriers as a whole will be electrically neutral. The absorption of the light will be a maximum near the surface, although in problems concerned with the behaviour of excess carriers the simplifying assumption of uniform generation is often used. Although equal numbers of carriers are generated the minority carrier density may increase by orders of magnitude whilst the percentage increase in majority density may be quite small. The excess minority carrier density governs the recombination process and we may now determine the mode of decay of the excess density formed by the single pulse of light.

As previously indicated, in an equilibrium state the thermal generation rate G of carriers is equal to the recombination rate R which is proportional to the product of the carrier densities. We may therefore put

$$R = knp = G, \qquad \ldots(8.55)$$

where k is a temperature-dependent constant for the semiconductor. If excess carrier densities Δn and Δp are produced by the pulse we have

$$n' = n + \Delta n$$

$$p' = p + \Delta p, \qquad \ldots(8.56)$$

where n' and p' are the non-equilibrium carrier densities. These will produce a new recombination rate R' of

$$R' = kn'p'. \qquad \ldots(8.57)$$

So that, using equations (8.55) and (8.56), the excess recombination rate over the generation rate is

$$R' - G = k(n\Delta p + p\Delta n + \Delta n\Delta p). \qquad \ldots(8.58)$$

This equation governs the rate at which the carrier densities return to their equilibrium values. Let us assume that the semiconductor is n-type. We then have both $p\Delta n$ and $\Delta n\Delta p$ much less than $n\Delta p$, i.e.

$$R' - G = kn\Delta p. \qquad \ldots(8.59)$$

Thus in the presence of a predominance of majority carriers the rate of approach to equilibrium is governed by the recombination of the *excess* minority carriers giving

$$\frac{\mathrm{d}}{\mathrm{d}t}(\Delta p) = -kn\Delta p = -\frac{\Delta p}{\tau_{\mathrm{p}}}, \qquad \ldots(8.60)$$

where τ_{p} is a characteristic constant for the semiconductor material termed the *minority carrier lifetime*. Integrating, we find an exponentially decaying excess carrier concentration

$$\Delta p = \Delta p_0 \exp\left(-\frac{t}{\tau_{\mathrm{p}}}\right) \qquad \ldots(8.61)$$

where Δp_0 is the initial excess minority carrier density. Charge neutrality requires that the majority carrier density also follows an identical decay pattern. Indeed, if instead of optical generation of excess carriers, we inject carriers of one type only into a semiconductor sample, equation (8.61) will still describe the process, since the electric field generated will draw in carriers of the opposite type and produce charge neutrality in a much shorter time than τ_p (see Section 9.2.3).

The approach to the equilibrium state of a semiconductor can be described by a simple model in which recombination and generation occur as single direct transitions from one band to another. However, during recombination it is a necessary condition that the electron loses both energy and momentum. The amount of momentum change as measured from the lowest conduction band minimum to the highest valence band maximum will depend on the details of the band structure for the particular semiconductor. For direct gap semiconductors this model describing direct recombination across the band gap applies and the energy is released in the form of light of appropriate wavelength. This forms the basis of the use of gallium arsenide as a semiconductor laser. For indirect gap semiconductors, momentum change must also occur so that the recombination energy is dissipated as that of lattice phonons. This requires the cooperation of several phonons and for a semiconductor with a moderate band gap the direct transition is a process of low probability which would yield carrier lifetimes much in excess of the experimentally observed values.

In order to account for recombination in these semiconductors a model on which recombination through localised levels near the centre of the band gap was developed by Hall and by Shockley and Read. This is illustrated in Fig. 8.16. Recombination is then considered as a process in which the centre consecutively captures an electron and a hole from the respective bands. Consecutive emission of an electron and a hole by the centre corresponds to the generation process.

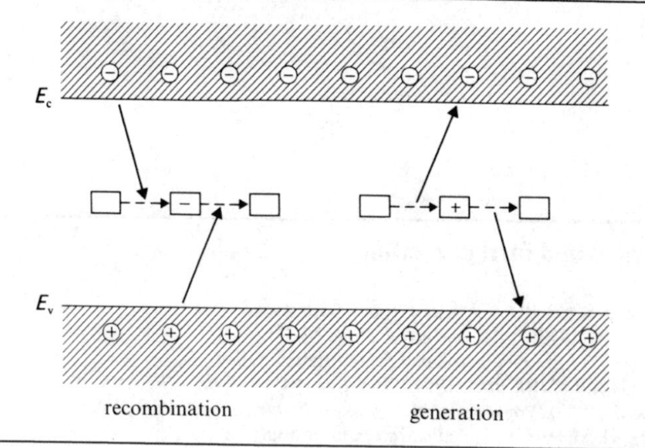

Fig. 8.16 The Shockley–Read model for transitions across the band gap via an intermediate level.

Certain impurity centres have a much higher probability of exchanging carriers with one or other of the bands, in which case the impurity centre will capture a carrier and release it to the same band. It is then termed a trap since it can temporarily immobilise carriers.

It has been shown that impurities, such as copper or nickel in germanium or gold in silicon, give rise to one or more of the localised levels near the middle of the band gaps in these semiconductors. The *lifetime* of the semiconductor material is determined by the concentration of these impurities and the concentration levels involved can be considerably lower than those of the donor or acceptor type impurities which control the conductivity. Thus copper at a concentration level of 10^{20} m^{-3} reduces the lifetime of intrinsic germanium from about 2 ms to 0.1 ms. There is also correlation between crystalline perfection and minority carrier lifetime, a high dislocation density producing a decrease in the observed lifetime of crystals.

Surface states can also give recombination centres so that the nature of the surface of a semiconductor, e.g. whether it is ground or electrolytically polished, may very appreciably affect the observed lifetime of a semiconductor sample. The reduced lifetime due to the presence of surface states leads to measured lifetimes which are functions of sample sizes, the smaller samples yielding the shorter lifetimes since in these the carriers are able to drift more easily to the surface to recombine.

The minority carrier lifetimes lie in the range 2 ms to 0.1 μs and may be determined by the decay of photoconductivity method. A semiconductor sample is illuminated with a short pulse of light and the transient increase in photoconductivity is monitored with a suitable electrical circuit and displayed. The bulk semiconductor lifetime is assessed from a series of measurements taken on samples of different cross-sectional areas and which have received different surface treatments.

References and further reading

Blakemore, J. S. (1969) *Solid State Physics*, Saunders.
Hart-Davis, A. (1975) *Solids, an Introduction*, McGraw-Hill.
Kittel, C. (1971) *Introduction to Solid State Physics*, Wiley.
Morin, F. T. and **Maita, J. P.** (1954) *Phys. Rev.*, **94**, 1525; *Phys. Rev.*, **96**, 28.
Pascoe, K. J. (1973) *Properties of Materials for Electrical Engineers*, Wiley.
Rosenberg, H. M. (1975) *The Solid State*, Clarendon.
Seymour, J. (1972) *Physical Electronics*, Pitman.
Smith, R. A. (1959) *Semiconductors*, Cambridge University Press.

Problems

(Data given in Table 8.1 should be used, as appropriate, for the solution of the following problems.)

8.1 Calculate the conductivity at room temperature of intrinsic germanium. What proportion of the atoms of germanium are giving rise to conduction electrons?

8.2 A sample of germanium contains a donor impurity concentration $N_d = 2.0 \times 10^{21}$ m^{-3}. Assuming that there is no acceptor impurity compensation determine the hole concentration p and the conductivity of the sample.

8.3 Determine the position of the Fermi level in germanium for $N_d : n_i$ equal to 10^2 and 10^4. Express the values obtained with respect to the position of the Fermi level in intrinsic germanium and assume room temperature condition. Would your method of calculation be valid if the ratio were 10^6?

8.4 The electron and hole effective masses in indium antimonide are 0.013 m_e and 0.18 m_e, respectively. Determine the position of the Fermi level, with respect to the middle of the band gap, in the intrinsic semiconductor at 300 K. The value obtained should be compared with that of 0.16 eV for the overall band gap of the semiconductor at the same temperature.

8.5 A sample of silicon was doped with arsenic at a level calculated to yield a resistivity of 0.03 Ω m. Due to the effects of compensation the sample obtained was of resistivity 0.04 Ω m and was n-type. Determine the donor and acceptor concentrations present.

8.6 Calculate an approximate value for the ionisation energy of donor type impurities in silicon and estimate the radial extension of the wave-function for the bound electron.

8.7 Intrinsic germanium has the following electron concentrations at the temperatures indicated.

T/K	200	300	400	500
n/m^{-3}	5.8×10^{15}	2.0×10^{19}	1.4×10^{21}	1.9×10^{22}
T/K	600	700	800	900
n/m^{-3}	1.2×10^{23}	4.3×10^{23}	1.4×10^{24}	2.7×10^{24}

Show that this data conforms to the form of equation (8.29) and determine the constants A and E_0 for germanium. Assuming appropriate values for m_e^* and m_h^* show that the energy gap for germanium at 300 K is 0.67 eV and that, for the temperature range concerned, the decrease of E_g with temperature amounts to 3.9×10^{-4} eV K^{-1}.

8.8 Using data from the previous question plot the variation of electron concentration and of conductivity with $1/T$ for the temperature range 300–900 K for a germanium sample which has $N_d = 10^{21}$ m^{-3}. Use the approximations given in the text for the transition region from extrinsic to intrinsic behaviour.

8.9 The electrical conductivity data of Morin and Maita on silicon lead to the representation

$$n_i = 3.87 \times 10^{22} T^{\frac{3}{2}} \exp\left(-\frac{1.21}{2kT}\right)$$

for the variation of intrinsic carrier concentration with temperature. Assuming constant values of carrier mobilities determine the carrier concentrations and conductivities for 300 K, 600 K and 900 K.

8.10 The multiplying constant in the equation for n_i in the previous question can be calculated from the experimentally determined values of m_e^*, m_h^* and α. Assuming that for silicon $\alpha = -2.4 \times 10^{-4}$ ev K^{-1}, by what percentage amount would the values of m_e^* and m_h^* given in Table 8.1 need to be modified so as to give a multiplying constant which agrees with that determined from the conductivity data.

8.11 Show that, at any given temperature, a semiconductor has a minimum conductivity σ_{min} given by

$$\sigma_{min} = 2en_i(\mu_h\mu_e)^{\frac{1}{2}}.$$

Determine the corresponding carrier concentrations in terms of the intrinsic concentration n_i.

8.12 Calculate the Hall constant and Hall mobility for the metal sodium using data given in Table 5.1.

Determine also the value of the Hall constant for a sample of p-type germanium of conductivity 100 Ω^{-1} m^{-1} and explain the differences between the value obtained and that for the metal.

8.13 Explain why the Hall constant R_i for an intrinsic semiconductor is not equal to zero. Derive, from equation (8.54) an expression for R_i in terms of the ratio $\mu_e:\mu_h$ and calculate the value for R_i for intrinsic germanium.

8.14 Denoting the ratio $\mu_e:\mu_h$ by b show that the Hall constant is zero if $p = b^2n$ and that when this condition holds the conductivity of the semiconductor sample is the same as that for intrinsic material.

Calculate the electron and hole concentrations for a germanium sample which has a zero Hall constant.

8.15 Determine the Hall constant for an n-type germanium sample of conductivity 50 Ω^{-1} m^{-1}. If the sample is of rectangular cross-section 1.5 cm × 1.5 mm determine the Hall voltage across the width of the sample if the current passing is 2 mA and the applied magnetic flux is 10^{-1} T.

9. The p-n junction

9.1 Introduction

In the last chapter the behaviour of bulk semiconductor material under equilibrium conditions was described. It was then assumed that the material was of uniform conductivity and it was shown that the mobile carrier concentrations in each band could be described in terms of the position of the Fermi level. In this chapter we will discuss the ways in which the band representation is modified, firstly when an electric field is applied to a sample and secondly when the conductivity of a sample varies from one point to another. This will be followed by a study of the properties of excess carriers and the behaviour of p-n junctions in equilibrium and with applied bias.

9.2 Electric fields in semiconductors

Let us first consider the effect of applying an electric field along an n-type rod sample of a semiconductor single crystal. As illustrated in Fig. 9.1 each electron, whatever its position in the band structure, has its energy influenced by the applied electric field. Electrons in the conduction band will experience a drift flow in the field which will be directed towards the position of lowest electron energy on the diagram; conversely, holes in the valence band move upwards on the band diagram although we note that in the n-type sample their concentration will be very small. Drift of both types of carrier will correspond to the flow of current in the conventional sense from the point of high electric potential to that of lower potential.

In accordance with Section 5.2 the electric field E_x at any point is given by

$$E_x = -\frac{\partial U}{\partial x}, \qquad \ldots(5.1)$$

where U is the electrostatic potential at that point. The band structure *as a whole* must follow the changes in applied potential, and converting equation (5.2) to *electron energy* we must therefore have

$$E_x = \frac{1}{e}\frac{\partial E_{\mathrm{j}}}{\partial x} \qquad \ldots(9.1)$$

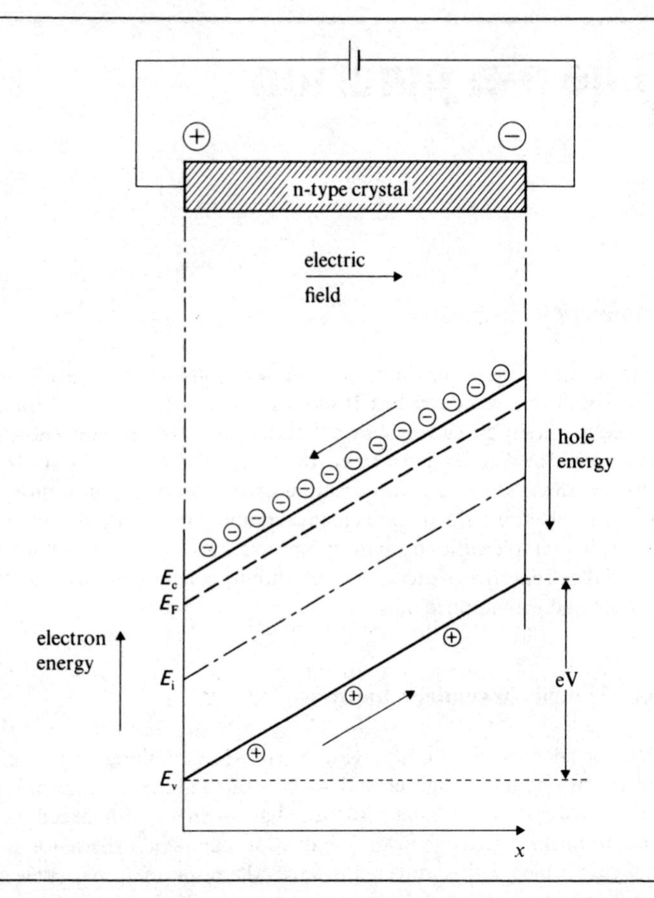

Fig. 9.1 The band diagram of a uniform n-type semiconductor rod sample in the presence of an applied external electric field.

where E_j may represent E_c, E_v, E_i or E_F. We note that the structure of the band, for example the band gap width E_g and the position of the Fermi level, is governed by factors such as crystal structure, temperature and sample impurity content, and is unaffected by the presence of the field.

We may next see how the band structure is represented for a rod sample of non-uniform conductivity. Let us, for example, consider as in Fig. 9.2 an n-type sample which has an impurity content varying continuously along its length. There is no external field applied in this case. For all positions along the sample the relative position of the Fermi level with respect to the band edges is determined by the impurity concentration. As previously discussed in Section 5.9, in the absence of an externally applied electric field, the Fermi level must be constant throughout, since states having the same probability of occupation

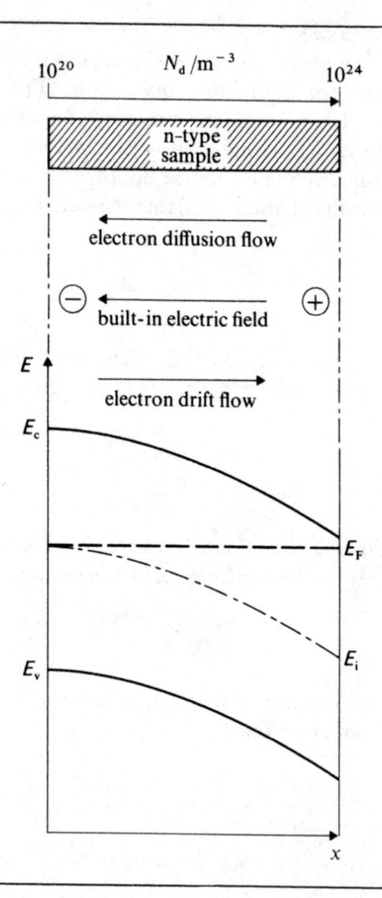

Fig. 9.2 The diffusion and drift flows of electrons in an inhomogeneous germanium rod sample. Note that the sample has a built-in electric field in the direction indicated.

must lie at the same energy value. It follows, therefore, that for this sample of variable conductivity, the band edges must vary in energy and that the band representation must be as shown in Fig. 9.2. In accordance with equation (9.1) there will therefore exist within the sample built in electric fields and comparing Figs. 9.1 and 9.2 these are such that the high conductivity end of the sample must be at a positive potential with respect to the near intrinsic end.

Although this internal field is present there cannot be a resultant flow of carriers since this would act in such a way as to destroy the field. The situation in the crystal can, however, be regarded as a dynamic equilibrium in which a drift flow of electrons under the action of the built-in field is exactly balanced by the diffusion flow of electrons due to the gradient in electron concentration. A similar type of balance exists independently for the mobile holes.

9.2.1 *The Einstein relationship*

The balancing of the drift and diffusion currents implies that there are interrelationships between the drift mobility and the diffusion coefficient for each type of mobile carrier. Let us consider, for example, an n-type sample which has a variable conductivity along its length. Provided that the donor concentration N_d at any point is not sufficiently high for the donors to be incompletely ionised we have, in accordance with equation (8.32) that the electron concentration n at any point is given by

$$n = N_d = N_c \exp\left(\frac{E_F - E_c}{kT}\right). \qquad \ldots(9.2)$$

In the absence of an externally applied electric field the Fermi level will be constant and by reference to Figs. 9.1 and 9.2 the electric field E_x at any point is seen to be given by

$$E_x = \frac{1}{e}\frac{\partial E_c}{\partial x}. \qquad \ldots(9.3)$$

For the conditions illustrated in Fig. 9.2 this field is in the negative direction of x and from equation (9.2) is given in terms of carrier concentration gradient by

$$E_x = -\frac{kT}{en}\frac{\partial n}{\partial x}. \qquad \ldots(9.4)$$

The resultant electron *drift current density*, which we may denote by J_1, is then given in accordance with equation (5.2) by

$$J_1 = E_x \sigma \qquad \ldots(9.5)$$
$$= E_x n e \mu_e$$

from equation (8.4). Substituting for E_x we obtain

$$J_1 = -\mu_e kT \frac{\partial n}{\partial x}. \qquad \ldots(9.6)$$

The electron *diffusion current density* is proportional to the concentration gradient of the mobile electrons and is given by

$$J_2 = eD_e \frac{\partial n}{\partial x}, \qquad \ldots(9.7)$$

where D_e is the diffusion coefficient for electrons. We note that J_1 and J_2 are *current* densities and that they are therefore in directions opposite to the electron flows indicated in Fig. 9.2. Applying the condition that, in the absence of an applied field, the total electron current must be zero it follows from equations (9.6) and (9.7) that

$$\mu_e = \frac{e}{kT} D_e. \qquad \ldots(9.8)$$

A similar consideration of the drift and diffusion hole currents yields

$$\mu_h = \frac{e}{kT} D_h, \qquad \ldots(9.9)$$

where D_h is the diffusion coefficient for holes. Equations (9.8) and (9.9) are known as Einstein's relations and may be combined to yield

$$\frac{D_e}{\mu_e} = \frac{D_h}{\mu_h} = \frac{kT}{e}. \qquad \ldots(9.10)$$

9.2.2 *Space charge*

The way in which internal fields arise in semiconductor samples may now be looked at in more detail. In a homogeneous sample with the uniform distribution of impurity there is no internal electric field. At each point in the sample there is charge neutrality so that denoting the ionised donor and acceptor concentrations by N_d^+ and N_a^- the space charge ρ is given by

$$\rho = (N_d^+ - N_a^- + p - n)e \qquad \ldots(9.11)$$

is everywhere zero. We neglect here ionised impurity centres other than acceptors or donors since they will normally be present at much lower concentration levels.

In an inhomogeneous sample in order to oppose the diffusive flow of carriers in the carrier concentration gradient it is necessary to have built-in electric fields. These fields are formed by a slight separation of the mobile carriers from the fixed ionised impurities. Charge neutrality will therefore not apply in these regions, although the sample as a whole will be electrically neutral.

For an inhomogeneous sample there are four parameters of interest which are interrelated, namely, field strength, electric potential, carrier density and space charge. Let us briefly examine the nature of these interrelationships. Firstly Gauss' law relates field strength $\boldsymbol{E}$ and space charge $\rho(\boldsymbol{r})$ in accordance with

$$\operatorname{div} \boldsymbol{E} = \frac{\partial E_x}{\partial x} + \frac{\partial E_y}{\partial y} + \frac{\partial E_z}{\partial z} = \frac{\rho(\boldsymbol{r})}{\varepsilon}, \qquad \ldots(9.12)$$

where E_x, E_y and E_z are the three components of the field. For a rod specimen this may be simplified to the form

$$\frac{\partial E_x}{\partial x} = \frac{\rho(x)}{\varepsilon}, \qquad \ldots(9.13)$$

where $\rho(x)$ is the space charge assumed to be a function of position along the rod. The field strength E_x is also, as we have seen, related to the carrier concentration in a sample and in terms of electron concentration is given by

$$E_x = -\frac{kT}{en}\frac{\partial n}{\partial x}. \qquad \ldots(9.4)$$

A similar type of interdependence is also found for the electrostatic potential. Thus by making use of equation (5.2) we obtain the alternative form to equation (9.13) in terms of potential, namely

$$\frac{\partial^2 U}{\partial x^2} = -\frac{\rho(x)}{\varepsilon} \qquad \ldots(9.14)$$

whilst the Boltzmann equations (8.32) and (8.33) give the potential in the form

$$U = \frac{E_F - E_i}{e} = \frac{kT}{e}\ln\frac{n}{n_i} = \frac{kT}{e}\ln\frac{p_i}{p}. \qquad \ldots(9.15)$$

Since $\rho(x)$ depends both on the impurity concentration and on the carrier concentration a simple solution for U cannot be obtained. We note, however, that a comparatively small space charge can set up an appreciable electric field and many problems involving non-uniform impurity distribution may be solved using the assumption that there is charge neutrality. This is equivalent to assuming that the carrier concentration follows exactly the variation in net ionised impurity concentration. Solutions obtained in this manner are valid for many problems. Alternatively we may assume that the density of mobile carriers is negligible compared with that of the fixed ionised atoms. This type of assumption is valid, for example, in the very high field region occurring near a p-n junction.

9.2.3 *Dielectric relaxation time*

In Section 8.5 the behaviour of excess carriers in a semiconductor sample was considered and the rate at which the carrier concentrations returned to their equilibrium values was described in terms of the lifetime of the minority carriers. Let us now consider the assumption then made that the return to *charge neutrality* on injection of carriers of one type only is much more rapid than the minority carrier decay time.

We may take as an example a homogeneous rod sample in which there is an instantaneous increase in charge density ρ within a small element given by

$$\rho = e(\Delta p - \Delta n). \qquad \ldots(9.16)$$

This charge density produces an electric field E_x and a current density J_x related by

$$J_x = \sigma E_x, \qquad \ldots(5.2)$$

where σ is the sample conductivity. Within a small element Δx of rod the charge $\rho\Delta x$ for unit cross-section will decay due to a net flow of current out of the element, i.e.

$$\frac{\partial}{\partial t}(\rho\Delta x) = -\frac{\partial J_x}{\partial x}\Delta x. \qquad \ldots(9.17)$$

Using equations (9.13) and (5.3) this yields

$$\frac{\partial \rho}{\partial t} = -\frac{\sigma}{\varepsilon}\rho. \qquad \ldots(9.18)$$

An instantaneous space charge density ρ_0 therefore decays exponentially to a value ρ_t at time t in accordance with

$$\rho_t = \rho_0 \exp(-t/\tau_d), \qquad \ldots(9.19)$$

where the quantity $\tau_d = \varepsilon/\sigma$ is termed the dielectric relaxation time. Substituting numerical values we find that for intrinsic germanium $\tau_d \sim 7 \times 10^{-11}$ s and this value should be compared with a typical value ~ 1 ms for the minority carrier lifetime in the material. Space charge therefore persists for only a very short time in a homogeneous semiconductor. In a sample of non-uniform conductivity diffusion effects need also to be taken into account and as we have seen space charges then persist and give rise to internal electric fields.

9.3 Behaviour of charge carriers

In Section 9.2 the way in which electric fields were produced in semiconductors by a variation in impurity concentration from point to point was described. In the present section the behaviour of excess carriers in semiconductor samples will be developed.

9.3.1 *Minority carrier injection*

Consider a uniform p-type semiconductor sample which has a potential difference applied across its ends by means of suitable contacts (Fig. 9.3). At a point in the sample well removed from these contacts the current will be carried predominantly by holes. Since the current in the connecting wires is carried by electrons it is clear that there must exist mechanisms producing the changeover in the current carriers at each end of the specimen close to the contacts. A similar situation must also arise at the boundary in a semiconducting crystal between regions which are of different conductivity type. The mechanisms involved in the example chosen are briefly as follows. Electrons from the negative contact B enter the semiconductor, and at points close to the contact recombine with holes. To replenish these holes and thus maintain charge neutrality a field is set up in the crystal which produces a hole flow into the region of the contact. This constitutes the current in the sample. The thermal generation of electron–hole pairs at the positive contact A provides the hole flow into the sample and the electron flow into the connecting lead. At the negative contact B electrons are drawn into the semiconductor from the metal contact and this is termed *minority carrier injection.* At the positive contact electrons are ejected into the metal contact. This is termed carrier extraction.

In the above discussion it has been assumed that electrons travel with equal

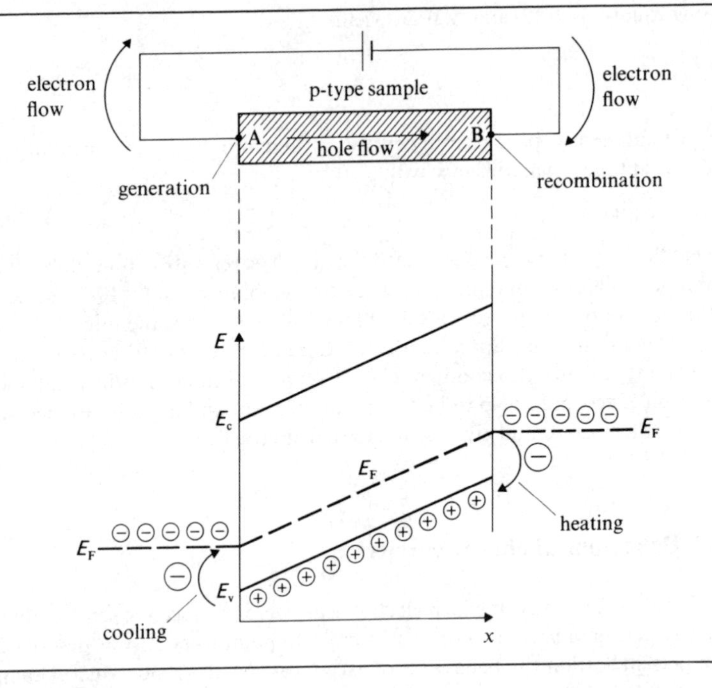

Fig. 9.3 The Peltier effect.

facility through each contact, i.e. there is no rectifying action and the contacts are then termed ohmic. On the band diagram representation electrons at contact B at the level E_F in the metal fill vacant levels near to E_v in the semiconductor. The electrons lose this energy to the semiconductor lattice and there is therefore a heating effect at this contact. Similarly at the contact A the generation of electron–hole pairs which are continually removed is accompanied by cooling. This is termed the Peltier effect. The production of a thermo-emf when either junction is heated is the inverse effect and is known as the Seebeck effect.

9.3.2 *The continuity equation*

Let us examine more closely the process of carrier injection by considering the effect in a long, uniform conductivity n-type semiconductor sample in which minority carriers are injected at a specific point. In Fig. 9.4 the hole injection is shown to take place at the point A, $x = 0$, and in order that the process may be continuous it is assumed that a small electric field is applied along the sample. The excess holes move into the semiconductor by a process of diffusion and simultaneously will be recombining with electrons. At a point sufficiently distant from A the carrier concentrations will therefore be essentially the same as those for a sample in the thermal equilibrium state. Additionally the fraction of the

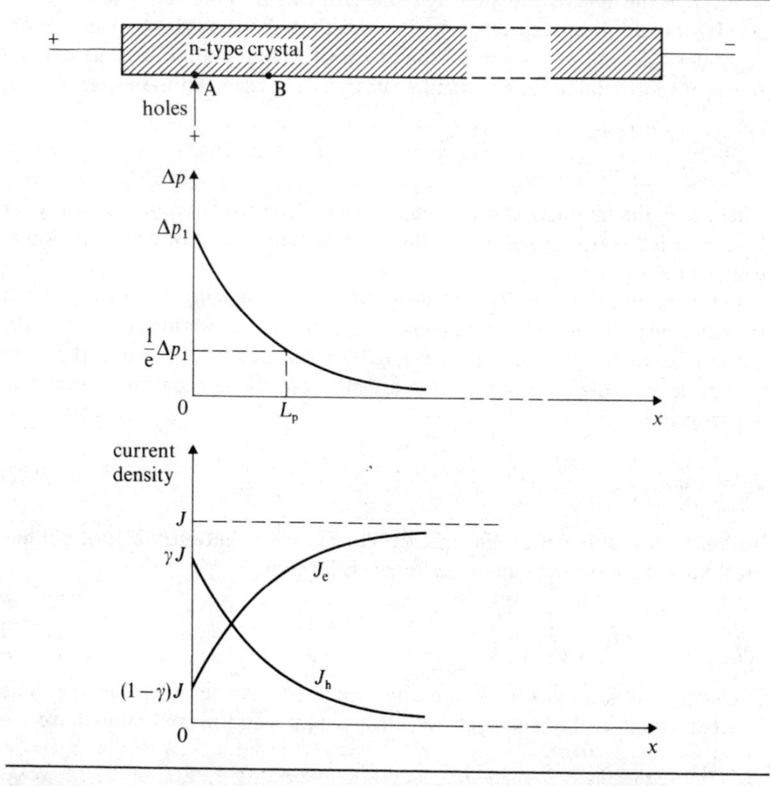

Fig. 9.4 Injection of holes into an n-type sample showing decay of excess minority carriers. L_p is the minority carrier diffusion length. The lower diagram illustrates the decay of hole current and the complementary growth of the electron current. Note that L_p is very small compared with the length of a rod sample.

current which at A is carried by the injected holes will at this distant point be carried by the majority electrons.

We may now determine how the excess hole concentration is dying away from the point A and how the magnitudes of the electron and hole currents will vary along the sample. Let p_1 and p represent the hole concentrations at A($x = 0$) and at any other point B at distance x. Let Δp_1 and Δp be the corresponding *excess* hole concentrations above the equilibrium value p_0. Considering an element of the sample at the point B the rate of change of hole concentration with time will be given by the sum of the excess generation rate G over the recombination rate R for electron–hole pairs together with the excess flow of holes inwards over the outward flow, i.e.

$$\frac{\partial p}{\partial t} = G - R - \frac{1}{e}\frac{\partial}{\partial x}(J_h), \qquad \ldots(9.20)$$

where J_h is the hole current density at the element. We note here that under the steady state conditions specified $\partial p/\partial t = 0$, although the equation given is more generally valid. As in Section 8.5 the excess of recombination over generation rate is proportional to the excess minority carrier concentration so that we have

$$\frac{\partial p}{\partial t} = -\frac{\Delta p}{\tau_p} - \frac{1}{e}\frac{\partial J_h}{\partial x}, \qquad \ldots(9.21)$$

where τ_p is the minority carrier lifetime. This is the equation of continuity for holes and is a statement of the condition that there cannot be accumulation of charge at any point.

Let us assume that the electric field applied along the sample is small and that the material is of fairly high conductivity. Under these conditions the part of the current due to drift of holes will be negligibly small and for J_h one may therefore substitute the hole diffusion current arising from the concentration gradient. This is given by

$$J_h = -eD_h\frac{\partial p}{\partial x}, \qquad \ldots(9.22)$$

where D_h is the diffusion coefficient for holes. Since in equation (9.21) *changes* in p are involved the substitution of Δp for p is valid giving

$$\frac{\partial \Delta p}{\partial t} = -\frac{\Delta p}{\tau_p} + D_h\frac{\partial^2 \Delta p}{\partial x^2}. \qquad \ldots(9.23)$$

Under the steady-state conditions specified in the problem the hole concentrations at all points must remain constant with time so that we have

$$D_h\frac{\partial^2 \Delta p}{\partial x^2} - \frac{\Delta p}{\tau_p} = 0. \qquad \ldots(9.24)$$

This has for its solution

$$\Delta p = \Delta p_1 \exp\left(-\frac{x}{L_p}\right), \qquad \ldots(9.25)$$

where

$$L_p = (D_h \tau_p)^{\frac{1}{2}} \qquad \ldots(9.26)$$

is the minority carrier diffusion length. It is the distance from the injecting source at which the excess concentration of minority carriers has decayed to $1/e$ of its value at the source. It is apparent that a corresponding diffusion length L_n can be defined in analogous manner for electrons in p-type material. Although minority carrier diffusion lengths are small, e.g. 7×10^{-4} m in n-type germanium having a minority carrier lifetime of 10^{-4} s, they often determine critical dimensions of semiconductor devices.

The total current density J at any point in the sample is the sum of the electron and hole current densities J_e and J_h. Denoting the hole current density at the source as J_{h_1}, the fraction of the total current carried by holes

$$\frac{J_{h_1}}{J} = \gamma \qquad \text{...(9.27)}$$

is termed the injection ratio and may vary between 0 and 1 depending on the type of contact. The variation in the hole diffusion current J_h from point to point along the sample follows directly from equation (9.25). Thus using equation (9.22) and noting that Δp can be substituted for p we have

$$\begin{aligned} J_h &= J_{h_1} \exp\left(-\frac{x}{L_p}\right) \\ &= \gamma J \exp\left(-\frac{x}{L_p}\right). \end{aligned} \qquad \text{...(9.28)}$$

The electron current density will clearly increase in a complementary manner according to

$$J_e = J\left[1-\gamma \exp\left(-\frac{x}{L_p}\right)\right]. \qquad \text{...(9.29)}$$

The variations in the current densities along the sample are illustrated in Fig. 9.4.

9.3.3 *Ambi-polar mobility and diffusivity*

In the last sub-section the variation in hole current was determined on the assumption that the drift current of holes could be neglected. Let this drift current now be included so that the total hole current is

$$J_h = pe\mu_h E_x - eD_h \frac{\partial p}{\partial x}, \qquad \text{...(9.30)}$$

where E_x is the electric field. On equating changes in p with those for Δp the continuity equation (9.23) becomes

$$\frac{\partial \Delta p}{\partial t} = -\frac{\Delta p}{\tau_p} - \mu_h p \frac{\partial E_x}{\partial x} - \mu_h E_x \frac{\partial \Delta p}{\partial x} + D_h \frac{\partial^2 \Delta p}{\partial x^2}. \qquad \text{...(9.31)}$$

If the applied field is small and the material is strongly extrinsic n-type ($n \gg p$) this equation reduces to the previous form discussed.

One may enquire at this point about the value of the electron density in the neighbourhood of the injected excess holes. In order to prevent large local electric fields being produced there must exist a corresponding excess density of electrons. For many problems it is assumed that there is charge neutrality, so that $\Delta n = \Delta p$. Corresponding to equation (9.31) for excess holes a similar equation therefore applies to excess electrons. Noting that the electron current density is given by

$$J_e = ne\mu_e E_x + eD_e \frac{\partial n}{\partial x}, \qquad \text{...(9.32)}$$

the continuity equation for electrons is

$$\frac{\partial}{\partial t}\Delta n = -\frac{\Delta n}{\tau_n} + \mu_e n \frac{\partial E_x}{\partial x} + \mu_e E_x \frac{\partial \Delta n}{\partial x} + D_e \frac{\partial^2 \Delta n}{\partial x^2}, \qquad \ldots(9.33)$$

where τ_n is a measure of the decay time for electrons in the material.

The behaviour of the excess carrier densities Δn and Δp may be put in a simplified form which is due to Van Roosbroeck. Charge neutrality is assumed to apply so that

$$\Delta n = \Delta p$$

$$\frac{\partial \Delta n}{\partial t} = \frac{\partial \Delta p}{\partial t} \qquad \ldots(9.34)$$

and

$$\tau_p = \tau_n. \qquad \ldots(9.35)$$

These assumptions are valid in equations (9.31) and (9.33) except for the terms $\partial E_x/\partial x$ describing the field variation produced by the space charge. These terms, however, may be eliminated from the two equations so as to give

$$\frac{\partial \Delta p}{\partial t} = -\frac{\Delta p}{\tau_p} - \mu^* E_x \frac{\partial \Delta p}{\partial x} + D^* \frac{\partial^2 \Delta p}{\partial x^2}, \qquad \ldots(9.36)$$

where μ^* and D^* are the ambi-polar mobility and diffusion coefficient given by

$$\mu^* = \frac{n-p}{\dfrac{n}{\mu_h} + \dfrac{p}{\mu_e}} \qquad \ldots(9.37)$$

and

$$D^* = \frac{n+p}{\dfrac{n}{D_h} + \dfrac{p}{D_e}} \qquad \ldots(9.38)$$

It is important to emphasise that the ambi-polar equation (9.36) describes the behaviour of excess carriers, which can be either in the form of continuous injection from a contact or as a single pulse. In the one case one would be interested in the decay of excess densities away from the contact and, in the other, both in a spatial and in a time decay. Charge neutrality imposes identical behaviour on both types of carrier and we simply write Δn for Δp in the equation in order to describe the behaviour of electrons.

Let us consider first minority carrier injection in n-type material, e.g. in the form of a single pulse. The behaviour of the excess holes and electrons within the pulse will be governed by equation (9.36) with respect to both position and time. If the material is strongly extrinsic ($n \gg p$), we have in accordance with equation (9.37) and (9.38) that $\mu^* = \mu_h$ and $D^* = D_h$. Thus excess carriers, of each type, will have the same mobility as that of the minority carriers in the material and in a small applied field the movement of the pulse will be governed by the drift of the excess holes. The excess majority carriers drawn into the pulse are constrained to move with the excess minority carriers in order to maintain charge neutrality. If the n-type material is near intrinsic, equation (9.37) shows that the effective mobility μ^* of the excess carriers becomes much reduced. The minority carriers may be considered to suffer a drag due to the majority carriers. This comes about by the external applied field bringing about a slight charge separation within the pulse giving a decreased effective value of the field within the pulse. It is exemplified in equation (9.37) as a decreased mobility.

A similar treatment is applicable for p-type material where we again find that the pulse movement is governed by the minority carrier parameters. Thus for $p \gg n$, $\mu^* = -\mu_e$ and the drift term in equation (9.36) becomes identical with that in equation (9.33) for electrons. In intrinsic material we have the interesting result that

$$\mu^* = 0$$

and

$$D^* = \frac{D_e D_h}{D_e + D_h}. \qquad \ldots(9.39)$$

There is then no overall drift of the pulse in an applied field and the pulse will decay by diffusion and recombination. A fuller account of the drift of a pulse of excess carriers is given by Jonscher (1960).

9.4 The p-n junction

A p-n junction is the boundary between a p-type and an n-type region in a single crystal of a semiconductor. The region over which the necessary changes in impurity concentration occurs is normally very small and for many problems the transition in impurity concentration can be regarded as being abrupt. Its exact form will depend on whether the junction has been manufactured by an alloying, a diffusion or an epitaxial deposition process. The necessary continuity in carrier flow across a junction cannot be achieved by having separate n- and p-type crystals in contact.

9.4.1 *The p-n junction in equilibrium*

The energy level diagram for an abrupt junction with no applied external field is shown in Fig. 9.5. As for the non-uniform semiconductor illustrated in Fig. 9.2 the band edges must vary in energy across the junction with the Fermi level remaining constant throughout. This leads to a potential energy difference eU_D between the two sides of the junction whose magnitude depends on the impurity

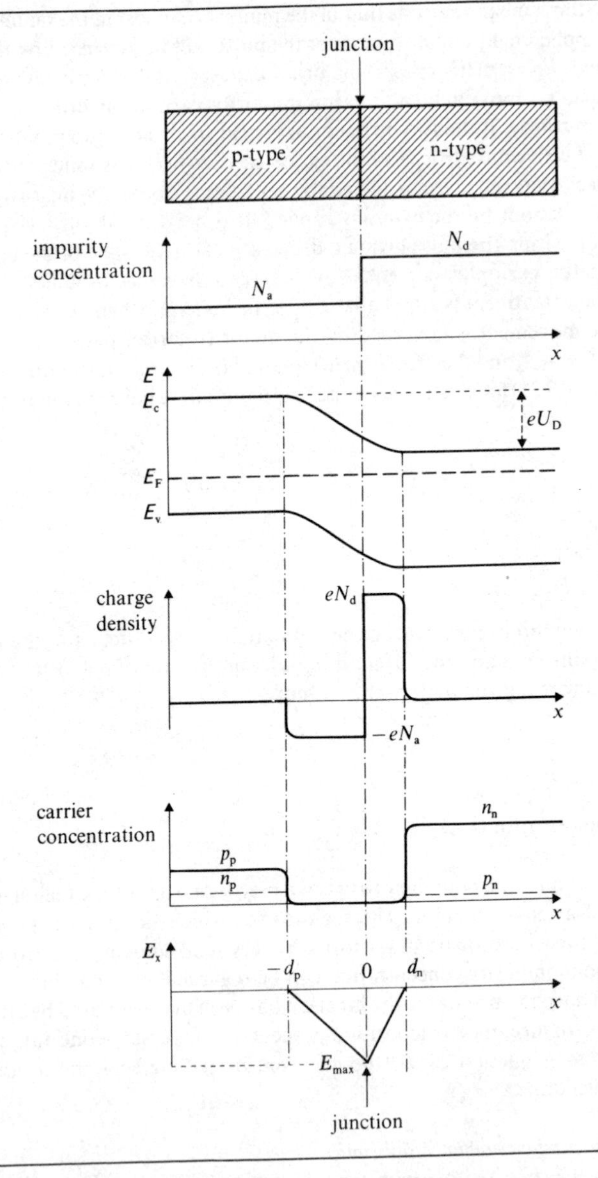

Fig. 9.5 The p–n junction in equilibrium.

concentrations in the p- and n-type regions of the crystal. The maximum value of this potential energy difference is approximately equal to the energy gap of the semiconductor.

As discussed in Sections 9.2 and 9.2.1 the presence of an electric field allows the diffusion flow of mobile carriers in a concentration gradient to be exactly balanced by an opposite drift flow in the field. The potential U_D exhibited across the junction is therefore termed the diffusion potential. We note also that this balance will exist separately for electron flow and for hole flow. The space charge necessary for the production of the potential barrier is formed in a very narrow region, extending for about 1 μm at the junction. This is termed the transition region. It consists of the layer of ionised donors and acceptors on either side of the junction, the strength of the internal field being such as to sweep the mobile carriers away from its immediate vicinity. For this reason it is often referred to as the depletion layer.

Let us now see how the diffusion potential U_D is related to the impurity concentrations on either side of the junction. Assuming a non-degenerate semiconductor we make use of the Boltzmann equilibrium conditions (8.32) and (8.33) together with the fundamental relationship

$$np = n_i^2. \qquad \ldots(8.31)$$

Denoting the carrier concentrations in the p- and n-regions with the corresponding subscripts we thus have for the p-region

$$E_{ip} - E_F = kT \ln \frac{n_i}{n_p} = kT \ln \frac{p_p}{p_i}; \qquad \ldots(9.40)$$

and for the n-region

$$E_F - E_{in} = kT \ln \frac{n_n}{n_i} = kT \ln \frac{p_i}{p_n}, \qquad \ldots(9.41)$$

E_{ip} and E_{in} denote the position of the intrinsic level on either side of the junction and we also note that E_F remains constant throughout. Adding equations (9.40) and (9.41) the change in the level of the band structure across the junction is then given by

$$\begin{aligned} E_{ip} - E_{in} &= kT \ln \frac{n_n p_p}{n_i^2} \\ &= kT \ln \frac{n_n}{n_p} \\ &= kT \ln \frac{p_p}{p_n}. \end{aligned} \qquad \ldots(9.42)$$

Assuming complete ionisation of the donors and acceptors and denoting their net concentrations in the n- and p-type regions of the crystal by N_d and N_a the diffusion potential will thus be

$$U_D = \frac{kT}{e} \ln \frac{N_d N_a}{n_i^2}. \qquad \ldots(9.43)$$

This diffusion potential, formed by the double layer of charge, is such that the p-side of the junction is at a negative electric potential with respect to the n-side. When both N_d and N_a are large the Fermi level will lie near the band edges on either side of the junction and the diffusion potential will then essentially correspond to the energy gap in the material. In most junctions, however, the value of the diffusion potential will be much less, e.g. 0.3–0.5 V for a junction in germanium.

The diffusion potential gives rise to a drift flow of minority carriers across the junction in both directions. The value of this current is limited by the thermal generation rate of electron–hole pairs in the semiconductor and it is termed a thermal generation current. Normally N_d and N_a are not equal and if, for example, we assume $N_d \gg N_a$ then $p_n \ll n_p$ and the thermal generation current is primarily that of minority electrons from the p-region. In the absence of an applied potential difference this minority electron flow is exactly balanced by the electron diffusion flow from the n-region.

In addition to the diffusion potential one is also interested in the variation of the field strength across the junction and in the width of the depletion layer. These will now be determined for the abrupt junction.

The requirement that overall space charge neutrality is maintained implies that for unit area of junction we must have

$$eN_d d_n = eN_a d_p, \qquad \ldots(9.44)$$

where d_n and d_p are the widths of the depletion region on the n- and p-sides of the junction, respectively. Again if it is assumed that one side of the junction has much higher conductivity than the other, the extent of the space charge region is much greater on the low conductivity side, e.g. assuming $N_d \gg N_a$ then $d_p \gg d_n$.

The variations in potential and field occur within the depletion layer and for the p-type region defined by $-d_p < x \leqslant 0$ will be given by equations (9.13) and (9.14) in the form

$$\frac{\partial E_x}{\partial x} = -\frac{\partial^2 U}{\partial x^2} = \frac{\rho(x)}{\varepsilon} = -\frac{eN_a}{\varepsilon}. \qquad \ldots(9.45)$$

Noting that the field $E_x = -\partial U/\partial x$ is zero at $x = -d_p$ and setting $U = 0$ at the junction at $x = 0$ this equation integrates to

$$E_x = -\frac{\partial U}{\partial x} = -\frac{eN_a}{\varepsilon}(x + d_p) \qquad \ldots(9.46)$$

and

$$U = \frac{eN_a}{\varepsilon}\left(\frac{x^2}{2} + d_p x\right). \qquad \ldots(9.47)$$

The potential at the edge of the depletion layer at $x = -d_p$ is therefore

$$U = -\frac{eN_a}{\varepsilon}\frac{d_p^2}{2} \qquad \ldots(9.48)$$

and it may readily be confirmed that a symmetrical expression is obtained for the potential at $x = d_n$ on the n-side of the junction. This latter potential is smaller by a factor N_d/N_a than that given by equation (9.48) so that with the approximation of $N_d \gg N_a$ it may be neglected. The potential given by equation (9.48) is thus the diffusion potential U_D so that

$$d_p = \left(\frac{2\varepsilon U_D}{eN_a}\right)^{\frac{1}{2}}. \qquad \ldots(9.49)$$

Since $d_n \ll d_p$ this gives the width of the depletion layer. The maximum field, in accordance with equation (9.46), is given by

$$E_{max} = -\frac{eN_a d_p}{\varepsilon} = \frac{2U_D}{d_p} \qquad \ldots(9.50)$$

and it occurs at the junction.

As an example we may take a typical junction in a germanium crystal at 300 K for which $N_d = 10^{24}$ m^{-3} and $N_a = 10^{21}$ m^{-3}. Taking $n_i^2 = 4.0 \times 10^{38}$ m^{-6} we find $U_D = 0.37$ V, $d_p = 8 \times 10^{-7}$ m and $E_{max} = 9 \times 10^5$ V m^{-1}. The potential drop is therefore occurring over a very narrow region and a large localised field occurs due to the depletion of carriers from the immediate region of the junction.

9.4.2 *The p-n junction with applied bias*

Before deriving the current–voltage characteristic of a p-n junction we may first look in a qualitative manner at the effect of applying a potential difference across a junction. First, let an applied potential be applied such as to make the p-region positive with respect to the n-region. This will reduce the barrier height for the flow of majority carriers across the junction and it is found that a current increasing exponentially with applied voltage is obtained. The junction is termed forward biased when in this condition. As for the equilibrium state one may deal with electron and hole currents separately since these are independent. Taking the hole flow, the decrease in barrier height allows an increased diffusion of holes from the p-region across to the n-region (Fig. 9.6). The hole flow in the opposite direction, which is limited by the thermal generation rate, remains unaltered, since the minority holes close to the junction on the n-side are still swept across by the junction field. Overall there is, therefore, a net flow of holes across the junction into the n-region. This is termed minority carrier injection. Within the n-type region the excess holes recombine with the majority electrons in the region adjacent to the depletion layer, electrons being continuously drawn in from the bulk material to maintain charge neutrality. The width of this recombination region is determined by the hole diffusion length L_p in the n-type region. We thus see that a hole current on one side of the junction is replaced by an equivalent electron current on the other.

A similar process occurs for electron flow across the junction and is now associated with recombination in the p-type region. We note that if the conductivity of the p-region is much greater than that of the n-region, i.e. if

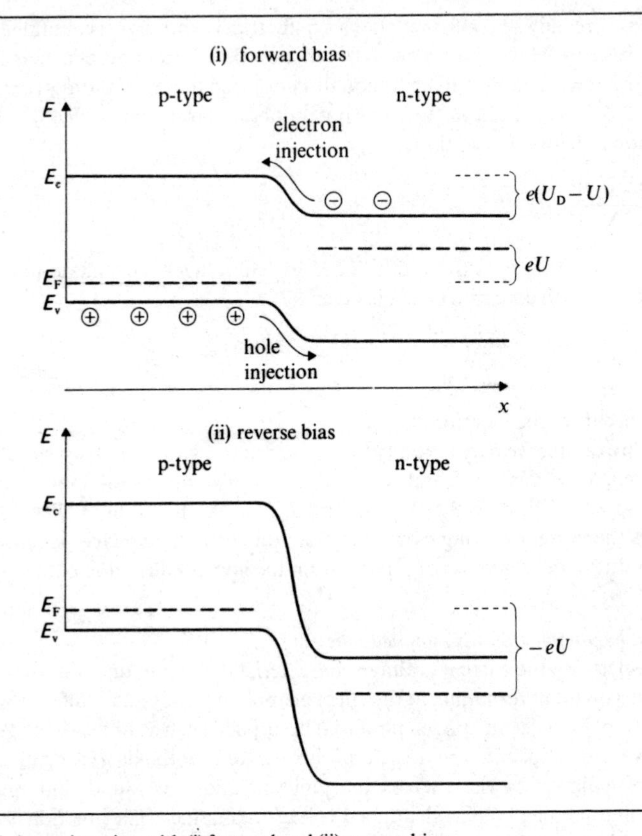

Fig. 9.6 A p–n junction with (i) forward and (ii) reverse bias.

$N_a > N_d$, the hole injection is predominant and the current flow across the junction is then primarily due to holes.

We have neglected here potential drops occurring in the bulk regions. These become of importance at moderately high injection levels when they may be comparable with the drop across the space charge region. The injected minority density may then approach the majority density and a modified exponential voltage–current relationship is found. At very high injection levels the effective potential across the junction may approach the diffusion potential U_D and the current flow is then governed by the conductivity of the bulk material. The exponential voltage–current relationship observed at the lower injection levels breaks down under such conditions.

An applied potential across a junction such as to make the p-side negative with respect to the n-region is termed a reverse bias. The barrier height is now increased and for a relatively small applied bias the current flow becomes constant and independent of bias. This reverse current consists of the drift flow of the minority carriers from the two regions and is limited by the thermal generation rate of the carriers. At a rather larger reverse voltage, whose value depends on the junction characteristics, breakdown occurs and large currents are then observed.

9.4.3 *The current–voltage characteristic of a p-n junction*

Under equilibrium conditions the carrier concentrations in a semiconductor are governed by the Boltzmann equilibrium equations and, in accordance with equation (9.43) a diffusion potential U_{D} given by

$$U_{\mathrm{D}} = \frac{kT}{e}\ln\left(\frac{n_{\mathrm{n}}p_{\mathrm{p}}}{n_{\mathrm{i}}^2}\right) \qquad \ldots(9.51)$$

exists across the junction. Using equation (8.31) with appropriate suffixes the minority carrier density in one region may be related to the majority carrier density in the other according to

$$\frac{p_{\mathrm{n}}}{p_{\mathrm{p}}} = \frac{n_{\mathrm{p}}}{n_{\mathrm{n}}} = \exp\left(-\frac{eU_{\mathrm{D}}}{kT}\right). \qquad \ldots(9.52)$$

Let us first see how the minority carrier densities are modified when a forward bias U is applied. The barrier height is reduced from U_{D} to $U_{\mathrm{D}}-U$ and, as illustrated in Fig. 9.7, injection leads to excess minority carrier densities which die away from the edges A and B of the depletion layer due to recombination. Charge neutrality requires also the presence of excess majority carriers and these will follow the same pattern of change as the excess minority carriers. Denoting the excess carrier densities at the planes A and B by Δ's and the appropriate subscript we therefore have

$$\Delta n_{\mathrm{A}} = \Delta p_{\mathrm{A}}$$

$$\Delta n_{\mathrm{B}} = \Delta p_{\mathrm{B}}. \qquad \ldots(9.53)$$

We now assume that a condition of the form of the equilibrium condition (9.52) is valid even in the presence of an applied potential U and obtain for the total hole density at B

$$p_{\mathrm{n}} + \Delta p_{\mathrm{B}} = (p_{\mathrm{p}} + \Delta p_{\mathrm{A}})\exp\left(-\frac{e}{kT}(U_{\mathrm{D}}-U)\right). \qquad \ldots(9.54)$$

This assumption may be shown to be valid for low injection levels. Using equation (9.52) and noting that $\Delta p_{\mathrm{A}} \ll p_{\mathrm{p}}$ this simplifies to

$$\Delta p_{\mathrm{B}} = p_{\mathrm{n}}\left[\exp\left(\frac{eU}{kT}\right) - 1\right]. \qquad \ldots(9.55)$$

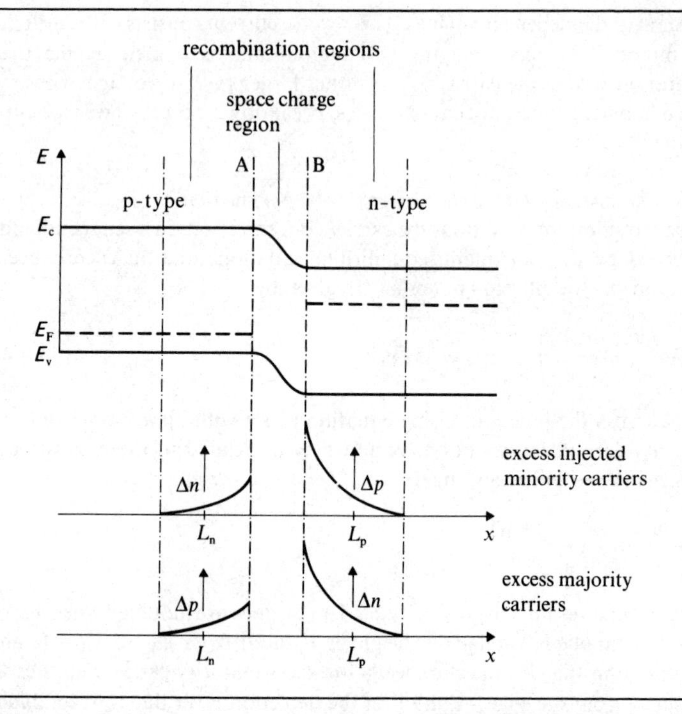

Fig. 9.7 Excess carrier densities in a forward biased p–n junction. Note that the excess injected minority carrier densities on either side of the junction are in the ratio of the majority carrier densities on the opposite sides of the junction. Charge neutrality in the recombination regions require the excess majority carrier densities to follow those for the minority carriers.

By a similar argument the symmetrical relationship

$$\Delta n_A = n_p\left[\exp\left(\frac{eU}{kT}\right) - 1\right] \qquad \ldots(9.56)$$

is also seen to be valid.

These relationships can now be used to derive the total current across the junction when the forward bias U is applied. Within the n-type region the excess hole concentration Δp will die away exponentially from the value Δp_B at the edge of the space charge region to give the equilibrium minority hole density in the bulk n-type region. This is achieved by electron–hole recombination and, in accordance with equation (9.25), the excess hole concentration will follow the decay

$$\Delta p = \Delta p_B \exp\left(-\frac{x}{L_p}\right), \qquad \ldots(9.57)$$

where x is the distance from the plane B and L_p the minority carrier diffusion length in the n-type region. The gradient in the minority carrier concentration is therefore achieved as a balance between the recombination mechanism and the diffusive flow of minority carriers within the recombination region. The hole current density within this region is given by

$$J_h = -eD_h \frac{\partial}{\partial x}(\Delta p)$$

$$= \frac{eD_h}{L_p} \Delta p_B \exp\left(-\frac{x}{L_p}\right) \quad \ldots(9.58)$$

This has a maximum value of

$$J_h = \frac{eD_h}{L_p} \Delta p_B \quad \ldots(9.59)$$

which occurs at the plane B and, in accordance with equation (9.55) can be put in the form

$$J_h = \frac{eD_h}{L_p} p_n \left[\exp\left(\frac{eU}{kT}\right) - 1\right]. \quad \ldots(9.60)$$

The maximum electron current density occurs at the plane A and in a similar manner is shown to be

$$J_e = \frac{eD_e}{L_n} n_p \left[\exp\left(\frac{eU}{kT}\right) - 1\right]. \quad \ldots(9.61)$$

Since the width of the space charge region is very narrow it is justifiable to neglect any recombination in this region so that the total current density J across the junction will be given by

$$J = J_h + J_e$$

$$= J_s \left[\exp\left(\frac{eU}{kT}\right) - 1\right], \quad \ldots(9.62)$$

where

$$J_s = \left[\frac{eD_h}{L_p} p_n + \frac{eD_e}{L_n} n_p\right]. \quad \ldots(9.63)$$

This relationship, sometimes referred to as the rectifier equation, is valid in this form for small injection levels. At a forward bias of a few tenths of a volt the injected carrier densities become equal to the majority carrier densities and a current density dependence on $\exp(eU'/kT)$ is then observed. U' here is the actual potential across the junction and may be much less than the applied bias since part of the voltage drop occurs in the bulk material. For larger applied voltages $U' \rightarrow U_D$ and the potential barrier disappears. The current–voltage relationship is then governed by the bulk material.

The proportion of the total forward current carried by holes defined by

$$\gamma = \frac{J_h}{J_e + J_h} \qquad \ldots(9.64)$$

is of interest in device applications. It is readily shown to be given by

$$\gamma = \frac{\sigma_p L_n}{\sigma_n L_p + \sigma_p L_n} \qquad \ldots(9.65)$$

where σ_p and σ_n are the conductivities of the p- and n-regions respectively. For $\sigma_p \gg \sigma_n$ the injection ratio γ becomes unity and the junction is then referred to as a p^+–n junction.

The behaviour of the junction under a reverse applied bias may be developed in a similar manner and is illustrated in Fig. 9.8. An increase in the barrier height now occurs such that the applied field is *extracting* minority carriers across the junction. The excess minority densities imposed at planes A and B are now negative and even for small reverse voltages will make the minority carrier

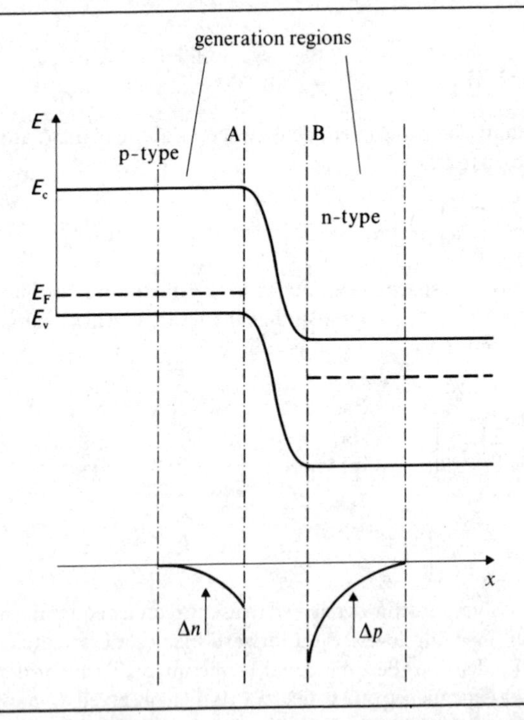

Fig. 9.8 The p–n junction with reverse bias. The minority carrier densities on either side of the junction are now less than the equilibrium values and even for small bias become zero, i.e. $\Delta n_A = -n_p$ and $\Delta p_B = -p_n$. The reverse current consists of minority carrier flow across the junction and is limited by the diffusion rate of the thermally generated carriers.

densities at these planes become zero. There is therefore a diffusion of minority carriers *towards* these planes, the carriers being thermally generated within the diffusion lengths L_p and L_n of the junction. A reverse saturation current, limited by the thermal generation rate, is thus obtained. Equations (9.55) and (9.56) remain valid but with Δp_B and Δn_A negative and equal to $-p_n$ and $-n_p$, respectively, for a wide range of negative values of U. The current density equation (9.62) thus also applies for a reverse biased junction and we identify J_s as the saturation reverse current density obtained for reverse voltages of magnitude greater than about $4kT/e$. The value of J_s is determined in accordance with equation (9.63) by the minority carrier densities on either side of the junction and the carrier lifetimes and diffusion lengths. It follows that if both the n- and p-regions are of rather low conductivity the reverse, or leakage current, tends to be high. On the other hand, in an assymetric junction, such as a p^+-n, the leakage current is governed only by the minority carrier flow from the low conductivity side.

In practice, the dimensions of a device may be such that either, or both, the semiconducting regions are shorter than the corresponding diffusion lengths L_p and L_n. In such cases L_p and L_n must be replaced in equation (9.63) by effective diffusion lengths. This leads to a higher leakage current but it also has the effect of reducing the forward voltage drop for a given current density.

The dependence of leakage current on minority carrier densities has an important consequence for the temperature dependence of the diode characteristic. In accordance with the concepts developed in the last chapter the majority densities n_n and p_p are unaffected by changes in temperature since both donors and acceptors remain completely ionised. The temperature dependence given by

$$np = N_c N_v \exp\left(-\frac{E_g}{kT}\right) \qquad \ldots(8.31)$$

which using equations (8.17) and (8.23) can be put in the form

$$np \propto T^3 \exp\left(-\frac{E_g}{kT}\right) \qquad \ldots(9.66)$$

will thus govern the minority carrier densities in the device and give the saturation reverse current

$$J_s \propto T^3 \exp\left(-\frac{E_g}{kT}\right). \qquad \ldots(9.67)$$

The fractional change in current density arising from a given temperature change is of most interest and is given by

$$\frac{dJ_s}{J_s} = \left(3 + \frac{E_g}{kT}\right)\frac{dT}{T}. \qquad \ldots(9.68)$$

Substituting appropriate numerical values it may be confirmed that a germanium diode at room temperature will have its leakage current doubled for every 9 °C rise in temperature. In any application where the forward bias is constrained to remain constant the forward current will also increase.

A typical experimental current–voltage characteristic for a diode is compared with the theoretical characteristic in Fig. 9.9. The reasons for the departures in

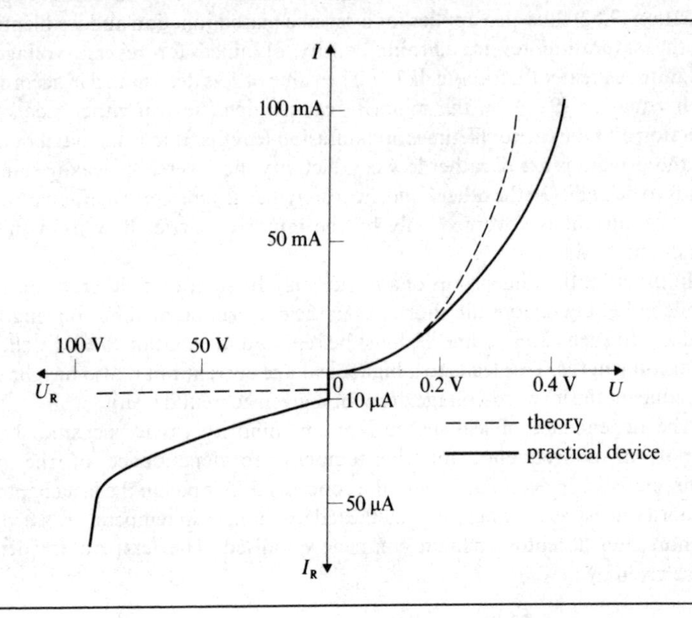

Fig. 9.9 Current–voltage characteristics of a p–n junction. Note that the scales for reverse applied voltage U_R and reverse current I_R are different from those used for the forward characteristic.

the forward characteristic have already been mentioned. For reverse bias, the departures at low reverse voltages are due primarily to surface leakage currents, whilst at large reverse voltages, e.g. at 100 V in the example chosen, the phenomenon of breakdown occurs. Provided that there is no overheating, which would cause physical damage to the device, the breakdown effect is reversible. At breakdown voltages the large junction field allows the carriers to attain very high kinetic energies within the depletion region. Such energies are sufficient to cause the production of electron–hole pairs by inelastic collision and an avalanche multiplication process similar to that in a gas discharge occurs. The minimum breakdown field in silicon is $\sim 5 \times 10^7$ V m^{-1} and is dependent on the impurity, or doping, levels in the device. Breakdown voltages as high as 1 KV in silicon and 500 V in germanium may be obtained by the use of material of low conductivity.

A different breakdown mechanism is operative in very heavily doped junctions in which the breakdown voltage is <5 V. The depletion layer is then very narrow and junction fields in excess of 10^8 V m^{-1} are obtained. In such cases a proportion of the atoms within the depletion region become ionised

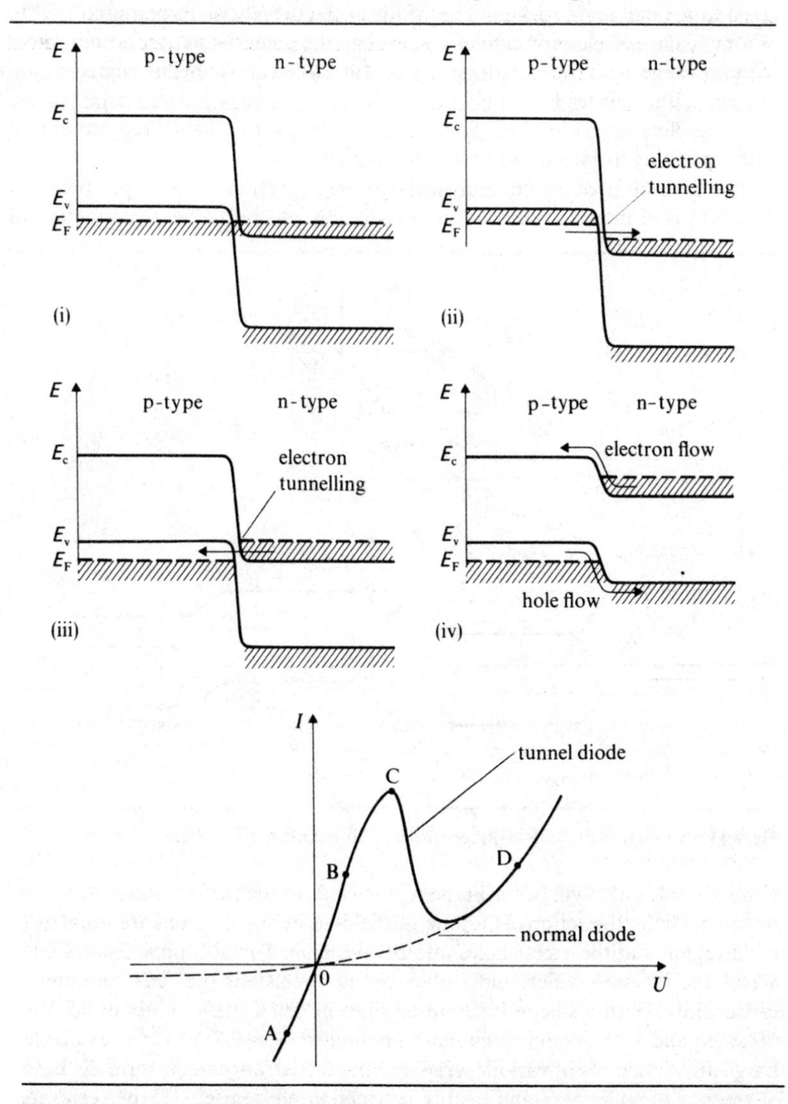

Fig. 9.10 Tunnel diode characteristic: (i) point O on characteristic. Note position of Fermi level lying within the respective bands due to high level of doping; (ii) point A. Large reverse current occurs due to electron tunnelling; (iii) point B. Electron tunnelling is occurring and reaches a maximum value for the positive bias corresponding to the point C; (iv) point D. Tunnelling has ceased and diffusion flow of carriers is dominant.

giving excitation of electrons into the conduction band. This is termed internal field emission or Zener breakdown.

When the impurity concentration is increased yet further, breakdown occurs even at a small forward bias. The width of the depletion region is reduced to about 10 nm and electron tunnelling through the potential barrier occurs. For a certain range of forward voltages both diffusion and tunnelling currents are obtained and this leads to a characteristic having a negative resistance region. This is illustrated in Fig. 9.10, together with the band representation corresponding to various points on the characteristic.

The behaviour of an illuminated large area junction of the type shown in Fig. 9.11 is of interest. Absorption of radiation and consequent production of

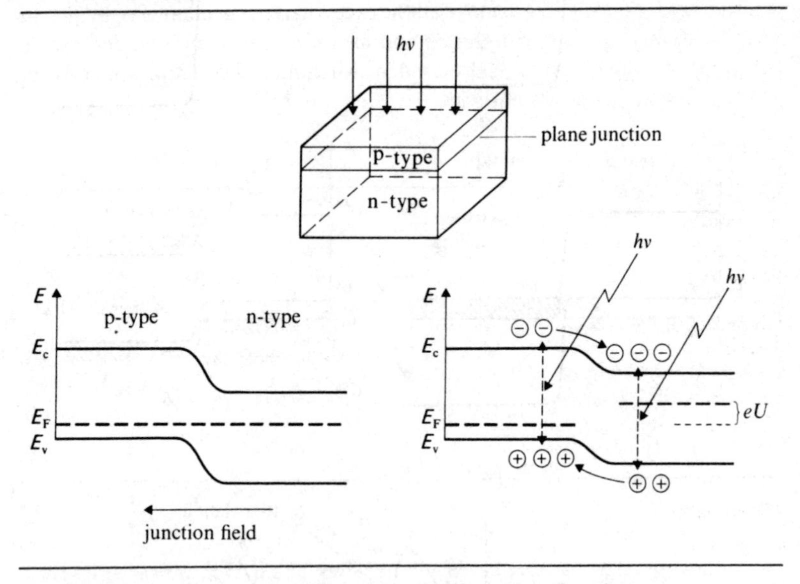

Fig. 9.11 Production of photo-voltage with an illuminated p–n junction.

electron–hole pairs will occur in the region close to the surface and near to the junction. Under the action of the junction field the excess electrons are swept into the n-region and the excess holes into the p-region. This sets up a *forward* bias across the junction which under open-circuit conditions prevents continuous carrier flow. With a silicon junction an open circuit voltage of about 0.5 V is obtained and with a suitable circuit a photocurrent of 0.3 A m^{-2} is available. Large area junctions in various series and parallel arrangements form the basis of the use of solar cells and energy conversion efficiencies ~ 15 per cent are possible.

The flow of excess carriers, formed by illumination, has the effect of increasing the leakage current when the junction is under reverse bias. Putting the total reverse current density as J_T we have

$$J_T = J_S + J_0 \qquad \ldots(9.69)$$

where J_0 is the photocurrent density and J_S is referred to as the dark current. If the optical power incident on unit area is P, and if a fraction η of the incident quanta yield electron-hole pairs then

$$J_0 = 2e\left(\frac{P}{h\nu}\right)\eta. \qquad \ldots(9.70)$$

A photocurrent proportional to the illumination level is thus obtained. This is nearly independent of reverse voltage since even at zero applied voltage the junction field is sufficient to remove all the excess carriers generated optically. In addition to their use in the detection and measurement of infrared and visible light, junctions of this type are also used in switching and counting applications and as detectors of nuclear particles.

References and further reading

Jonscher, A. K. (1960) *Principles of Semiconductor Device Operation*, Bell.
Lynch, P. and **Nicolaides, A.** (1972) *Worked Examples in Physical Electronics*, Harrap.
Morant, M. J. (1964) *Introduction to Semiconductor Devices*, Harrap.
Pascoe, K. J. (1972) *Properties of Materials for Electrical Engineers*, Wiley.
Seymour, J. (1972) *Physical Electronics*, Pitman.
Smith, R. A. (1959) *Semiconductors*, Cambridge University Press.

Problems

9.1 If the potential difference between two points A and B 0.4 m apart is 50 V what is the magnitude of a uniform electric field at angle 30° to the line AB which gives this potential difference?

9.2 Show by means of Gauss' theorem that if there is no space charge in a semiconductor sample the electric field is everywhere constant. Hence show that such a sample in the form of a rod must have an exponential distribution of impurity along its length. In what way does the electric field then arise?

9.3 Derive the Einstein relationship

$$D_h = \frac{kT}{e}\mu_h$$

for holes in a semiconductor.

9.4 Assuming values of electron and hole mobilities for germanium from Table 8.1 determine the corresponding diffusion coefficients D_e and D_h for a temperature of 300 K.

Explain the term minority carrier diffusion length. If this has a value of 5×10^{-3} m for a p-type sample of germanium what is the minority carrier lifetime?

9.5 Derive an expression for the potential difference between two points in an inhomogeneous n-type semiconductor for which the impurity concentrations are n_1 and n_2.

One end of an n-type germanium sample has a room temperature conductivity of $3.7 \times 10^4\ \Omega^{-1}\ \mathrm{m}^{-1}$. What is the conductivity of the other end of the sample given that it is 0.2 V lower in potential? There is no external field applied.

9.6 Show by means of Gauss' Law that the magnitude of the electric field at a distance a from the centre of a dielectric sphere having a uniform volume density of charge ρ is

$$E = \frac{\rho a}{3\varepsilon},$$

where ε is the permittivity of the dielectric.

9.7 Excess holes are injected into an n-type semiconductor rod sample and a small electric field is applied. Determine the relationship between the hole current density and the excess hole concentration at any other point in the sample.

9.8 Confirm the form of equation (9.36) and the values of the ambipolar diffusion coefficient D^* and mobility μ^*.

9.9 Show that the injection ratio γ for a p-n junction is given by

$$\gamma = \frac{\sigma_p L_n}{\sigma_n L_p + \sigma_p L_n}, \qquad \ldots(9.65)$$

where the symbols have the meanings assigned to them in the text (Section 9.4.3).

9.10 Derive for the conditions illustrated in Fig. 9.4 the distance from the injecting source at which the electron and hole currents are equal. Assuming $\gamma = 1$ what is this distance?

9.11 The room temperature conductivities on the n- and p-sides of an abrupt junction in silicon are respectively $10^4\ \Omega^{-1}\ \mathrm{m}^{-1}$ and $0.77\ \Omega^{-1}\ \mathrm{m}^{-1}$. Determine the diffusion potential across the junction, the width of the depletion layer and the value of the maximum field at the junction. Given that the minority carrier lifetime for the p-region is 100 μs determine the minority carrier diffusion length in this region.

9.12 Show that the excess injected minority carrier densities at the edges of the depletion layers on either side of a p-n junction are in the ratio of the majority carrier densities on the opposite sides of the junction.

9.13 Determine the fractional change $\mathrm{d}J_s/J_s$ in the saturation reverse current for a unit degree rise of temperature for a germanium and for a silicon diode. Assume room temperature operation.

9.14 Explain why the forward bias U required to sustain any given forward diode current is decreased when the temperature of the device increases. Show that the required change in forward bias with temperature is given by

$$\frac{\mathrm{d}U}{\mathrm{d}t} = \frac{U}{T} - \frac{k}{e}\left(3 + \frac{E_g}{kT}\right)$$

and that this amounts to $-2.5\ \mathrm{mV\ K^{-1}}$ for a silicon diode operated at a forward bias of 0.5 V.

9.15 Show that the incremental resistance R of a forward biased p-n junction defined by $\mathrm{d}U/\mathrm{d}I$ is given by

$$R = \frac{\mathrm{d}U}{\mathrm{d}I} = \frac{kT}{e(I_0 + I)},$$

where I_0 is the saturation reverse current.

Given that a diode has a saturation reverse current of 10 μA determine the incremental resistance for forward currents of 20 μA, 0.5 mA and 10 mA. Assume $kT = 0.025$ eV.

Index